The Ocean Manifesto

ISBN 978-2-36568-115-5

The Ocean Manifesto

Curated by Caroline Ha Thuc
with the participation of the Metis Fund

METIS

JBE BOOKS

Unless otherwise stated, all artist quotes in this book come from interviews conducted between May 2024 and January 2025 by Caroline Ha Thuc.

Passion and Reason Are in Attendance Here

Tiphaine de Mombynes,
Director of the Metis Arts and Development Fund

I am certain of at least one thing, which is that our actions and decisions are far from being the result of a precise chain of causes and effects that are rational, considered, deliberate, and consciously weighted.

Remove our hearts, our emotions, our senses, our feelings, our intuitions . . . society would look very different. For better or for worse? What do you think?

I'll say it loud and clear: our sensitivities are vital. They are not sufficient, but they are vital. They inform our choices, our positions, our visions. They are multiple, infinite. Isn't it therefore useful to nurture them as a means to grasp and build the world more equitably? A world that reflects this richness, intrinsic to each and every one of us, but also intrinsic to every community and every group.

Who currently takes the time to nurture them, to let them express themselves, to listen to them? What place do we leave for informed emotions? Technology is everywhere. Highly regarded, massively used. So very useful! But if technology advances alone . . . Aren't boredom, emptiness, and error increasingly lurking to suffocate us?

This conviction led me to launch the Metis Arts and Development Fund in 2019. A bit of a crazy gamble! Metis, the Greek goddess of cunning and wisdom. She who gives Zeus his unrivaled strength. Supported by the French Development Agency and committed private partners, this program invests in our emotional intelligence. What could be more commendable?

The Metis Fund facilitates the creation of ambitious artistic works "for" and "with" people in isolated and vulnerable areas. It supports the opening of new spaces for connections, exchanges, partnerships, and construction. Painting, sculpture, photography, music, dance, theater, circus, poetry, street art, slam . . . the arts unite and re-unite us. They transcend social, cultural, religious, and political differences. The arts heal, emancipate, and educate. They stir our will and our capacity to learn and set us in motion.

Beyond these local creations that transform all the actors involved, Metis also hopes to exert its influence on a larger scale . . . show, demonstrate how relevant it is to trust our senses and support the production of artists. What distinguishes us is that we strive to combine works of art, link them to technical level actions that are being undertaken to develop and improve our living conditions. And even more so for those who are the most disadvantaged. Yes, we can think of infrastructure construction (water, energy, etc.) and artistic creation. Yes, we can think of health, education, and artistic creation. Yes, we can think about agricultural and artistic creation.

Think about the fight against climate change and the protection of biodiversity and artistic creation. Yes, we can think of the fight against poverty artistic creation. Of course we can. Think and act. Let's not forget that this artistic expression dates back to the earliest times of humanity. It evolves, shapes, and is shaped by the reality of our societies. It is everywhere. Even where you least expect it.

By establishing a program dedicated to the oceans, we have therefore chosen to use the artistic lever locally in a dozen countries to encourage changes conducive to a better understanding of the issues that characterize these oceans. These works of art, what they tell us, we choose to present them to as many people as possible in a comprehensive reference work . . . The perfect opportunity to convene not only the artists who have worked with us, but also the many others around the world who create with, on, for, and around oceans.

An insatiable curiosity is at the origin of this work: who are these artists? What are they telling us? How will their works make us feel–you, me, us? We have decided to entrust this work to Caroline Ha Thuc, an accomplished arts/science academic and curator. With her, we set off in search of these artists. And we found them. In addition to their statements, we wanted to include in this book the voices of a few "non-artists." To get them conversing. Six leading figures do us the honor of lending their voices to our adventure.

Who is this book for? For you, for me, for all, for us . . . yesterday, today and tomorrow. Thank you, thank you to all those who collaborated on this manifesto, the result of a sensitive and well thought out approach. Passion and reason are very much in attendance here.

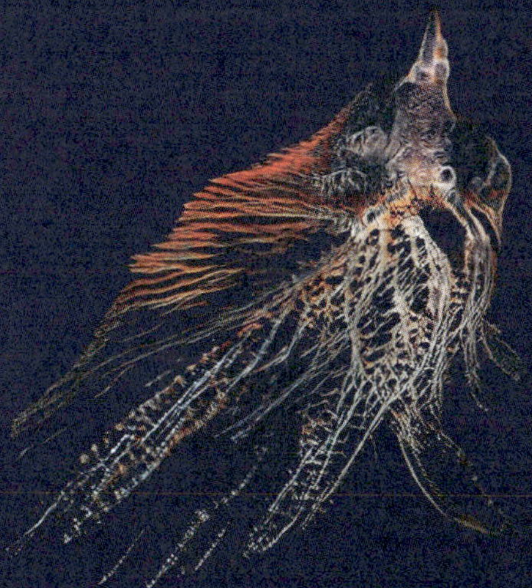

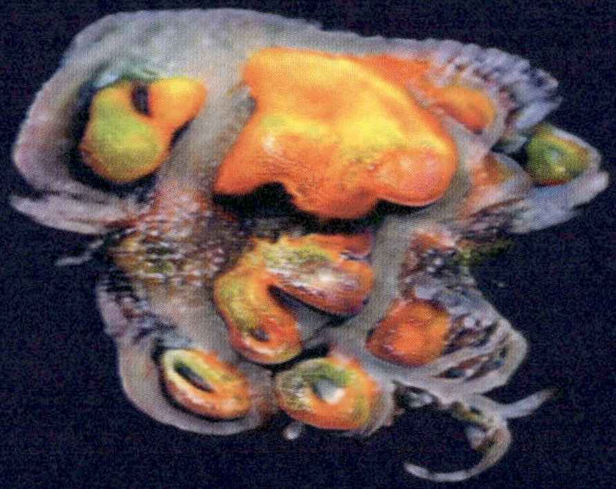

Oceanic Perspectives: A Cultural Shift

Caroline Ha Thuc,
Art Historian and Curator

Artists have always played a fundamental role in making visible that which eludes the human eye. Due to the inherent limitations of direct observation, oceanic knowledge and collective representations of the ocean have predominantly emerged through their imaginative descriptions and narratives. The science of oceanography derives its name from Oceanos, the Greek god of the sea, as depicted by the poet Homer. Many scientific terminologies, such as "plankton," denoting "the wanderer," "nekton," referring to "the swimmer," and "benthos," signifying "the organism residing on the seabed," have their origins in the epic narratives of the Iliad and the Odyssey. Artists were on board the first research marine expeditions, diving with early submersible to sketch the underwater world. To this day, our understanding of the ocean remains heavily reliant on visual representations facilitated by both technological instruments and artistic interpretations. Scientific research vessels such as the schooner Tara and Falkor from the Schmidt Ocean Institute continue to invite artists to complement their scientific perspectives.

In contemporary contexts, this traditional role has become increasingly critical. Artists are not only tasked with capturing the invisibility of the underwater world but also with testifying of the multiple impacts of climate change and anthropogenic activities on marine environments. The ramifications of global warming–manifesting as rising sea levels, glacial melt, and coral bleaching–alongside issues such as ocean acidification, fish depletion, seabed extraction, habitat loss, biodiversity degradation, and various forms of pollution (industrial, plastic, chemical, acoustic, etc.)–constitute a lengthy and well-documented list. However, aside from plastic pollution and the rapid rise in sea levels, many of these phenomena are challenging to conceptualize and visualize. Moreover, the interconnectedness of these effects and the deep global relationships linking the ocean, our environment, and humanity are complex and elusive to grasp.

Until the 1970s and what was called the "oceanic turn," the ocean was primarily viewed as an endless body of water, capable of absorbing all our excesses, providing routes for ships, and offering resources for humanity. Its own agency was denied, and its inner life, limits, and principles went unnoticed. It is hard to believe that it was only in 1993 that international treaties were signed to prohibit the ocean disposal of radioactive waste.

The ecological crisis facing human society today is unprecedented in its scale and severity. It calls for new paradigms, new ways of existing in the world, and innovative ideas. In essence, it demands creativity to develop and share new models. The recognition of the ocean as a living entity, with its own local specificities, is one of the challenges in front of us. However, changing our perceptions and collective representations of the sea is a long-term endeavor that requires engagement from all disciplines across every level of society. As concepts like "blue humanities," "hydrocene," "jellyfish aesthetics," and "aquatocene" proliferate, what does it truly mean and imply to "think with" the ocean? Beyond theories and discourses, how do we interact with and relate to a living sea, driven by its own dynamics and rules? From which perspective should we observe it: the shores, the surface, the depths, or the abyss? Artists emerge as ideal mediators in this dialogue; they connect with both scientists and communities, bridging the global with the local, and transforming abstract ideas into concrete, sensory experiences.

This book serves as a manifesto for this artistic mediation. It features nearly 100 artists from around the globe, working in diverse mediums and representing various backgrounds and generations. This compilation of artistic practices, voices, and testimonies seeks to foster new commons that transcend geographical and cultural boundaries. It is conceived as a collection of shared ideas and aims at inspiring action and behavioral change.

For these artists, the goal is to make tangible, accessible to our senses, the increasingly complex reality of human relationships with the ocean.

Despite their differing objectives, methods, and mediums, they are united in their commitment to deepening our understanding of and connection to the ocean. The book is mainly based on a series of conversations I had with them over the last couples of months. Together, we questioned the capacity of art to contribute to the radical transformation necessary in how we perceive and engage with the marine environment. One of the central concerns that emerged was how to navigate the pervasive sense of powerlessness in the face of ecological crises, to avoid spreading fears and eco-anxiety, and how to meaningfully embody the vast scope of complex issues that often feel both overwhelming and abstract.

Each artist supplied their own observations of how climate change has impacted their respective environments. These localized perspectives enrich our understanding of specific contexts and ground the manifesto into tangible realities. They inject complexity in what are, too often, simplified and over-homogenized views. At the same time, participation in this project helped to mitigate feelings of isolation as common threads and shared concerns emerged, fostering a sense of connection and collective purpose.

In addition to the artists' voices, the manifesto incorporates contributions from experts in other fields, including scientists, policy makers and anthropologists. This interdisciplinary approach underscores the interconnectedness of ecological, social, political, historical and cultural issues, highlighting how these domains often overlap and inform one another.

Finally, the manifesto emphasizes the essential role of poetry within this endeavor. Language itself must be reconsidered as part of this paradigm shift, as conventional rational discourse often proves insufficient in capturing the depth and complexity of ecological issues. Poetry, with its openness and capacity for ambiguity, allows us to reconnect with imagination and creativity.

The manifesto is divided into eleven chapters that often overlap, inviting readers to approach this categorization with fluidity and openness.

The first chapter showcases artworks that depict the ocean as the origin of life. From this primeval soup, hybrid living beings emerge, with the sea frequently associated with a maternal figure. The second chapter takes a historical perspective, presenting the ocean as a collective archive and a site of memory. In the third chapter, the focus shifts to the sea's ambivalent nature—portrayed as a liminal space embodying both freedom and isolation, fluidity and boundaries, serenity and violence. These artworks primarily address the issue of migration while also grappling with the inherent contradictions of the oceanic realm.

In the fourth chapter, the sea becomes a subject in its own right, asserting its agency and principles. It transforms landscapes, erodes coasts, and is often personified as a spiritual entity. Chapter five examines the many excesses of human activity and their devastating impact on marine ecosystems. Following this, chapter six adopts a more political perspective, addressing legal loopholes and political mismanagement related to the ocean, which was defined by the United Nations in 1970 as our "Common Heritage."

As no day passes without extraordinary discoveries in the marine environment, chapter seven is dedicated to this sense of awe, celebrating the beauty and treasures of the ocean as a shared heritage. Increasingly, we learn to interpret the sea and listen to coastal communities that have long lived in harmony with their marine ecosystems. Chapter eight highlights these processes of knowledge-sharing, with artists engaging directly from the field, attentive to voices and territories that have been historically overlooked.

Chapter nine invites us to radically change our perspective, adopting an oceanic worldview that dissolves individual identities and challenges social norms. From these paradigm shifts and metamorphoses, new possibilities arise: in chapter ten, which looks toward the future, artists explore symbiotic living forms, innovative renewable resources, and reimagined ways of inhabiting the world in harmony with the natural environment.

Finally, the eleventh chapter presents ten projects supported by the Metis Fund. These initiatives combine community-based development efforts with collective artistic practices, demonstrating, once more, the transformative potential of transdisciplinary artistic collaborations in the context of climate change.

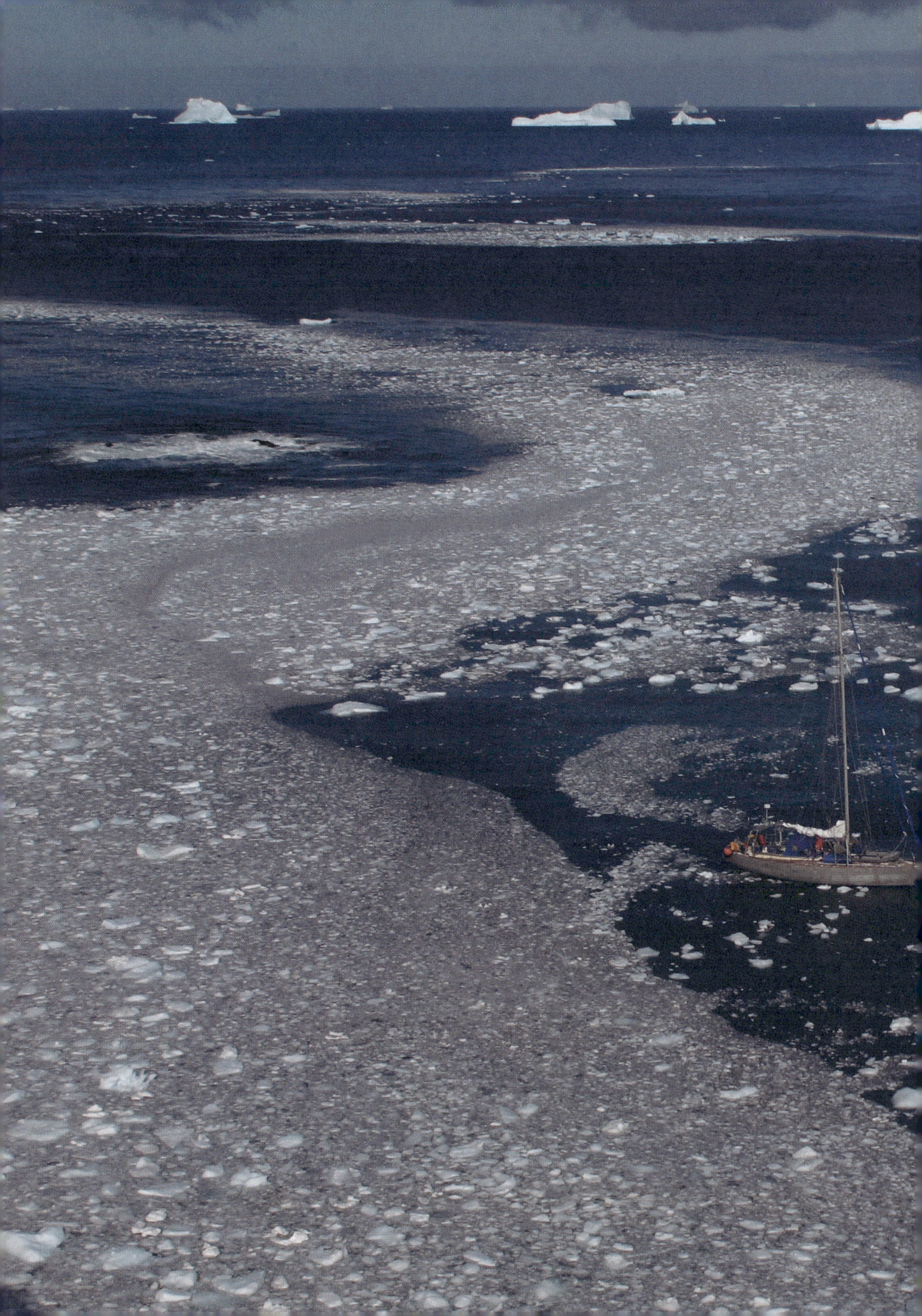

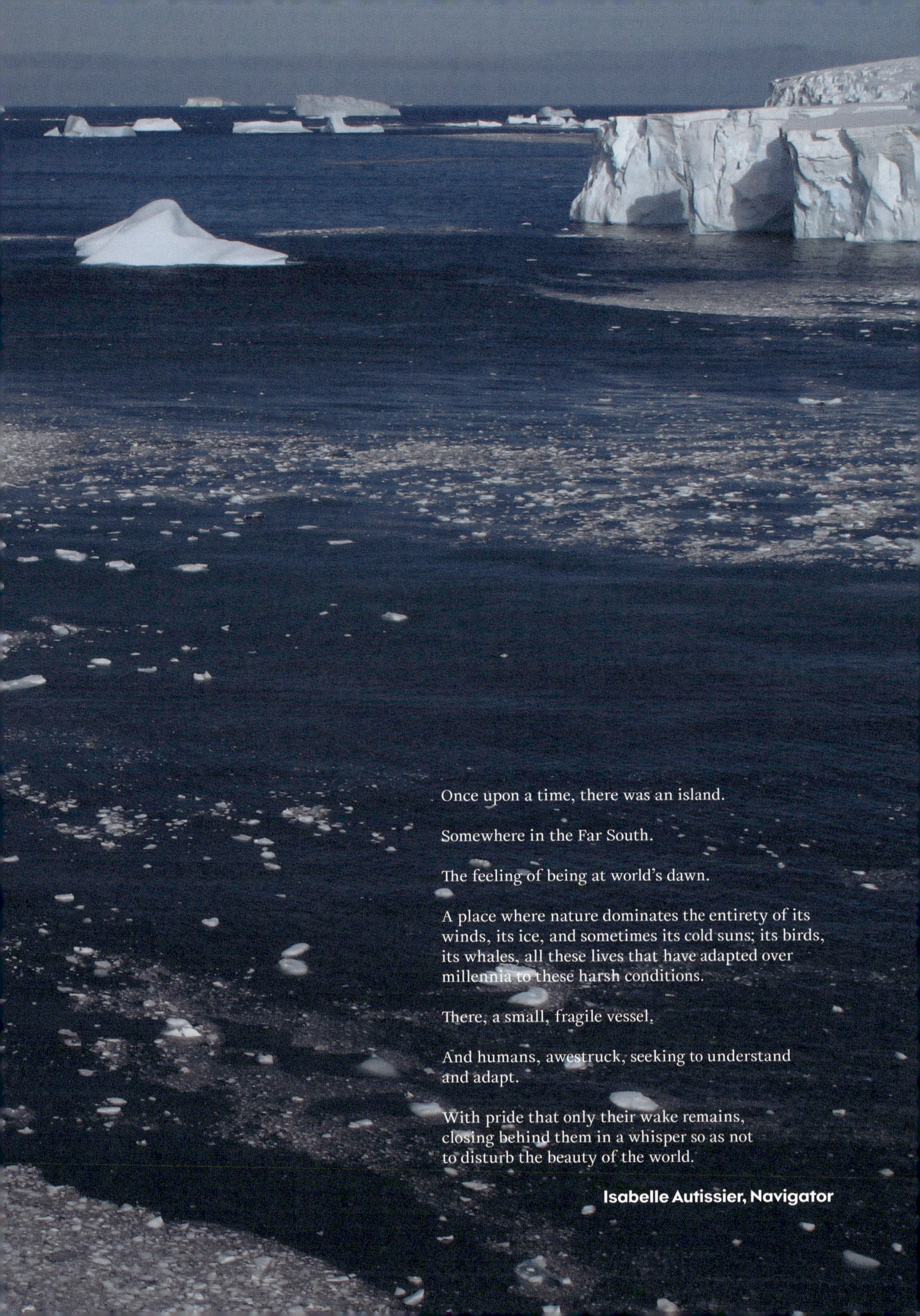

Once upon a time, there was an island.

Somewhere in the Far South.

The feeling of being at world's dawn.

A place where nature dominates the entirety of its winds, its ice, and sometimes its cold suns; its birds, its whales, all these lives that have adapted over millennia to these harsh conditions.

There, a small, fragile vessel.

And humans, awestruck, seeking to understand and adapt.

With pride that only their wake remains, closing behind them in a whisper so as not to disturb the beauty of the world.

Isabelle Autissier, Navigator

The Origin of Life

For all at last return to the sea—to Oceanus, the ocean river, like the ever-flowing stream of time, the beginning and the end.

Rachel Carson

In the beginning there was no sun, no moon, no stars. All was dark and everywhere there was only water.

Legend of the Maidu Native American Indians, Northern California

1 Shuyi Cao
Born in China in 1990

2 Arnold Fokam
Born in Cameroon in 1996

3 Claudia Müller Montes
Born in Chile in 1983

4 Maki Ohkojima
Born in Japan in 1987

5 Nyaba Léon Ouedraogo
Born in Burkina Faso in 1978

6 Ghizlane Sahli
Born in Morocco in 1973

7 Katherine Takpannie
Born in Canada in 1989

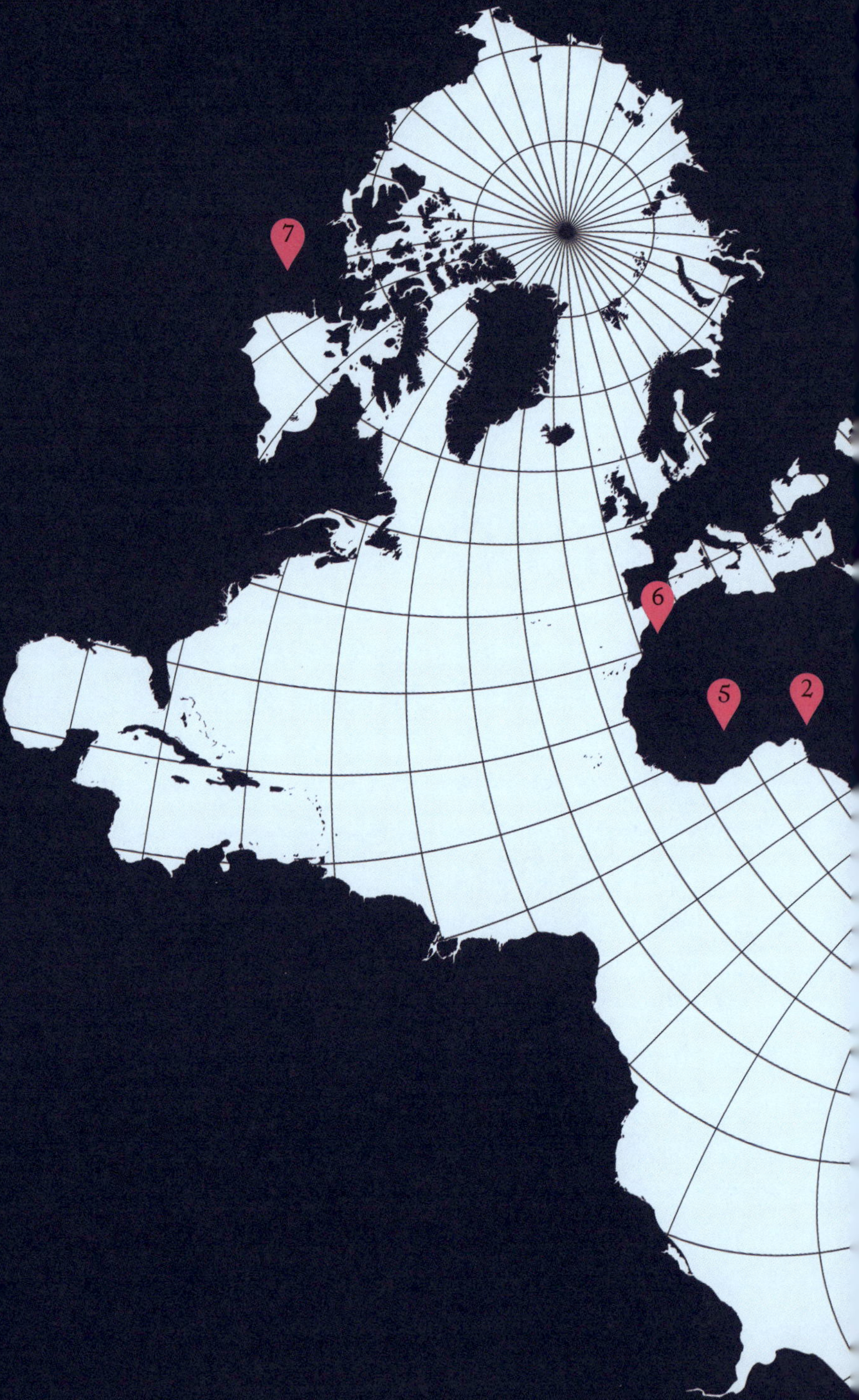

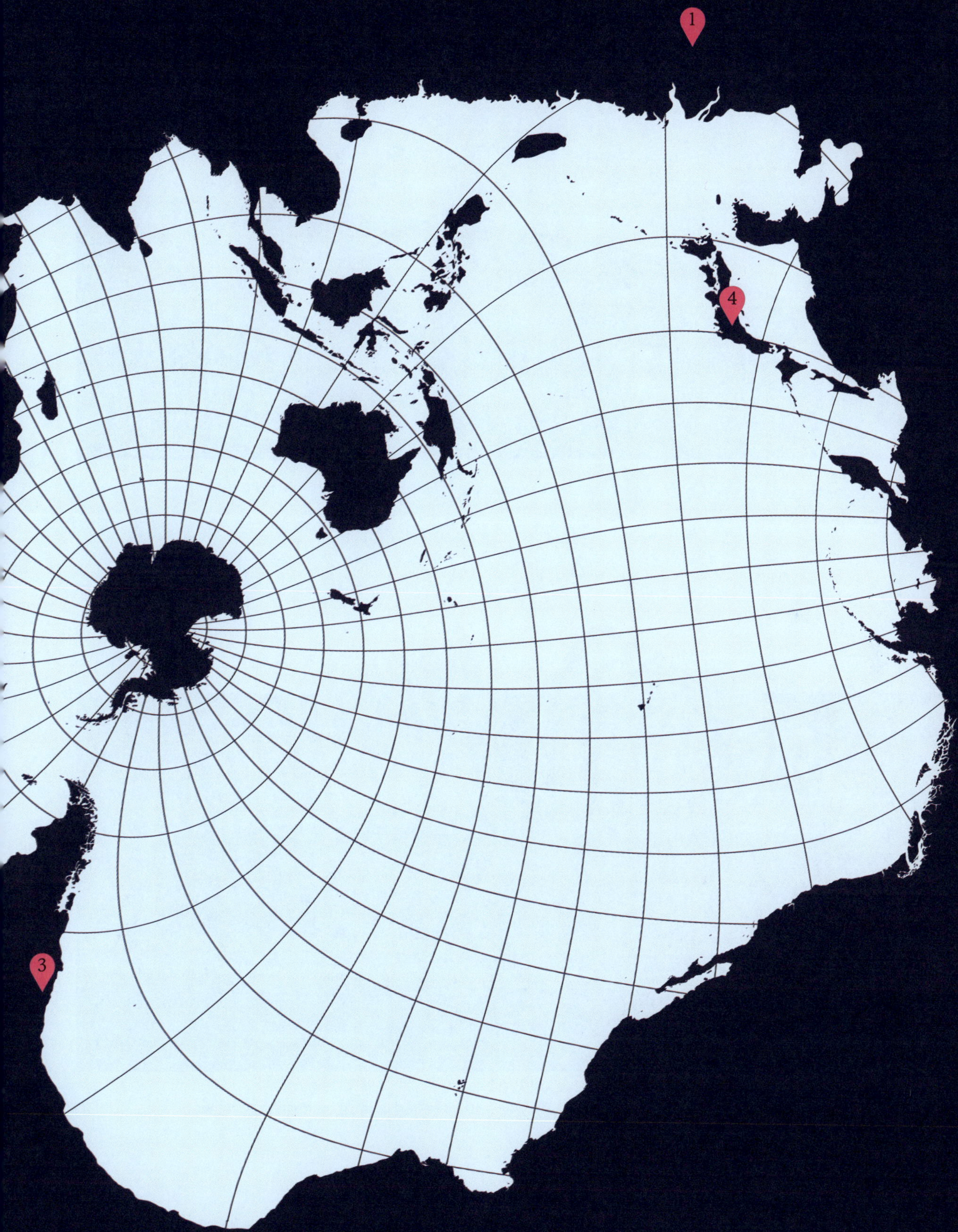
1
4
3

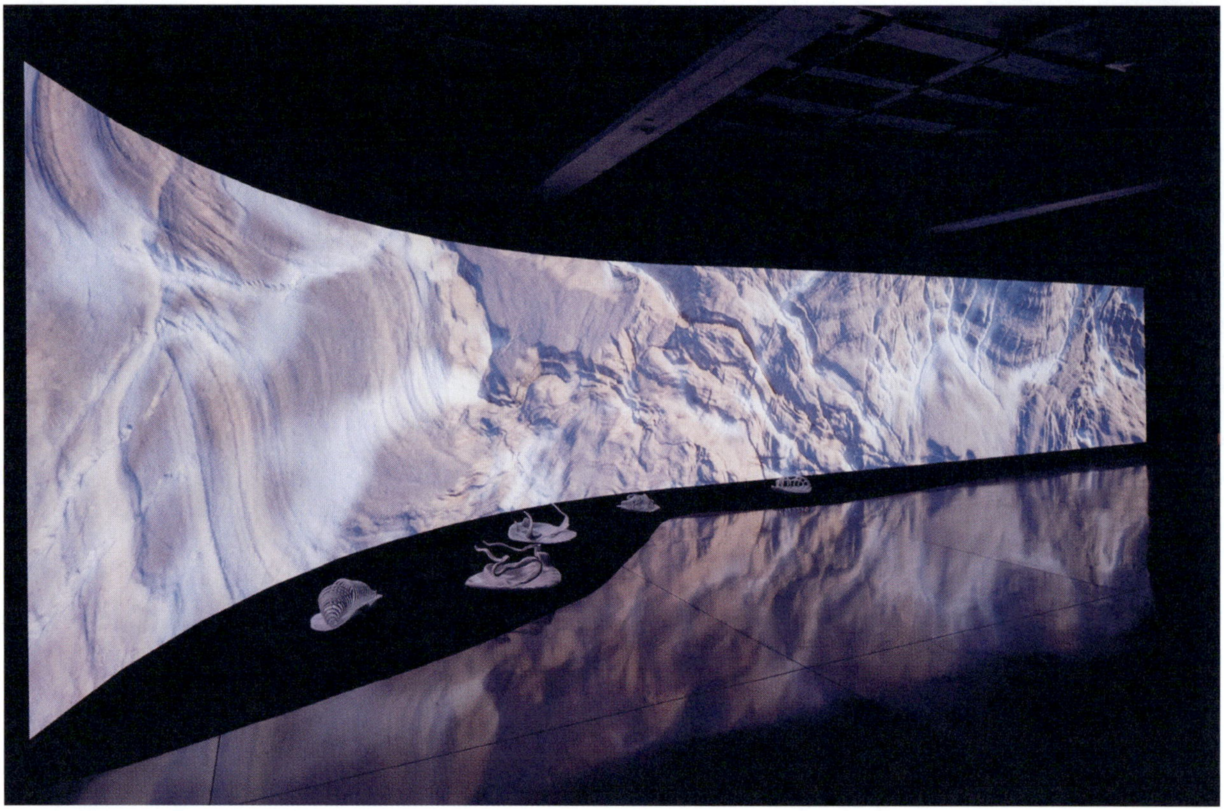

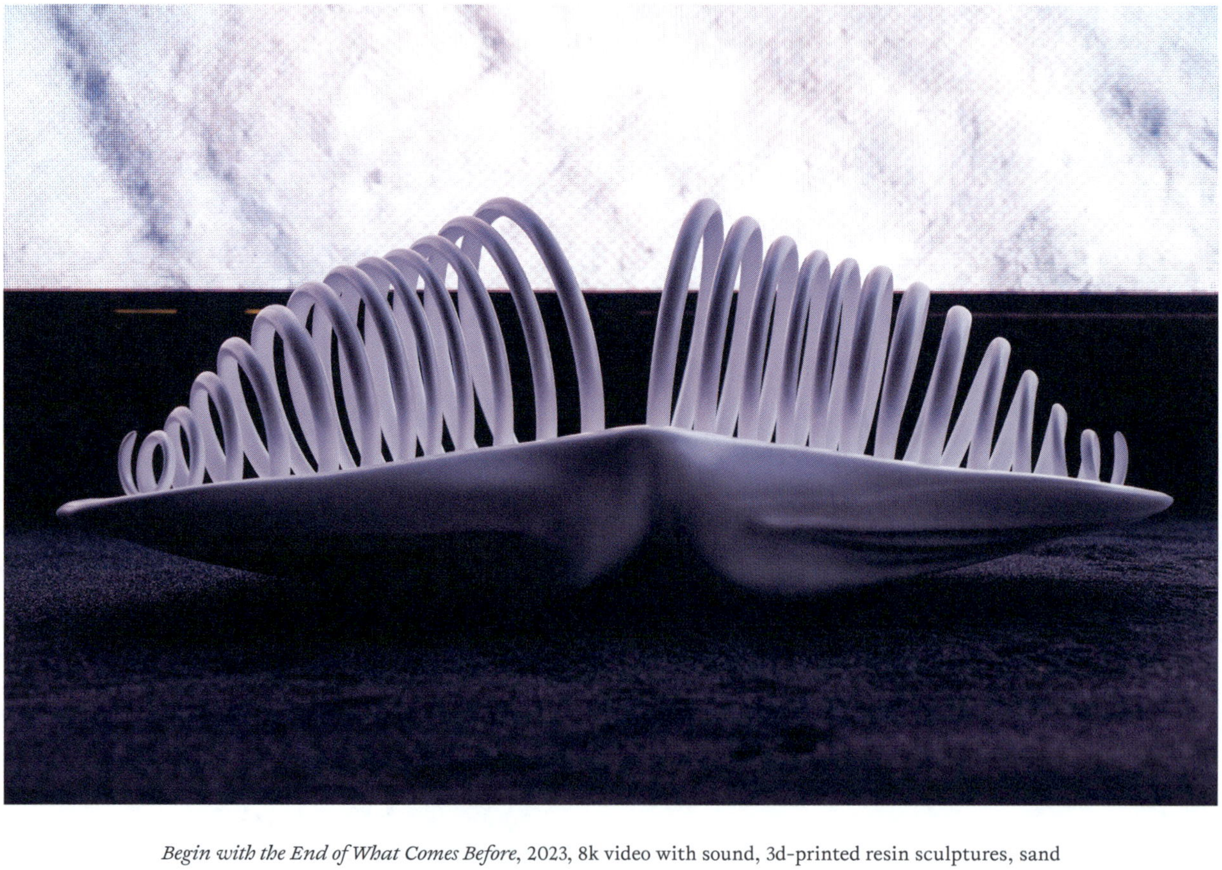

Begin with the End of What Comes Before, 2023, 8k video with sound, 3d-printed resin sculptures, sand

Shuyi Cao

Born in China in 1990

"To me, the ocean is an archive of life primeval and life to come. Many of my series contemplate such transcorporeal cycles across species, symbiotic becomings, and evolutionary times."

Shuyi Cao grew up in Guangzhou, in the southern part of China. At that time, during the 1990s, the sea was considered a primitive place. The focus was on ideas of modernization and urban development: "Our identity has never been shaped around the sea or surrounding islands. Chinese culture appears to prioritize the centrality of land while marginalizing the ocean." Cao's interests in the sea began when she was investigating landscapes that were previously submerged by the ocean. The geology of these sites is often still deeply marked by their oceanic history, allowing her to read past stories about climate and ecological changes. Inspired by the ever-changing relation between the sea and the land over long periods of time, her practice moves from a microscopic to a geological perspective as she evokes collective stories of the evolution of specific species.

"As an artist who was born and raised in a coastal city (Guangzhou) and has since migrated to another coastal city (New York City), over the past few years, tracing the stories 'offshore' has become my ongoing ritual of researching, re-searching, and re-positioning in erratic climates and discontinuous landscapes."

The *Foram* series (2023), for instance, deals with the single-celled marine planktonic organism foraminifera, one of the most primitive marine organisms that has been living in the open ocean since the Cambrian period. Forams secrete calcium and agglutinate sediment particles to form shells, producing thick layers of chalk, limestone, and reefs that function as markers of our evolution. Inspired by their multiple organic mutations, Cao has created digitally modelled sculptures based on machine learning-generated hybrid foram species, which she then sculpted by hand.

"In my works, the human and more-than-human beings are interconnected bodies of water that are under constant infiltration, exchange, and transmutation. This includes all forms of biological life–from animals, plants, fungi, and single-celled organisms to non-biological entities like marine pollutants such as microplastics and chemicals."

Cao's sculptures and sculptural installations offer a poetic, often ambivalent way to look at matter, living organisms, and the evolution of species. For her, the sea is both a primeval site for all forms of life and a site of constant regeneration and perpetual creation. It is, for instance, by studying various fossil forms in Canada and China that she began delving into the complex process of phosphatization, by which organic matter is turned into fossil. Her research always leads her back to the sea, as evidenced in her video *Begin with the End of What Comes Before* (2023), which addresses the deep transmutation between biological and geological matter through the circulation of phosphorus, from the transformations of marine organisms to today's issues of mining, extraction, and disposal of phosphorus.

Cao's practice is very much embedded in her own relationship with material, as she involves her body in her creative process. Although she bases her work on scientific research, fieldwork, and literary review, she increasingly involves herself in these embodied conversations, letting the materiel speaks for itself. "My belief lies in the matter itself, in our encounter." For *Of Shape Offshore* (2023), a series of sculptures made from diverse organic material such as seashell fragments, barnacles, sand, gravel, and fish bones, the artist plays with the idea of fossils to reconstruct a site of decaying organisms, inviting us to ponder the current evolution of living forms and our impact as human beings. Rather than praising any form of activism, she wishes to open up our imagination and to shed light on the complexity and multiplicity of the living.

Arnold Fokam

Born in Cameroon in 1996

Although he did not grow up near the sea, Arnold Fokam has always felt a profound attraction to water; it captivates and calls to him. This affinity was one of the reasons he chose to relocate to Douala, seeking proximity to the ocean. However, climate change has inflicted damage in this region that extends beyond visible effects. For him, its impact is palpable, affecting both their physical environment and collective psyche: "What was once a vibrant vacation destination in my childhood memories has been transformed into a desolate and unwelcoming place. The mangrove forests, which previously served as a natural barrier against severe weather and tumultuous sea waves, are now deteriorating or being exploited unsustainably. As these vital ecosystems decline, a sense of spiritual desolation emerges, leaving behind an emptiness and a deep-seated yearning to escape from home."

In Fokam's paintings, men and women are often depicted in shades of blue, their bodies filled with small bubbles, jellyfish, and corals. Although they appear to originate from an alien world, they represent real human beings engaged in their daily lives. This transparency, for the artist, serves as a metaphor for their inner lives and highlights the deep connection between humanity and nature, particularly the marine world.

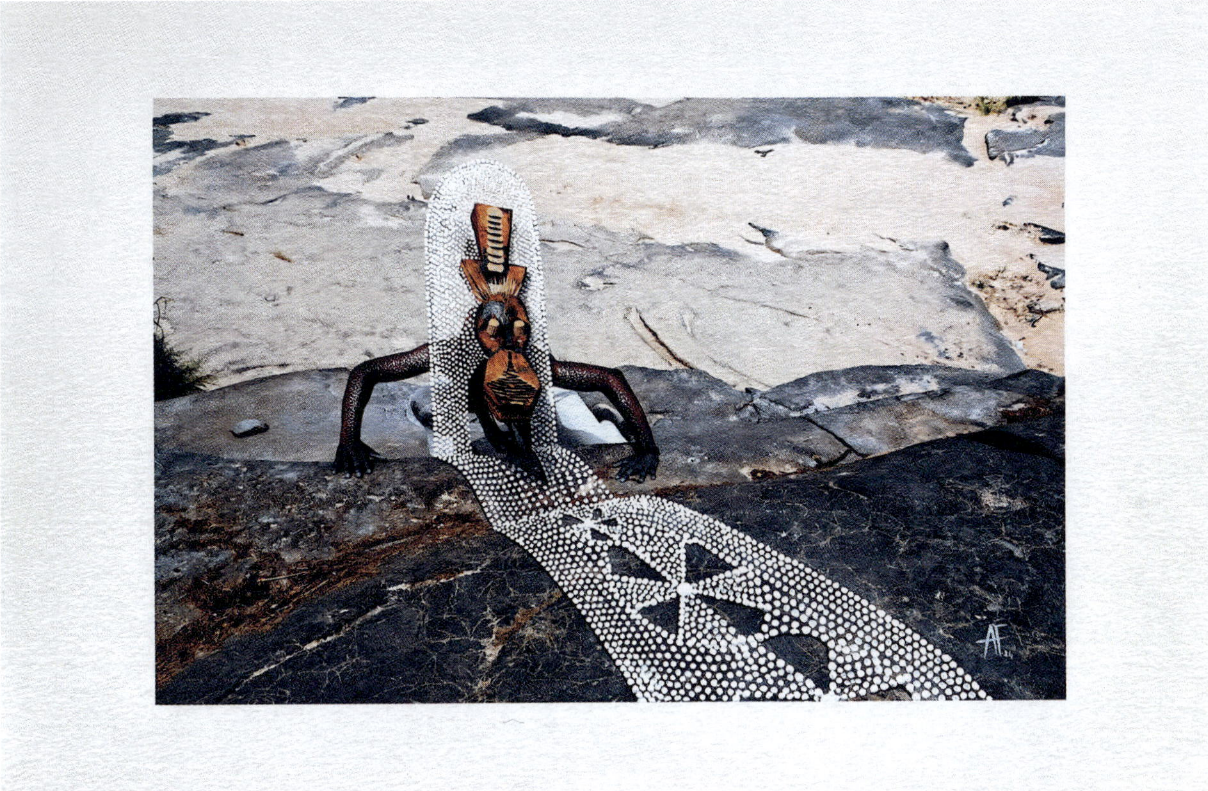

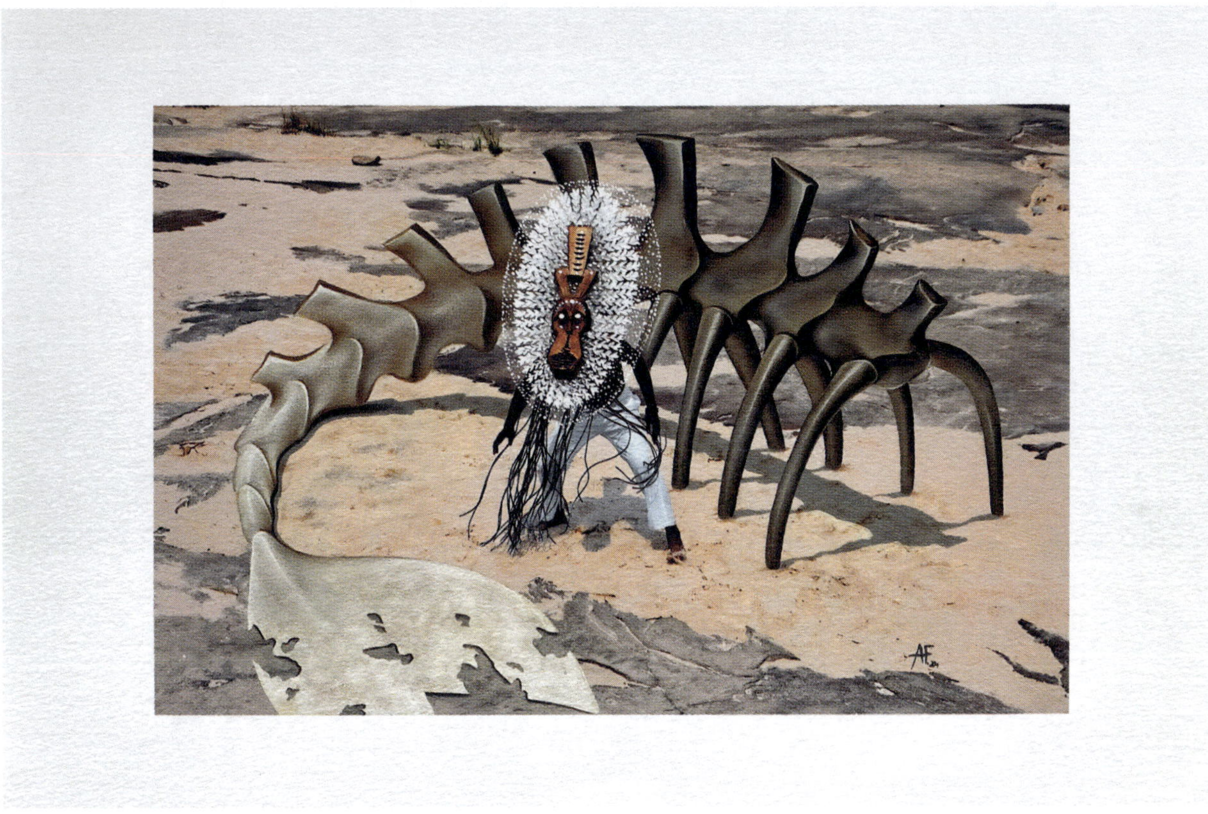

Processions Series, 2024 , Acrylic painting on inkjet prints, 30 × 40 cm each

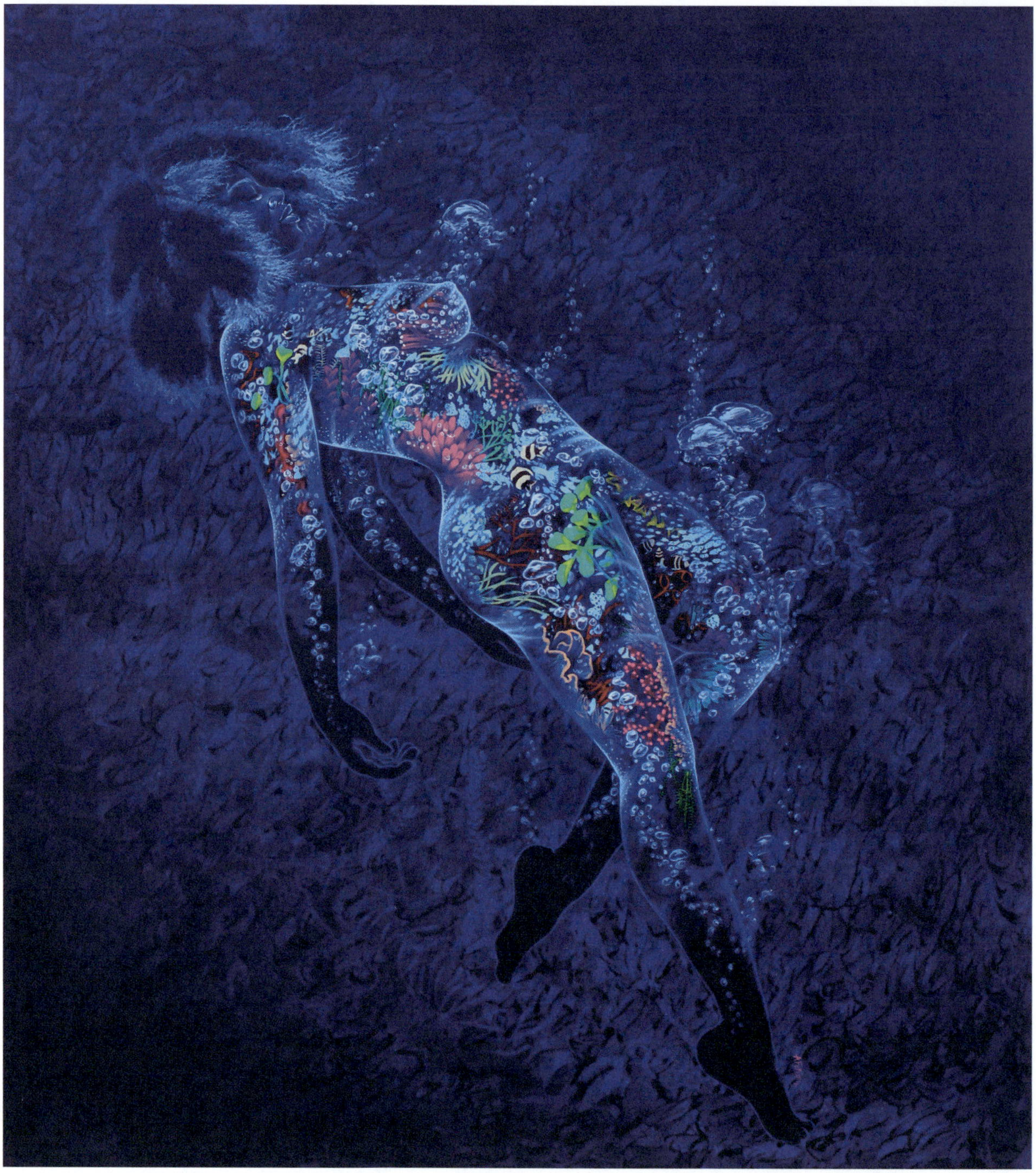

Immersion, 2020, Acrylic on canvas, 160 × 140 cm

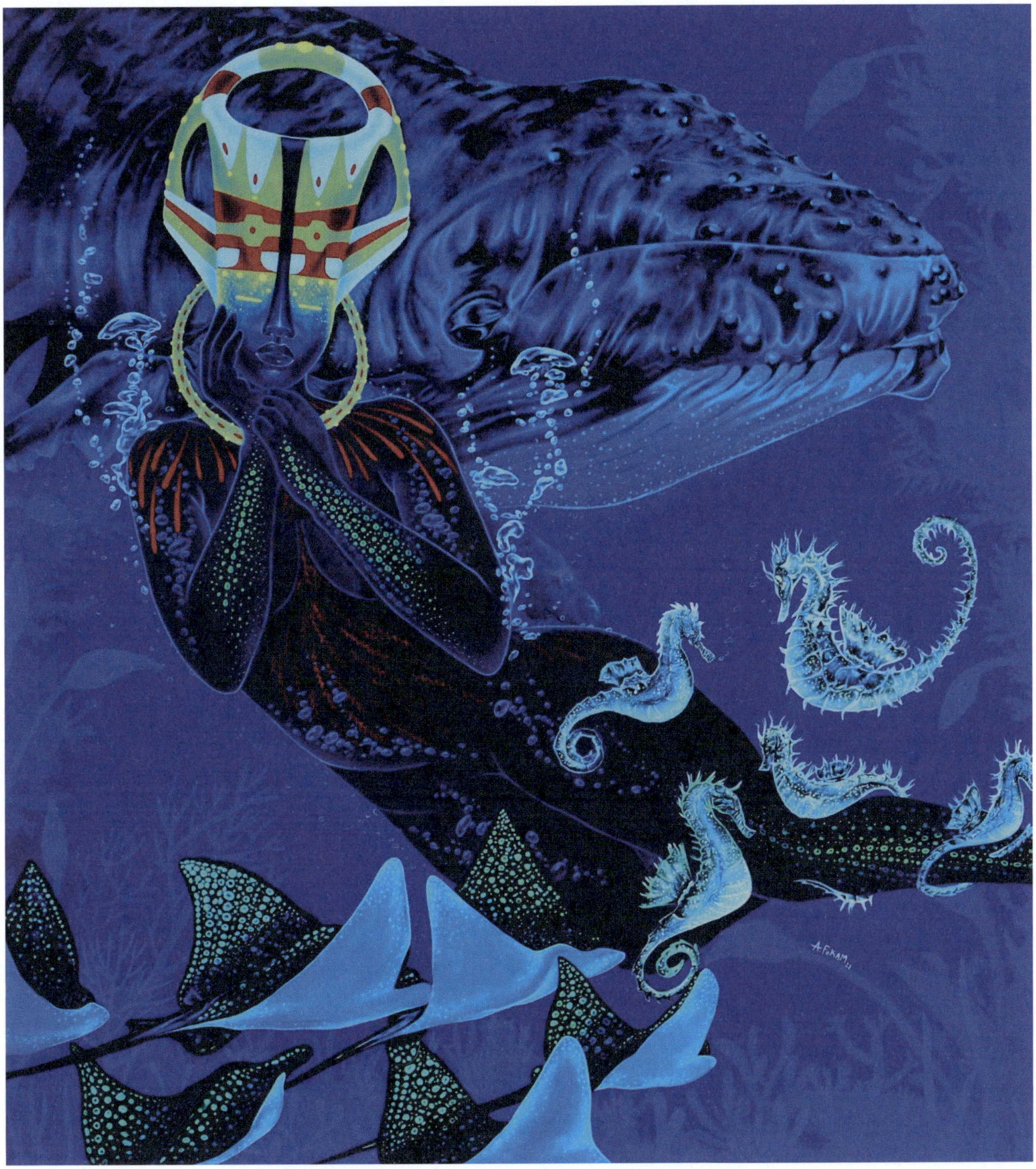

The Goddess of the Deep Blue, 2022, Acrylic on canvas, 160 × 140 cm

Not only is the human body composed of approximately seventy percent water, but all life is believed to have originated from the sea. Beyond the physical form, Fokam seeks to trace a shared genealogy from the primordial fish swimming in ancient oceans to the first creatures that walked on land. For him, the color blue also evokes "ndop", a traditional fabric from western Cameroon, used by his native people, the Bamileke. This textile, classified as national heritage, is typically dyed in blue hues, symbolizing life.

In addition to his cultural heritage, Arnold Fokam's interest in water was sparked by his discovery of video works by the American artist Bill Viola, whose characters are often submerged in water, with the element flowing and surging in mesmerizing forms. This encounter inspired Fokam to explore the relationship between the human body and water, from not only to a physical but also a spiritual perspective. Over time, he developed the belief that the soul of the deceased might reside in the ocean, deepening his connection to this element. "Water is that part of me that I still don't fully know, a vehicle guiding my artistic explorations. Opened to the reality of the sea, this entity becomes for me an ancestral reflection of our deeper, forgotten selves–those parts of us yet to be discovered. We, like all living things, human and non-human, are bound to water."

In parallel, Fokam has produced artworks addressing the issue of access to clean drinking water in various African regions, particularly in Cameroon. In his paintings, individuals are frequently portrayed amidst scenes of flooding, clinging tightly to large jerry cans of drinking water as if they were life buoys.

The *Processions* series (2024) originated from an art residency the artist undertook in Brazzaville, along the shores of the Congo River, where he researched water mythologies in the context of today's climate crisis. Inspired by the spirit of Mami Wata, the local water spirit, Fokam created a crocodile mask, which he wore during a series of performances on the beach. Constructed from cardboard and recycled materials, the mask references both the spirit of nature and the contemporary issues of hyper-consumption and waste. These performances, documented in photographs, serve as a foundation for Fokam's creative process. Back in his studio, he reworks these images with paint, transforming reality by infusing it with his personal poetry and imagination. While his earlier works often depicted figures standing or submerged in the ocean, Fokam's recent paintings increasingly portray them marching in protest. Today, his figures emerge from the water to reclaim the rights and respect owed to nature.

"When I am wearing the crocodile mask, I aim to embody the spirit of the ocean associated with that mask. Through my performance, I seek to facilitate a connection to the essence of creation–the ocean–a realm where every pure soul resides before and after life. Mami Wata is integral to this expansive energy of creation. I offer my body as a vessel to the world and to the deities, allowing them to communicate through me."

Claudia Müller Montes

Born in Chile in 1983

Chilean artist Claudia Müller draws on her daily observation of natural forces to question their history, materiality, and associations, which she transforms through photography, video, and installation. Her practice aims at exploring how art can embody the dynamics of nature, mainly water, and the interconnectedness between all living beings.

Caroline Ha Thuc *What is your own relation with the sea, and when did you start paying attention to it?*

Claudia Müller Montes The sea and the ocean have always seemed like exceptional places on this planet to me; I think these bodies of water allow access to a new dimension different from the terrestrial one. Since childhood, I have been in contact with nature and water. I spent a lot of time in the mountains in Chile and travelled along the Pacific coast from northern Chile (border with Peru) to the south, Magallanes. Then, in my adolescence, I learned to dive by chance, informally. There, I discovered the different temporalities that exist underwater and how it is a fluid just like air but thicker, with a different gravity and a different time/space. I was able to observe how the algae move in the direction of the ocean currents and purify the waters, and also how the respiratory rhythm and heartbeat change when underwater. I could understand gravity through my own body weight and observed some fish whose colors mimic their environment. I was lucky to observe the ocean's flora and fauna from a free place. This exploration led me to want to delve deeper into this world. Water, its forms, and manifestations seem infinite to me, like a spiral that cyclically leads us into a dimension that transcends that of humans and connects us with the Universe.

CHT *The video* Liquid Border *(2021) takes us on a cosmological journey to the beginning of the universe. How did you get the idea of focusing on the history of water on Earth?*

CMM I have always wondered about the origin of water, how old the water we drink from the tap is, how old the drops of water in a lake, river, or sea are. I like to think about how the water that emerges from us ends up in the clouds through sweat. Water seems to me to be a dynamic entity, both terrestrial and extraterrestrial; thinking about its origin is also thinking about the origin of humanity. We, as people, are also dynamic bodies of water composed of the same minerals and elements.

CHT *Are there any specific Chilean mythologies about human beings and the ocean?*

CMM One of the myths I have worked on and researched is Lethe, the river of forgetfulness. According to Greek mythology, when people die, their souls drink from this river to forget and be reincarnated into a new life. →

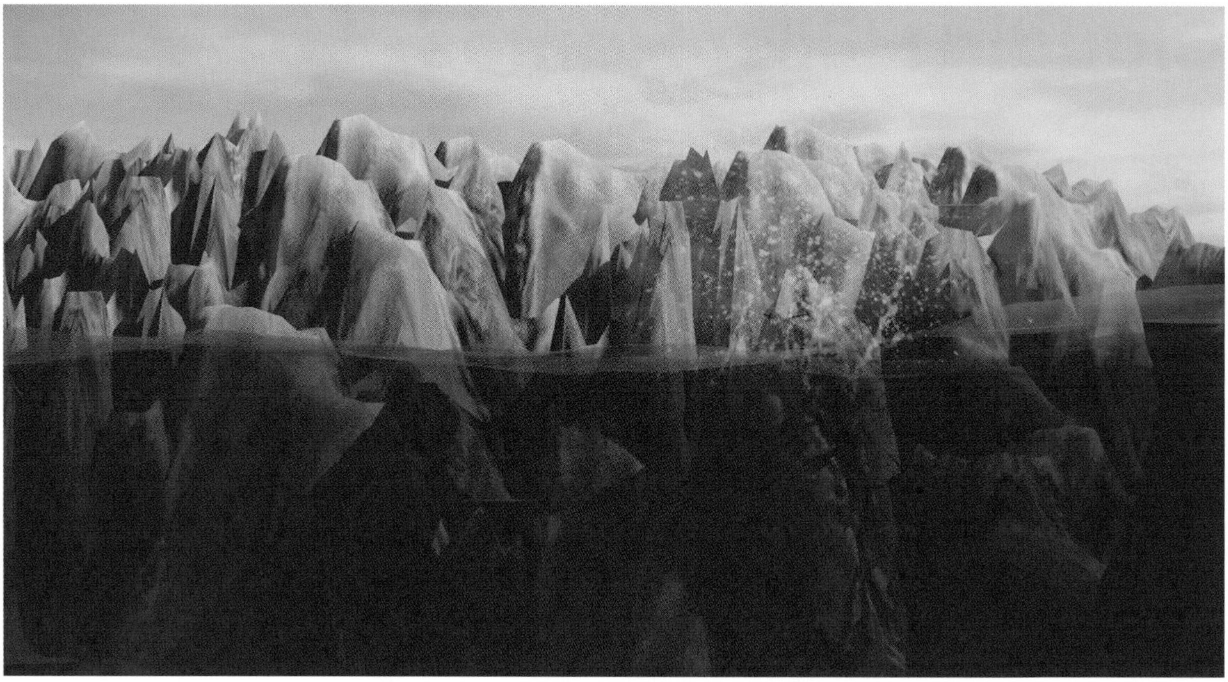

Hexa, 2021, Video, 3D, stop motion, waterproof image, 9 min 03

Similarly, Andean mythology treats bodies of water as beings capable of opening a portal to a spiritual world or the underworld. Healers use the rainbow to mediate between the world of the dead and the world of the living. Water is a gateway to the spiritual world and a source of healing.

In Bolivia, there is the myth of the goddess Jichi, an anaconda that inhabits freshwater lagoons. It was said that she was the spiritual guardian of the waters, and if people wasted or polluted the water, Jichi would leave the lake or lagoon, bringing famine and disaster to that population. With the arrival of the Spanish conquest and the Catholic Church, the anaconda came to be seen as a symbol of evil, and in a short time, their numbers dwindled significantly, leading to their extinction.

CHT *The images of your video are very poetic and take us underwater as if we were floating. Is this a way to bring us back to our fetal time?*

CMM It is a beautiful concept that through water we can return to the primal realm. We begin life breathing in water and emerge into the terrestrial world, crossing another boundary and portal: our mother's body. In *Liquid Border*, the temporality of water is depicted in the filming of a submerged forest, which continues its life in an aquatic manner. The forest is transformed underwater to continue its existence among mosses and plants that grow within and thanks to the

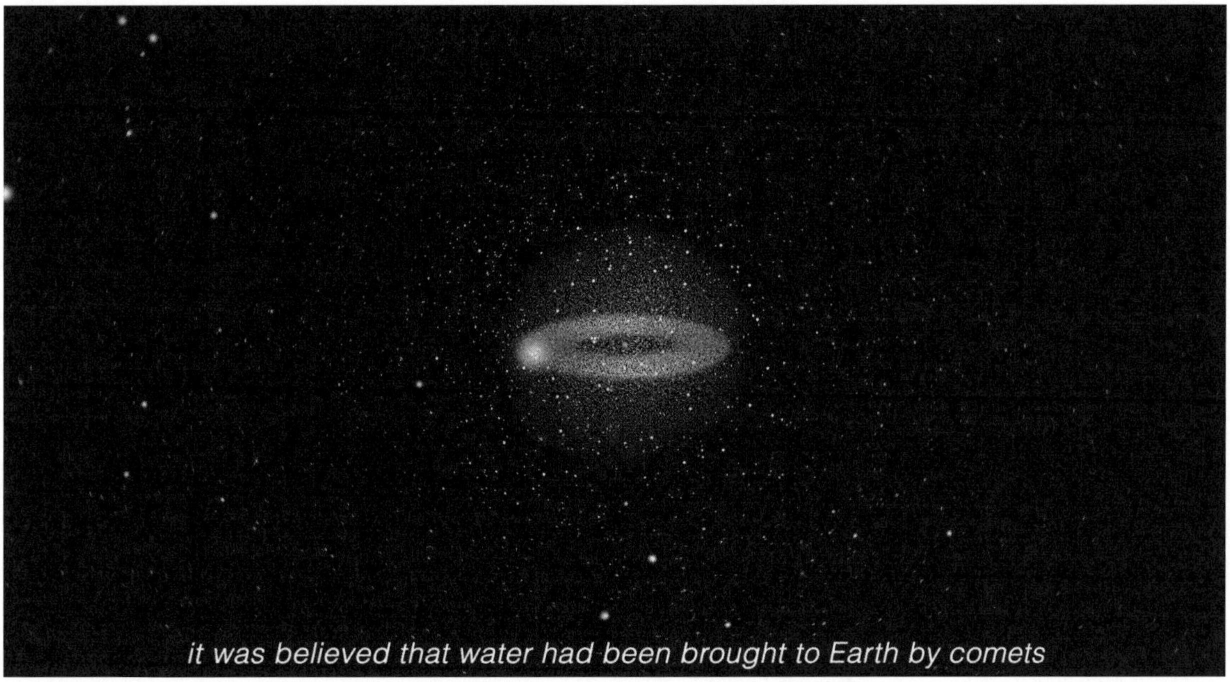

Liquid Border, 3D video, underwater filming, 12 min

water. In this hidden lagoon, there is a small flow of water that allows a gentle movement of the beings found there, traveling through the current like visitors in a still, silent forest, suspended in time.

CHT *Your video* Hexa *(2021) focuses on the melting of the Schiaparelli glacier, in the sub-Antarctic region of Magallanes, Chile. The sound plays an essential role, with heartbeats that suggest that the glacier is alive. Do you wish to play on a feeling of fear?*

CMM The idea behind *Hexa*'s pulse is to demonstrate that the glacier is a dynamic entity in constant advance and retreat and that these glaciers might become the dinosaurs of the future. The sensation I aim to convey is not fear, but rather the finite nature of time due to climate change, to raise awareness. We are a relatively young species compared to the age of the planet, and we have accelerated the circumstances of climate change to the point that, in a few generations, life as we know it may need to be transformed or perhaps even face extinction.

Maki Ohkojima

Born in Japan in 1987

I was dreaming.
In the sea, impregnated by the sea.
My form was like a fish. But the next day, my form was like a lizard.
Was it a dream? No, it was not just a dream. It was a memory.
I remember the landscapes I saw on my far-flung journey of life's evolution.
I know the remnants of our ancestors who have been with us for 3.5 billion years, generation after generation.
Those who impregnate the sea.

Initially trained as a painter, Maki Ohkojima has expanded her experimentation of artistic mediums to better explore human interactions with the natural environment and express their mutual connections. In 2017, she embarked on an extensive research journey on board the *Tara*, navigating through the Pacific Ocean. This expedition had a great impact on her practice. Since then, she has been researching oceanic ecosystems, with the whale emerging as a central symbolic figure in her work.

Ohkojima's 2019 installation, *Cave Mural as Narration, and the Story of a Whale Being Reborn as Oxygen*, exemplifies the artist's ability to translate scientific inquiry into artistic expression. During her time aboard the *Tara*, she witnessed a striking scene: the carcass of a whale floating in the ocean, surrounded by a thriving ecosystem of fish and birds feeding on the remains. Even after decomposition, the whale's bones continued to provide sustenance, forming a habitat for marine life. To Ohkojima, this spectacle encapsulated the cyclical nature of life, where all elements are interconnected and interdependent. She refers to the ocean as "the soup of life," a dynamic space where living organisms continuously transform, dissolve, and recompose within a symbiotic system.

For the creation of this installation, Ohkojima collaborated with the residents of Awashima, Japan, where the work was produced, and with three artists from the Warli ethnic minority in India, known for their expertise in traditional wall paintings. Together, they constructed a cave-like environment where the earth and ocean realms converge. The murals within this space narrate the story of humanity, with the skeletal remains of a whale serving as a metaphorical messenger, traversing temporal and spatial boundaries.

As she worked on the whale sculpture, Ohkojima reflected on the symbolic potential of the whale's bones, considering them as representations of the ocean in its solid form. She proposed that the bones store essential minerals that living organisms require for survival on land. This conceptual framework also underpins her 2022 installation, *Amniotic Sea*. The piece features a video projection onto a bed

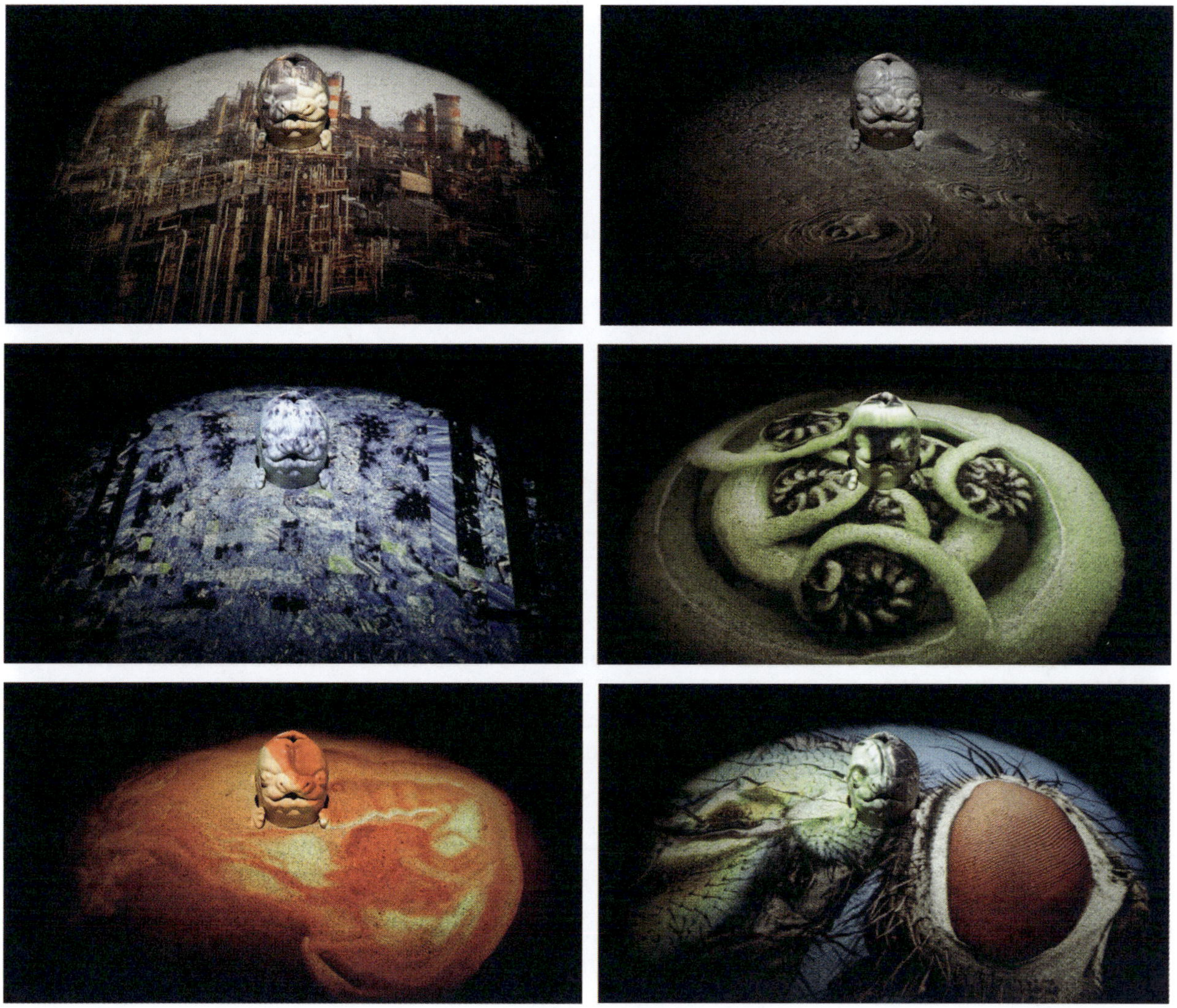

Amniotic sea, 2022, Ceramic, sand, projection, Video, 3 min 07

of sand, in which a ceramic fish head is placed. The projected imagery creates the sensation that the fish is undergoing constant transformation within what Ohkojima refers to as an "amniotic sea."

"I have just given birth this year and through childbirth I feel a renewed connection to the sea. The composition of amniotic fluid in mammals is very similar to that of ancient seawater. The baby was born through the sea in the womb."

For Ohkojima, the ocean is the primordial source of all life, including human beings. She draws upon the work of anatomist Shigeo Miki, who argues that the composition of mammalian amniotic fluid closely resembles that of ancient seawater. According to Miki, in order for mammals to transition from aquatic to terrestrial environments, they had to carry seawater–the "soup of life"–with them into the womb. He posits that the mammalian embryo, during its developmental stages, recapitulates evolutionary history, as it is transformed from fish to amphibian to reptile within the aquatic environment of the womb. "We carry the memory of the ancient oceans within us, along with the seawater itself," she posits.

Since 2023, the artist has been working in collaboration with Yosuke Tsuji.

Nyaba Léon Ouedraogo

Born in Burkina Faso in 1978

Nyaba Ouedraogo was raised in Ouagadougou, Burkina Faso, a landlocked nation in West Africa. For Ouedraogo, the first ocean is metaphorically the one within our mother's womb. Water is a primordial element; despite his native city's distance from the sea, he has always felt a profound connection to the ocean. "I am liquid," he says, emphasizing that his relationship with water is fundamentally physical. His reflections on the nature of water began during his work in Congo in 2011, where he recognized its deep spiritual significance, transcending the tangible realm. "All our cultures originate from water," he asserts. "Water, particularly the ocean, is associated with both life and death, serving as a source of fear as well as sustenance. In both contexts, it possesses a magical dimension."

Ouedraogo's experiences in Congo inspired his series *Phantoms of the Congo River* (2011), wherein he examined the connections between invisible realms, a theme he further explored through his reappropriation of traditional masks in *The Visible and the Invisible* (2020-2021). He also engaged in two years of research to investigate the mythological figure of Mame Coumba Bang, a water goddess revered by the people of Saint-Louis, a city situated at the confluence of the Senegal River and the Atlantic Ocean on Senegal's northwest coast.

The spirit and power attributed to Mame Coumba Bang are both formidable and elusive. Ouedraogo aimed to explore the goddess's imaginary world, seeking to understand contemporary perceptions of her. His research involved diverse interlocutors, including an anthropologist, a local Muslim theologian, and fishermen. One fisherman, named Samba, related, that the inhabitants of Saint-Louis

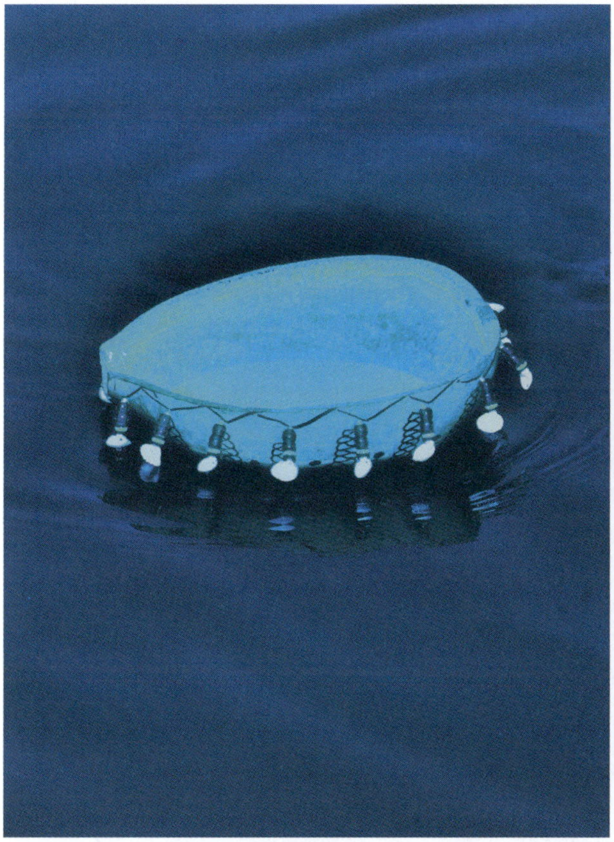

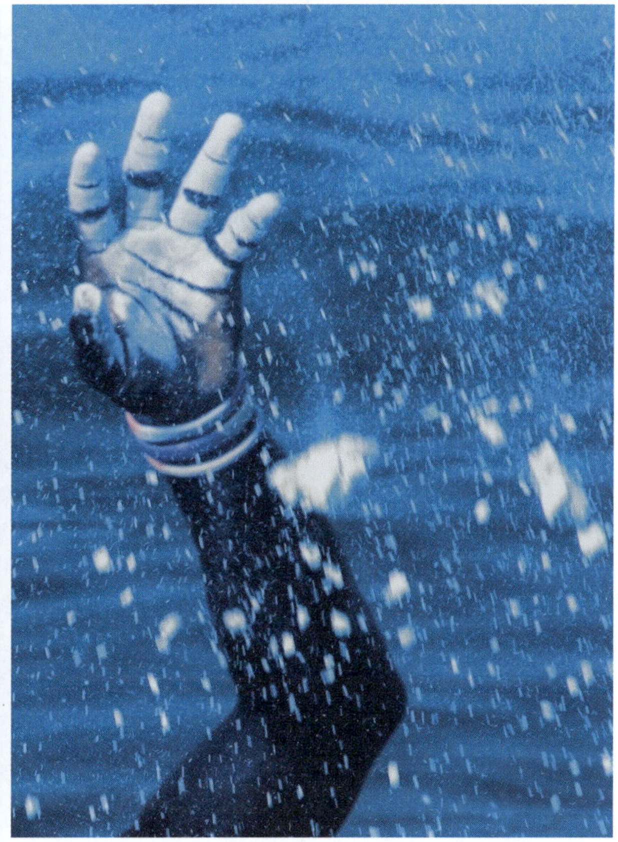

La calebasse & *La main aimante*, 2022, Baryta fine art printing, pigment inkjet, 70 × 50 cm

experience her presence within their own bodies. Perceived as omnipresent, she is able to manifest herself through any individual. For many, she symbolizes the spirit of Saint-Louis and embodies the collective narrative of its people. Gradually, Ouedraogo began to weave together these fragmented narratives in his pursuit of understanding the elusive figure of Mame Coumba Bang. Nevertheless, he acknowledges that she will always evade any attempts to fully capture her essence, suggesting that this exploration may indeed be a lifelong endeavor.

Two series emerged from Ouedraogo's research. The first series features fishermen from Guet N'Dar, a long island situated opposite the city of Saint-Louis in the Atlantic Ocean. The second series depicts a ritual performed by the eighty-year-old dancer Germaine Acogny as she offers curd milk to the water goddess on the banks of the Senegal River, near its estuary. The milk is presented in a calabash, a traditional vessel considered a divine gift due to its origin from a plant.

Acogny's gesture and offering resonate with various animist traditions

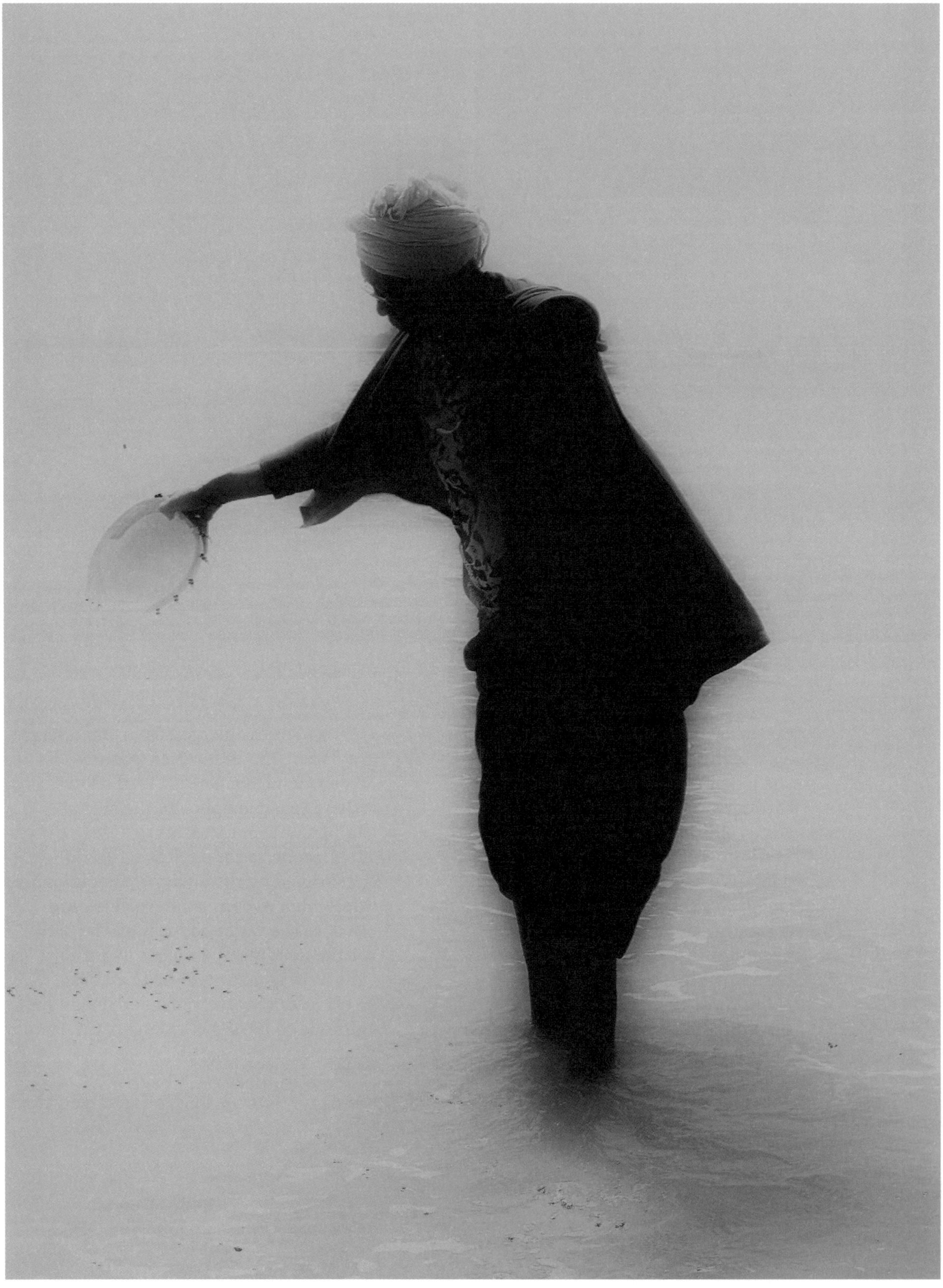

Germaine Acogny à l'embouchure de Saint-Louis, 2022, Baryta fine art printing, pigment inkjet, 70 × 50 cm

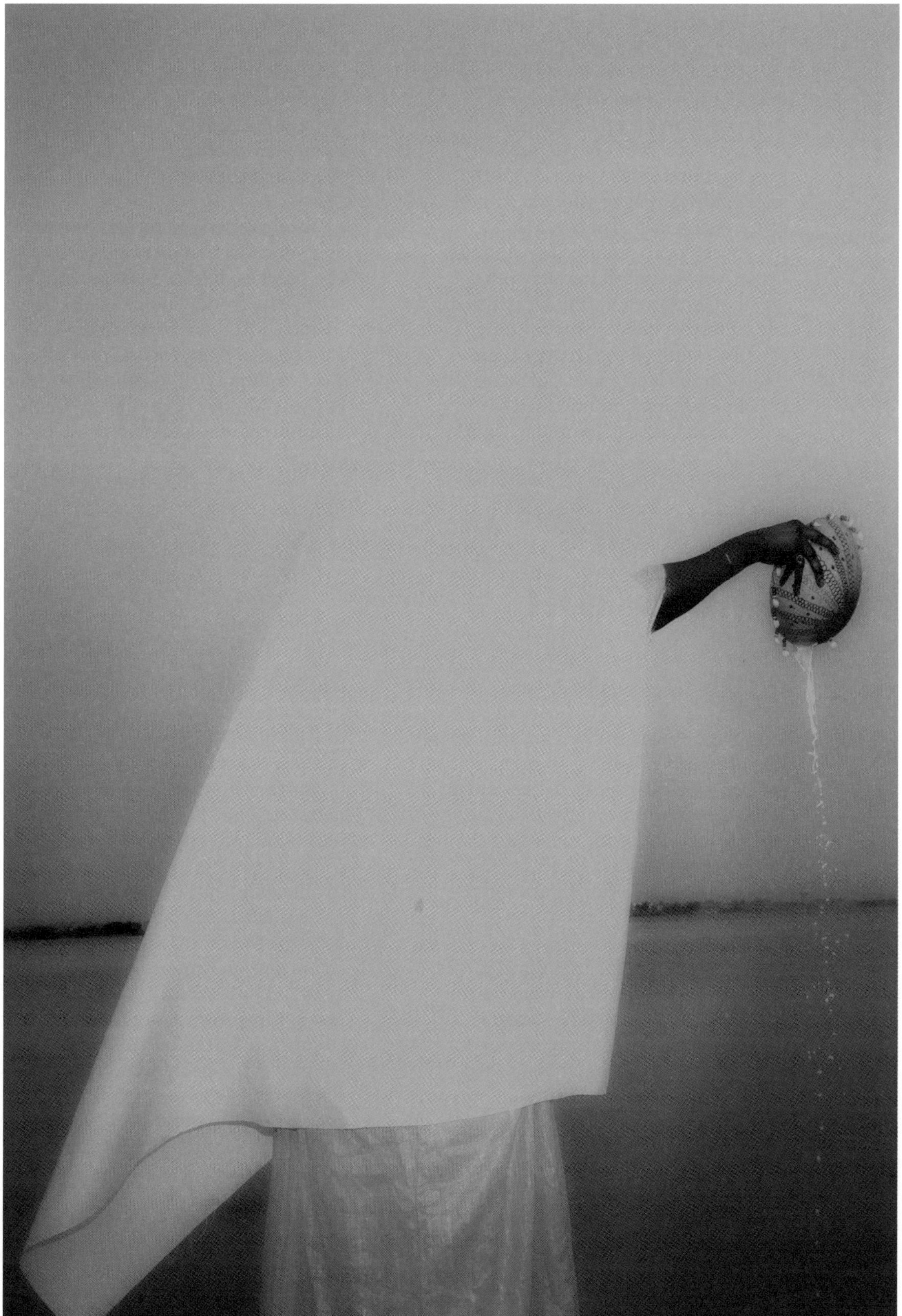

Offrande pour la déesse Mame Coumba Bang, 2022, Baryta fine art printing, pigment inkjet, 70 × 50 cm

in West Africa, a subject of significant interest to Ouedraogo. Having grown up in a family where animism was prevalent, he possesses an innate connection to his environment. In his artistic practice, he seeks to reflect on these mysteries and ancestral systems of thought, while also interrogating their relevance in contemporary contexts. Although he actively searches for answers, Ouedraogo does not aim to demystify the collective beliefs surrounding Mame Coumba Bang. Instead, his photographs lend a sense of tangibility to oral traditions, imbued with a poetic and evocative quality that defies any definitive form. The series is characterized by blue tones that enhance the atmosphere of mystery. Ouedraogo's photographic process itself remains enigmatic, operating as a kind of magical alchemy that expands the boundaries of reality.

The artist characterizes himself as a "griot of modern times." His practice begins with an exploration of reality, from which he departs to delve into imaginary realms. He perceives his work as part of a continuum with his ancestors, aiming to transmit and revitalize myths from a contemporary perspective.

Ghizlane Sahli

Born in Morocco in 1973

"I grew up in Agadir, by the sea, and the sea has always been present in my life. As an artist, I am interested in metaphors related to women, particularly the liquids of the female body. The connection to the ocean emerged naturally for me, as I perceive the ocean as a matrix–a nurturing force akin to a mother figure for all of humanity."

Sahli's first major installation was an artistic work created from waste materials. For the artist, waste is a particularly compelling medium. It is universal, and each discarded object carries its own unique history. Who drank from this plastic bottle? Where has it traveled, and for how long? We know that all oceans are interconnected, and it is possible to find plastic debris in one part of the world that was discarded in another. By working with such materials, Sahli brings together these individual narratives to form a new, collective story. The process also involves the gathering of energies. Each piece is composed of accumulated bottle tops, functioning as an accretion–an organic formation that grows and takes on a life of its own. Her sculptures are organic and circular in form, inspired by the varying shapes of the bottles she repurposes. Each bottle becomes a cell, a unit, that enhances the overall structure. "It's like looking at matter under

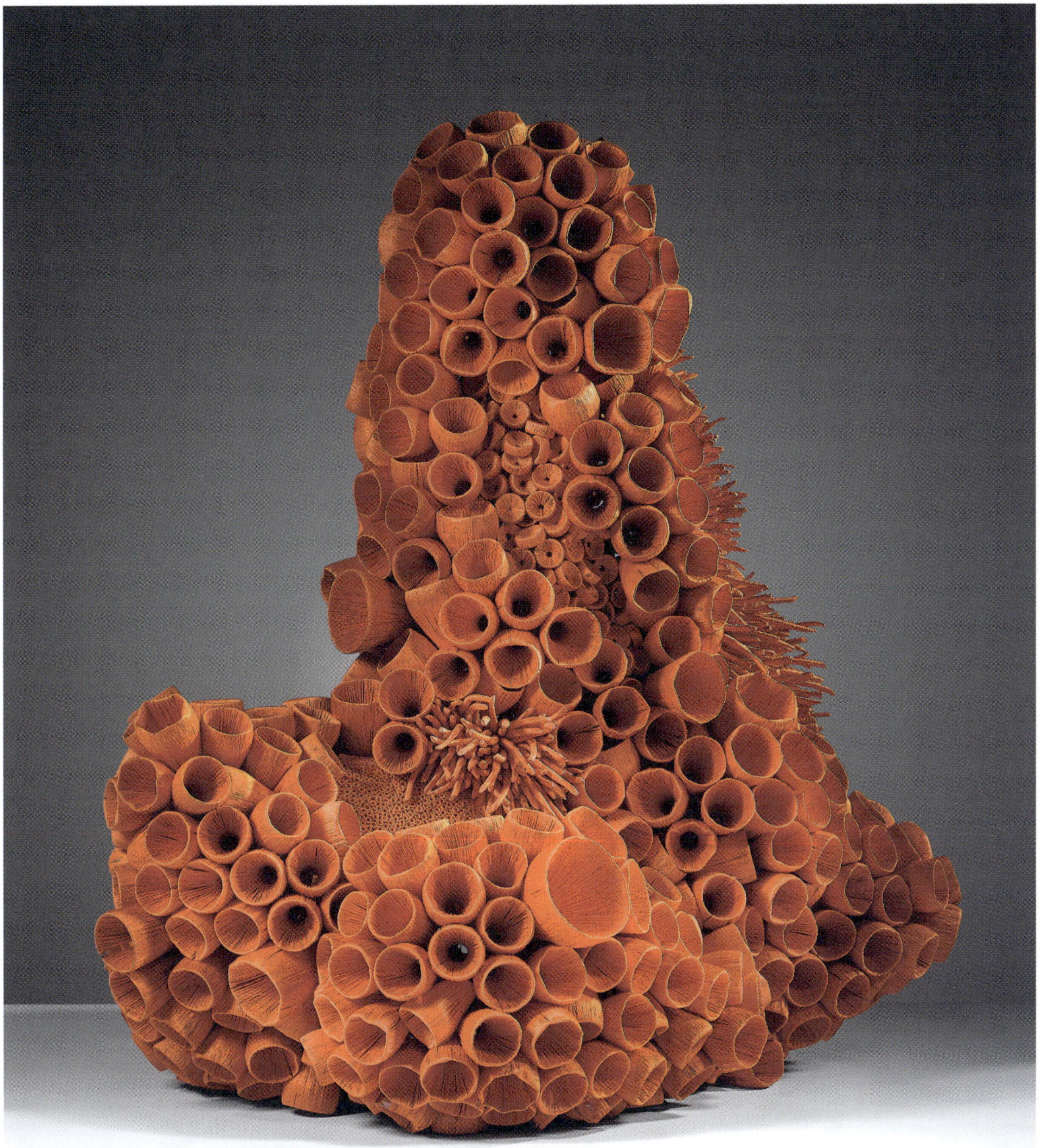

La Mer(e), Origine du monde, MOMS 002, 2020, Silk threads on plastic waste, 140 × 110 × 90 cm

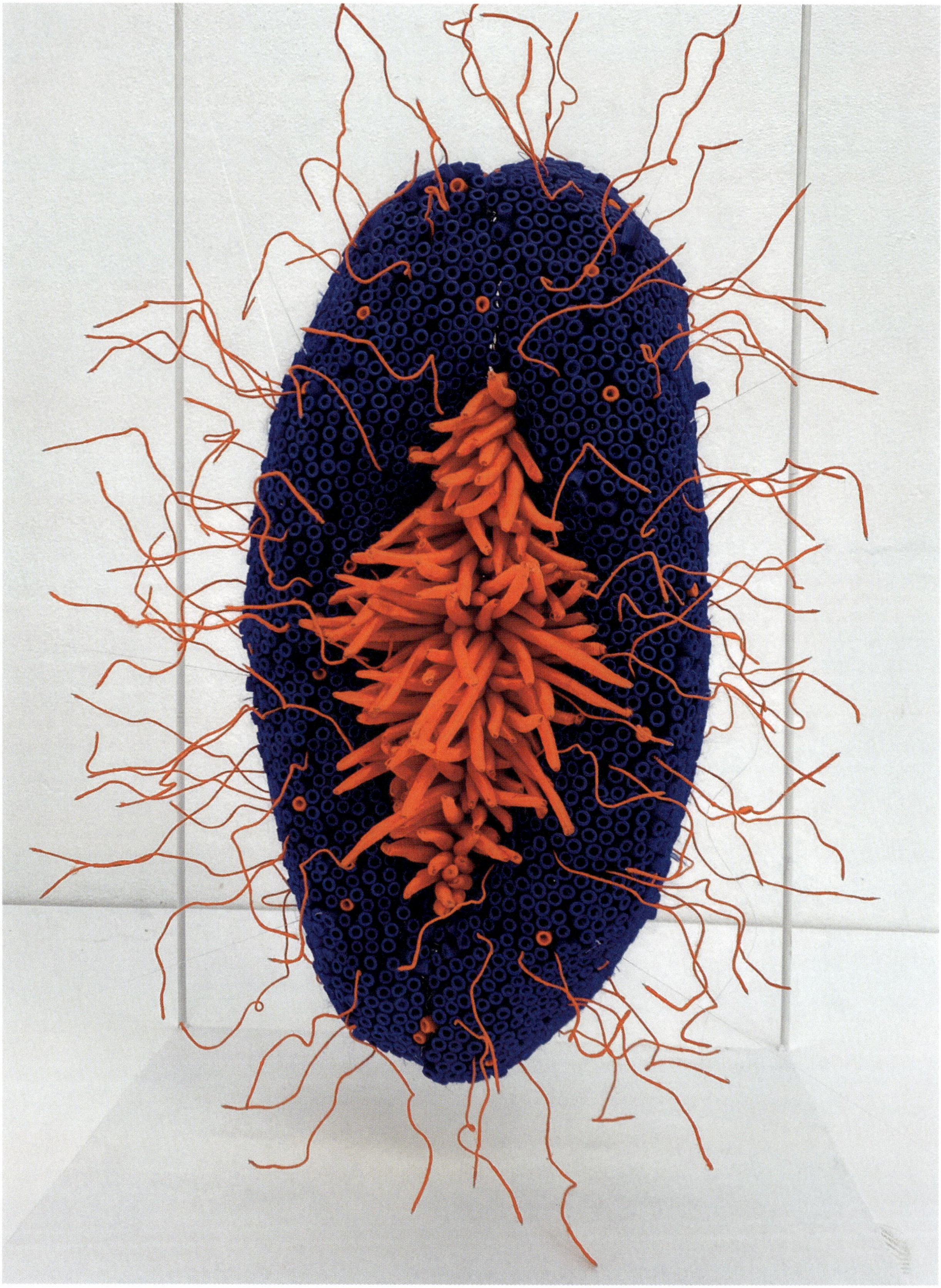

La Mer(e), Origine du monde, MOMS 001, 2020, Silk threads on plastic waste, 60 × 30 × 25 cm

a microscope," she explains. "You begin to recognize the unity of all things—the human body and the ocean. Everything is interconnected."

Sahli's work also incorporates a strong spiritual dimension. She views the energies within these materials as souls that come together and are reanimated. Moreover, the slow and meticulous process of covering each bottle with silk threads is inherently meditative. For Sahli, the repetitive nature of this task resembles the recitation of a mantra. In reconnecting with her African heritage, Sahli discovered the goddess Mami Wata, whose influence resonates across the African continent. This cultural syncretism is a driving force in her work, which is enriched by a diverse range of cultural inputs. Sahli's extensive travels and experiences across multiple continents, along with her background as an architect and her integration of artisanal techniques, further enhance the depth and complexity of her artistic practice.

Although Sahli initially focused on embroidery, it was in 2016 that she fully embraced her identity as a visual artist. Nevertheless, she continues to collaborate with women from her community in the creation of her works. She has also mobilized those around her to collect plastic bottles, forming a collective known as the "Zbel Manifesto" (with *zbel* meaning "waste" in Arabic).

Her series *La Mer(e): Origine du Monde* (2020) is a play on words, as the French words for "sea" (*la mer*) and "mother" (*la mère*) are homophones. This series of sculptures evokes dreamlike, primordial forms of life—creatures that might be imagined lying at the bottom of the ocean. The color palette is simple yet striking, composed of bright blue, orange, and white tones that evoke a phosphorescent quality, as if glowing in the dark depths of the oceanic abyss. Through these creations, Sahli challenges our conception of human beings—not as isolated individuals, but as entities deeply connected to the natural elements and the environment.

Katherine Takpannie

Born in Canada (Inuk) in 1989

A self-taught photographer, Katherine Takpannie's practice is deeply grounded in her Inuk culture, with her family originating from Nunavut. Her name, "Takpannie" means "up there." "My grandfather used to say, 'we're from up there' (meaning, the Arctic)." Photography and performances allow her to explore her own identity as an urban Inuk and to document the history, culture, language, and know-how of urban Inuit life. In particular, she wishes to shed light on the current issues that Indigenous Canadians face daily, including the impact of climate change and increased local industrial mining.

Sedna | ᓴᓐᓇ (2021) features a siren, half woman, half fish, lying on the ice crust and on the shore of a desert place. She looks at a distant horizon and seems to wait for something or someone to come. Below her body, the ice breaks and the snow melts. She embodies the Goddess of the Sea, known as the mother of all marine mammals in the Inuit culture, from Greenland to Alaska. The series was created as a response to the artist's question: "What will Sedna do when she hears the seismic testing?" In these embodied representations, Takpannie draw on the epistemology of Inuit societal values to raise awareness about the violent disruption of her Inuit homelands through resource extraction.

With *Sedna* | ᓴᓐᓇ (2023), the artist continues to explore the complex relationships between Inuit culture and the land/sea. "Within Inuit cosmologies and epistemologies, uncertainty, unpredictability, and change are foundational. The world is understood to be ambiguous and to elude full comprehension and thus change is both expected and accepted." However, the artist adds, this philosophy of life does not imply the changes triggered by man. The term "climate change" in Inuktut

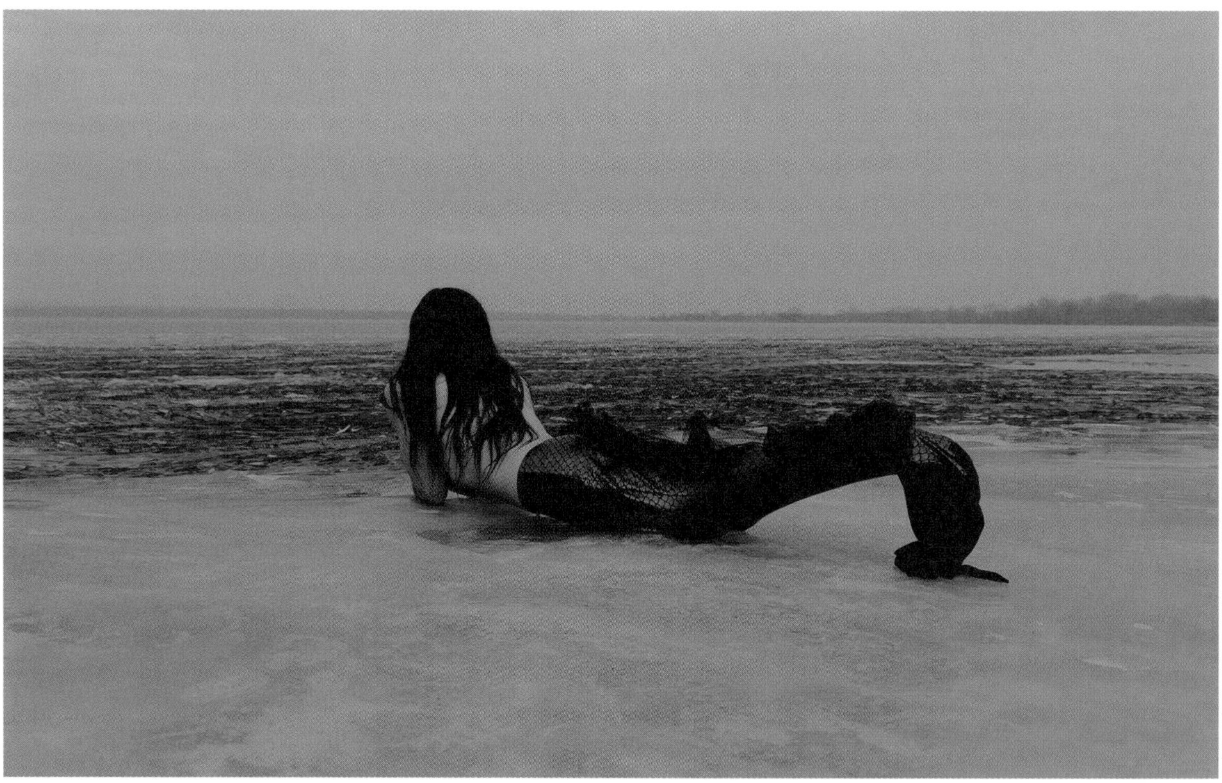

Sedna / ᓴᓐᓇ *: On Thin Ice, #8*, 2023, Archival pigment ink print, 60 × 90 cm

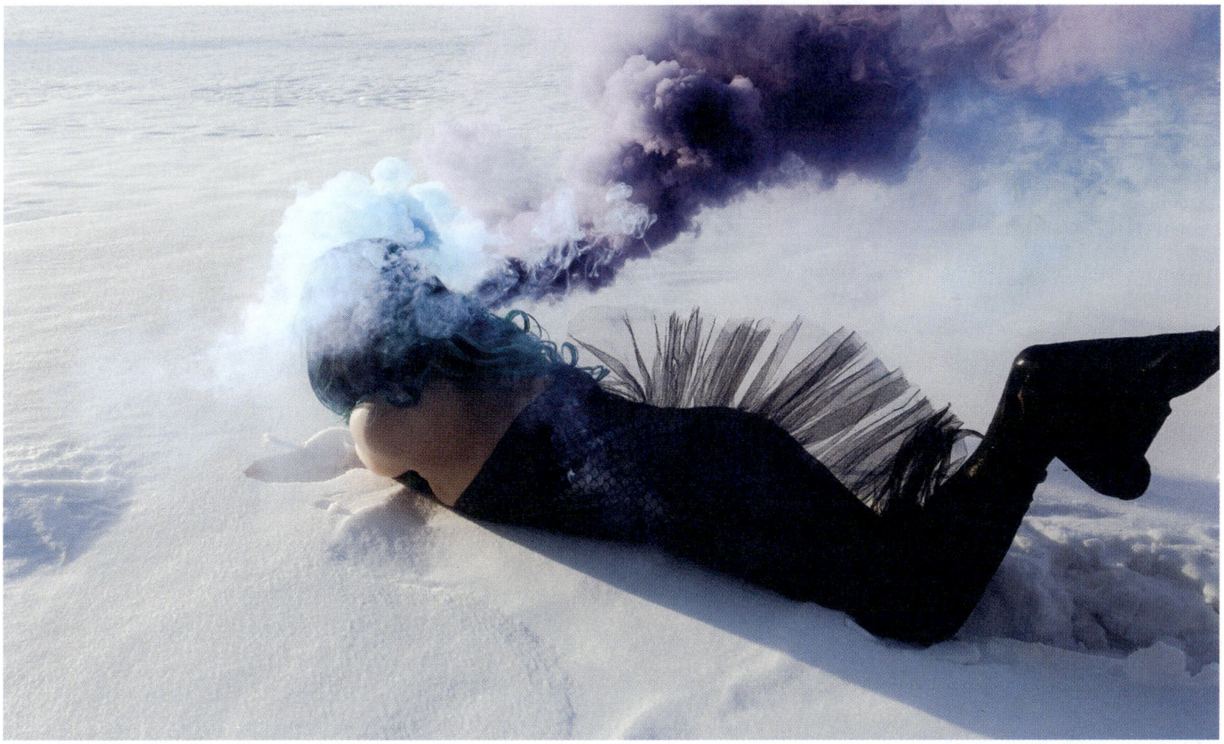

Sedna / ᓴᓐᓇ #7, 2021, Archival pigment ink print, 60 × 90 cm

is *silaup asijjiqpallianinga*, which can be translated to "Sila being made to change." In the Inuit culture, Sila is a fundamental concept that pervades the daily life of the people. It englobes the air, the wind, the atmosphere and weather, and it is what dominates the life of the community.

"For centuries, the Inuit have maintained a close relationship with ice (*siku*), land (*nuna*), sky (*qilak*), and wildlife (*uumajut*). We have cared, respected and honoured all life. Until we couldn't. Our new neighbors didn't value Sila and Nuna as we did. So, in the midst of humanity's collective climate crisis, Inuit lives in the Arctic are increasingly vulnerable.

"Reoccurring observations from Inuit elders, hunters, and community members show us that our *aniuvat* (permanent snow patches) are decreasing in size. Sea ice conditions have changed. The temperature has increased. The length and timing of the traditional Inuit seasons have changed. Inuit perspectives on climate change are built on lifetimes of observing the environment. For us, sea ice is a fundamental source of food, learning, memories, knowledge and wisdom. Climate change impacts the way we live, interact, and know the world."

Sedna warns: "We are on thin ice."

Olivier Poivre d'Arvor

Special correspondent of the President of the French Republic for the United Nations Ocean Conference, French Ambassador for Poles and Oceans (Paris)

A place of infinite marvels, an inspiration for magnificent works, a realm of mystery and adventure, the Ocean is now in danger. It is insufficiently protected, subject to rival conquest strategies, the victim of overexploitation by humans oblivious to the need for its preservation, it nevertheless plays an irreplaceable role in regulating the warming process that increasingly threatens global ecosystems.

Is the awareness of its vulnerability and its importance sufficiently shared? Admittedly scientists, governments, and international organizations are certainly working to spread the message and find solutions, but for people, in all their diversity and in their different living environments, appropriate them, artists from all disciplines have a preeminent responsibility.

Their ability to create dreams and to showcase the vital force of marine realms reveals their beauty, their rarity, the risks they face, but also the forcefulness of the movement that campaigns for their urgent preservation.

Visual arts, dance, theater, music, cinema, photography... all cultural initiatives must be mobilized to this end. The artists whose works are presented here, and their contributions, illustrate a vision of the Ocean that is not reduced to a technical dimension, but which stimulates our imagination, with an invitation to share the messages as widely as possible.

The voices of the artists and their perspectives are necessary as they persuade us that despair is not the appropriate response, nor is the feeling of failure or of a catastrophe foretold. The prospects for increased protection, and sustainable resource management, are within reach: concerted international action is needed in the short term; we owe it to future generations and to the marine ecosystem.

To this end, several lines of work are being pursued simultaneously at the international level and some appear to be chief among them.

Governments and international organizations must be galvanized to strengthen political, individual, and collective commitment and the undertaking of ambitious, rapid actions to support the implementation of the 2030 Agenda for Sustainable

Development, in particular SDG14. This entails defining monitoring mechanisms to ensure the effectiveness of the normative processes adopted on a voluntary basis.

Global scientific cooperation certainly exists, but the knowledge available must be better shared, [better disseminated,] and better used. New, effective tools, such as indicators for measuring the health of the Ocean, a platform to pool experiences, and exploratory missions will facilitate the adoption and development of scientific knowledge, which is known to play a major role in the protection of the sea.

No viable project can be considered without appropriate funding therefore the international community must increase investment in ocean research and strengthening marine technologies. Public funding and contributions from companies are needed to support a blue economy, the more desirable it is, the more sustainable it will be, and the "return on investment" will appear to be extremely positive in its medium- and long-term impact.

Major international events, such as the United Nations Ocean Conference, organized by France and Costa Rica to be held in Nice in June 2025, are beneficent occasions to formalize commitments and raise funds, highlight research results, contribute to strengthening the networks created by civil society, which are rightly concerned about the dangers that threaten, blow the whistle, and spur on decision-making processes.

Here again, the voices of the artists must be heard alongside political, scientific, and financial discussions as they resonate all the more strongly carrying a message of hope and call for widespread mobilization in support of the Ocean.

Such international events strengthen our belief that the conservation and defense of the Ocean compel us. They also remind us that the power of dreams and the abundant imagination of artistic creation often draw their origin from the unfathomable depths of the sea.

The Sea as a Collective Memory

The ocean separates us, us, our lonely islands, scattered in the Indian Ocean.
Woman, scarred island, sister island, woman with gills, Zazavavindrano. Where are you from? Are you truly my sister? I hunger for your journey, your story fascinates me.
You left, marooned.
I dream of catching a glimpse of you beyond the broken blue line.
I watch, I peer, and I imagine your hair undulating all the way to the shore of my broken body ... feet, pebble, thighs, ravine, hips, mountains, sex, spring, arms, river, head, peak.
Then the sea roars, black, oily, and the scarred ocean opens its belly and sings.
I am the woman of the colonized, of the colonizer, of capitalism, of patriarchy, of assimilation, of power, and of conflict. And yet I come from a long lineage, from a submerged continent populated by mirages and countries.
Mozambique, Tanzania, Madagascar, Comoros, Juzu Al Quamar, the islands of the Moon, Réunion, Mauritius, Seychelles, and then India, Pakistan, and then ...
My liquid roots anchor themselves to the navels of many lands.

Magalie Grondin

1 Dineo Seshee Bopape
Born in South Africa in 1981

2 Marina Gioti
Born in Greece in 1972

3 Magalie Grondin
Born in La Réunion, France in 1972

4 Sithabile Mlotshwa
Born in Zimbabwe in 1975

5 Enrique Ramírez
Born in Chile in 1979

6 Lisa Reihana
Born in New Zealand in 1964

7 Bao Vuong
Born in Vietnam in 1978

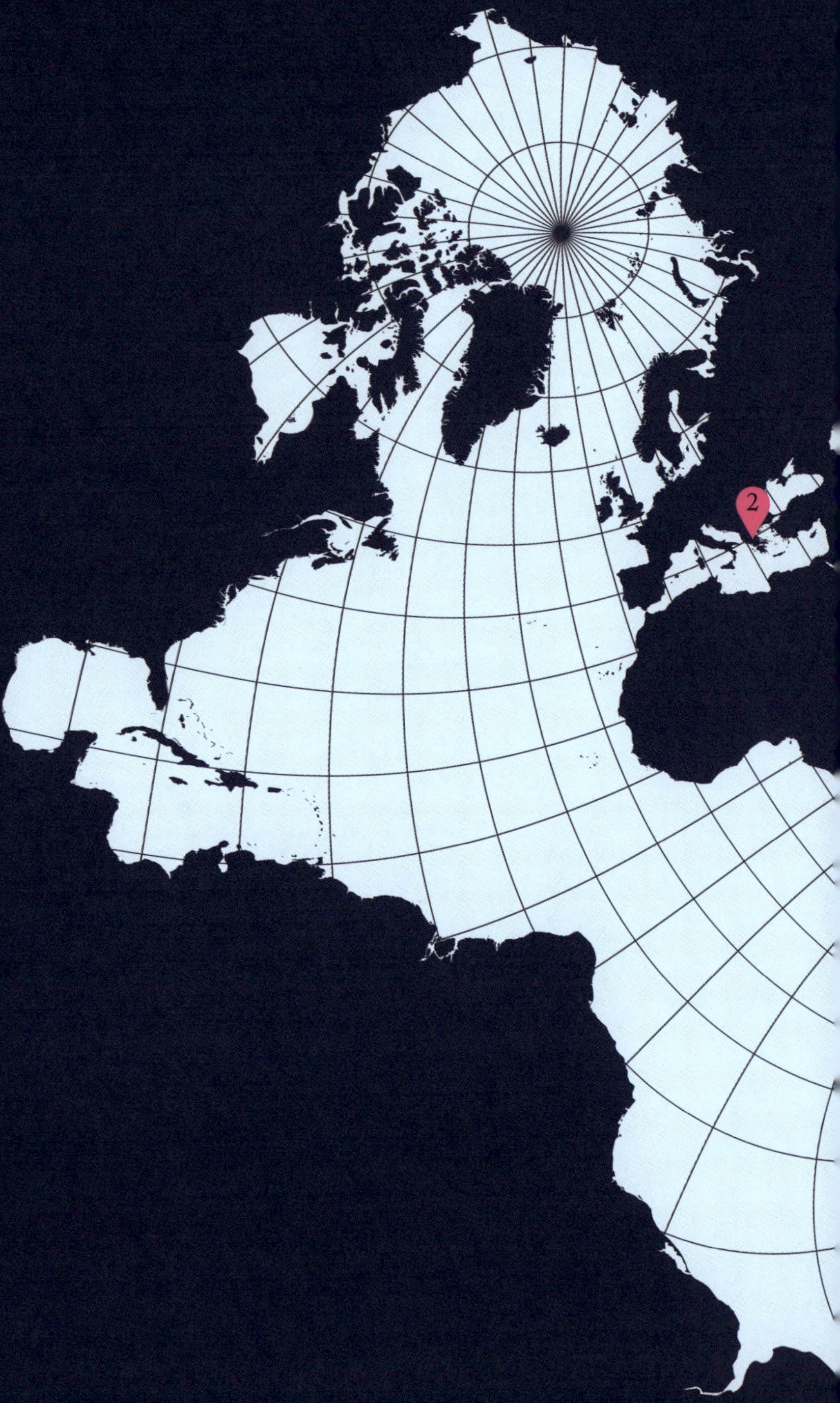

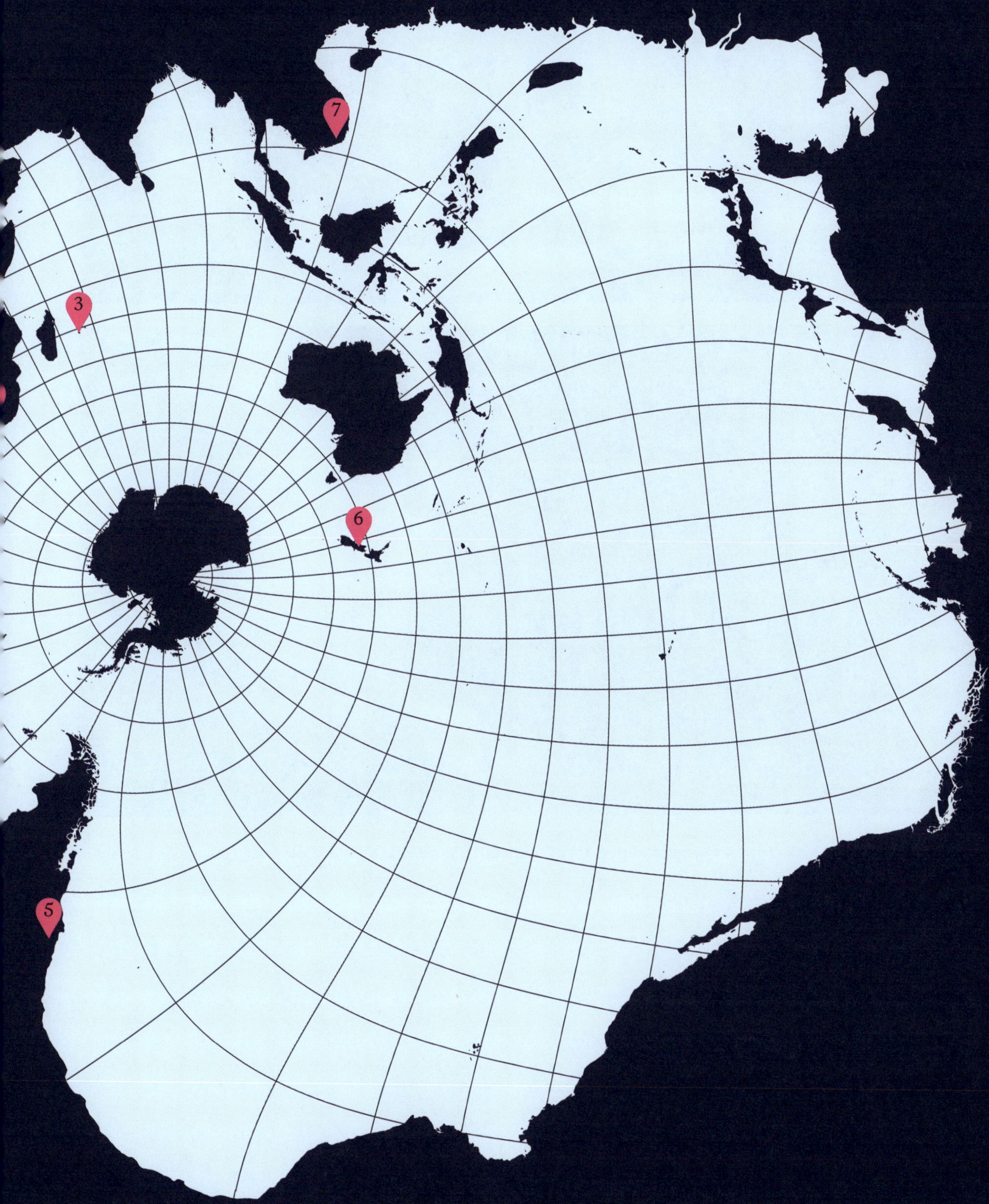

7
3
6
5

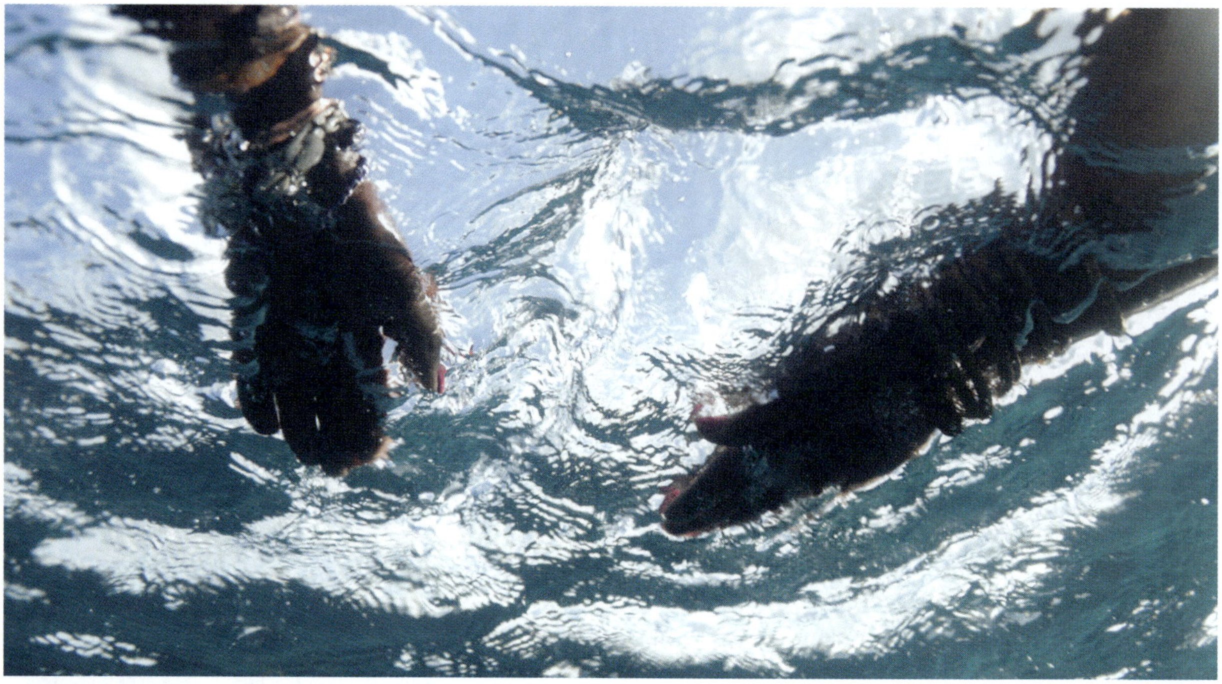

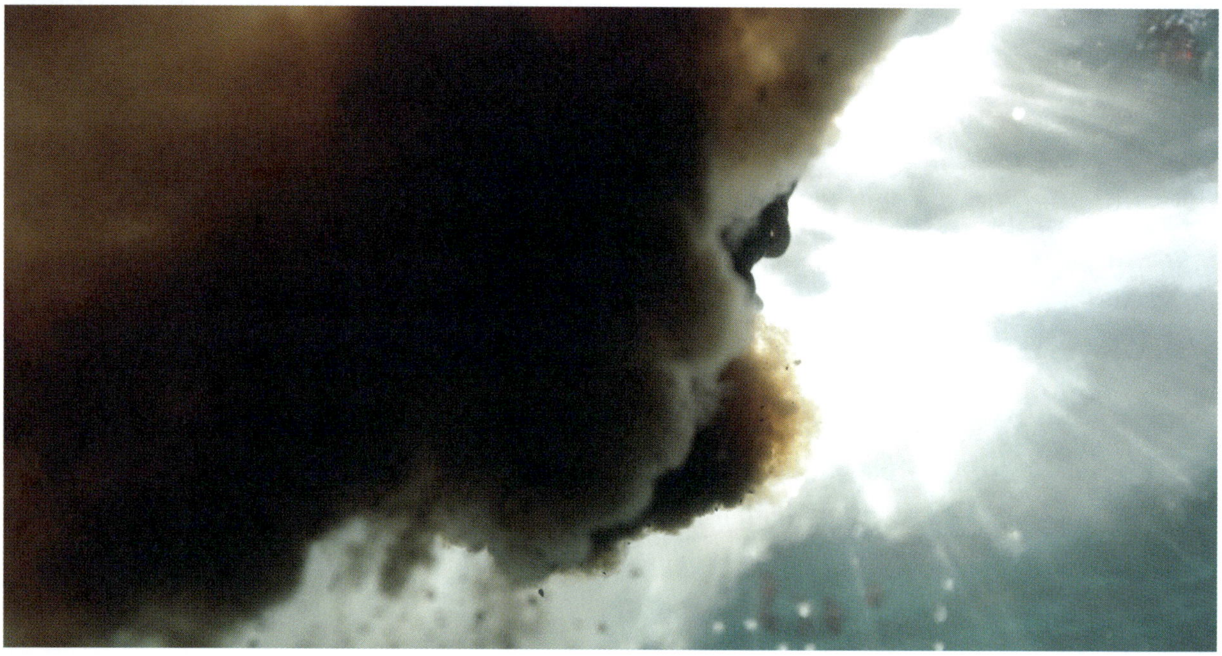

lerato laka le a phela le a phela le a phela/My love is alive, is alive, is alive, 2022, Three-channel video installation, color, sound, 17 min

Dineo Seshee Bopape

Born in South Africa in 1981

When she introduces herself, Dineo Seshee Bopape notes that "during the same year of her birth, there were perhaps twenty-two recorded Atlantic Ocean hurricanes and four Indian Ocean cyclones close to Mozambique." The practice of the South African artist is grounded in her environment, drawing strong connections to natural elements. Commonly, she incorporates materials such as soil, water, and stones into her installations. Her artworks frequently explore themes of collective memory, mythologies, and archetypes, reflecting on South African history and extending beyond its borders. A key concern in her work is the experiences of the African diaspora and the psycho-spiritual aspects of the transatlantic slave trade, as she investigates the enduring presence of historical traces in contemporary contexts.

For Bopape, soil and water serve as foundational elements in the construction of identity. In her works, the ocean emerges as a repository of collective memory, its resonance continuing to be felt on the shores. The sound and fluidity of water symbolically link the past to the present: contemporary climate disasters and population displacement echo the forced migrations and labor of colonial histories. In *(Seithi) The Rest, as They Used to Say, Is Story* (2021), the waves metaphorically connect past and present as they wash ashore, simultaneously bringing history to the fore while erasing traces in the sand. The video component of the installation is projected onto a floor covered with marble dust, creating a space where reality, imagination, and memory converge and engage in dialogue.

The camera focuses on the artist's hand as it gently touches the surface of the water, exploring its texture and sound. The work draws inspiration from the story of Peter, an escaped American slave whose image—his back scarred from extensive whippings—became emblematic of the brutal conditions endured by slaves, contributing to the abolitionist movement in the United States. The simple yet intimate gesture of the hand caressing the water evokes a sense of healing and care, as if the artist were soothing Peter's scars. At the same time, this gesture connects the artist's body to a broader history of trauma, diaspora, and suffering. Progressively, her hand is giving offerings to and through the ocean, fruits and flowers, as an homage to the millions of enslaved people who crossed the Atlantic and to the ones who drowned in the passage.

"Think about the water itself and my relationship with it: the water of the ocean, the water in my body, as well as the water in our collective body or collective consciousness, us all in relation to each other and to the water, and what love has to do with that, where love sits in that connection." For Bopape, water is holding everything together, memory and oblivion, resilience and hope, sufferings and love.

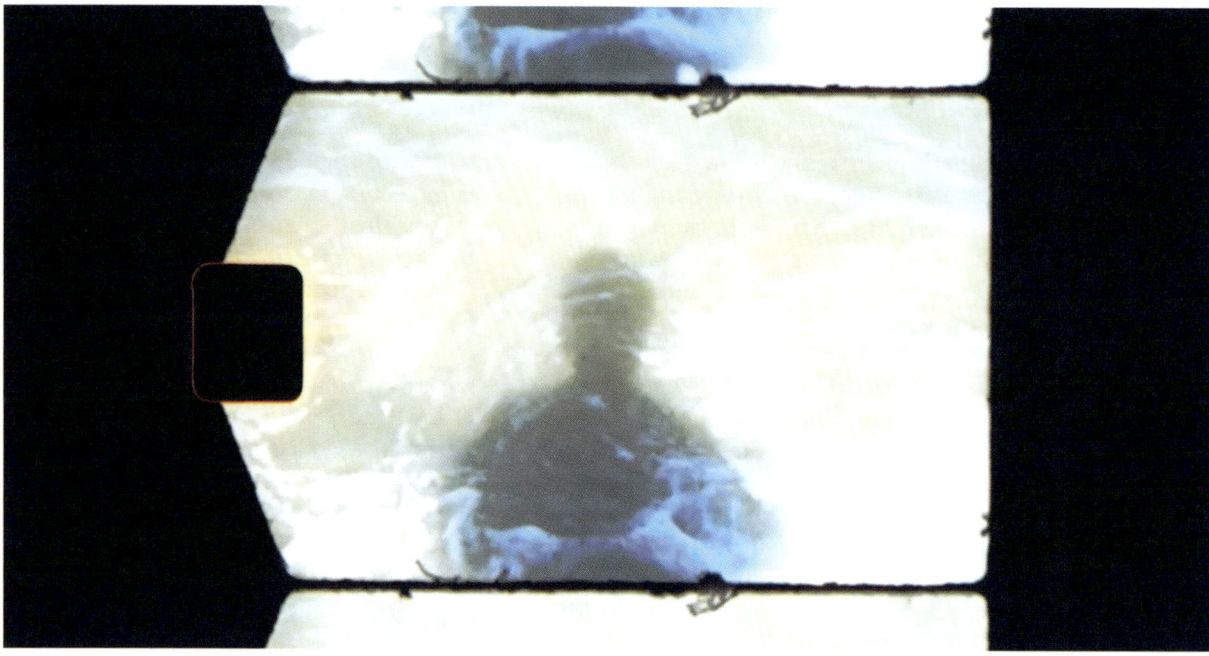

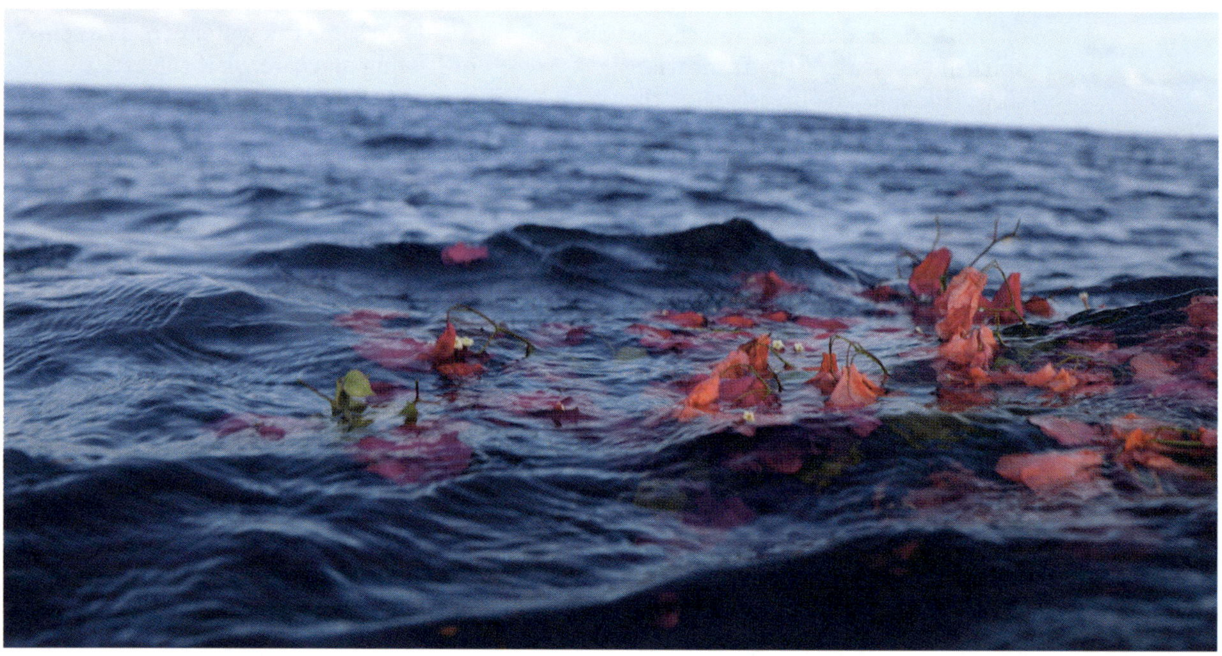

lerato laka le a phela le a phela le a phela/My love is alive, is alive, is alive, 2022, Three-channel video installation, color, sound, 17 min

Marina Gioti

Born in Greece in 1972

"The sea is full of hidden, overlooked stories that I am attempting to salvage and bring to the surface."

Marina Gioti is a Greek visual artist based in Athens. With a background in chemical engineering, environmental management, and filmmaking, her multimedia and research-based practice is enriched by her multidisciplinary training. Through her work, she aims to question established narratives and historiography by presenting new perspectives on the past.

Gioti's recent investigation and long-term project focus on locating and identifying the many shipwrecks that lie at the bottom of the sea. This endeavor began with *Launching Ceremony* (2021), a video based on archival footage of ceremonial vessel launches. For Gioti, the idea is to trace the journey from the birth of a ship to its demise, working backwards. These shipwrecks carry pieces of the past and symbolize what she perceives as human failures. Examining them allows her to reflect on the human condition and question our relationship with materiality. What should we consider waste, and what should we regard as heritage?

"The leading global heritage organization, UNESCO, protects wrecks that have sunk over one hundred years ago; therefore, recently wrecked ships are currently in a state of limbo, depending on governments and regional policies towards their removal and safe decommissioning." (Marina Gioti's statement for her exhibition at Galerie Dominique Fiat, Paris, May 2024.) Some ships are abandoned due to corruption, while others are left adrift due to financial or functional problems. Each vessel has its own story that reflects our times, and many, although classified as dangerous shipwrecks, remain in the waters. When tankers sink and cause oil spills, they are simply sealed off but not removed. Moreover, legislation regarding these ships remains unclear. In Greece, for instance, the artist discovered she was not allowed to name the ships in her artwork. Why? Who owns them now, and who is responsible for their decaying hulks?

Gioti's investigation reminds us that shipwrecks represent the third-largest source of ocean pollution. In total, in 2023, there are an estimated three million sunken and abandoned vessels in the ocean, with over 8,500 classified as "potentially polluting wrecks" dating mostly back to World War I and World War II. These wrecks contain harmful chemical pollutants, unexploded munitions, and tons of heavy fuel oil. Researchers warn that the effects of climate change are likely to accelerate the deterioration of these wrecks, increasing the risk of environmental contamination. However, there is still a lack of data and of international cooperation regarding the topic.

Initially, Marina Gioti focused on the Gulf of Elefsina, near Athens. The bay

City of Mykonos shipwreck, 2023, Multi beam echosounder video still

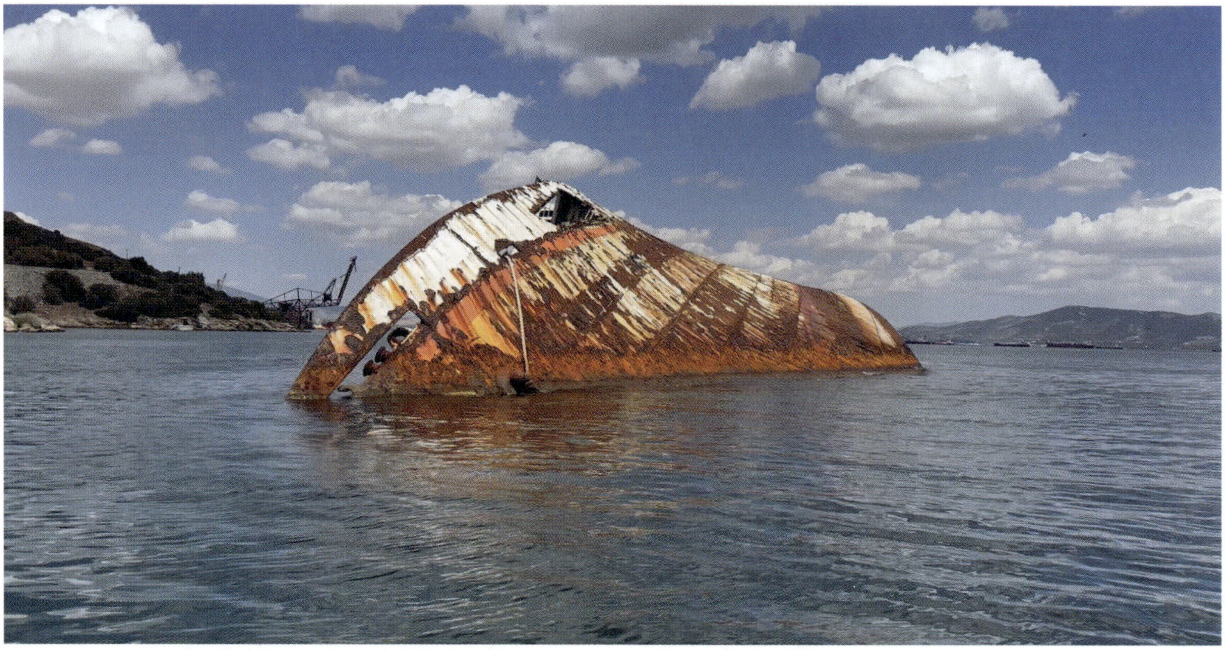

Ship Graveyard of Eleusis, 2023, Photographic print

is infamous for the high number of ship graveyards lying at its bottom. It is also a historical and sacred place, where the Eleusinian Mysteries took place. This ritual, referring to the Greek myth of Persephone, can be seen as a myth of regeneration. This concept is key for the artist, who questions the material traces and environmental impact of human activities. How might we envision regeneration, and what can be recovered from these long-term damages?

Marina Gioti collaborated with a team of marine researchers to record and scan the shipwrecks in the Gulf using specific acoustic sonars. "The sea is so loud," says the artist. "We are just deaf to its voice." Recognizing the importance of sharing her research in a sensory way, she explores various modes of presenting her findings. For instance, she used sonar images to create prints displayed within lightboxes and to produce an intriguing video. Observing the images of the sea bottom feels like looking into one's own body because they resemble medical ultrasound. This perspective highlights the fact that our bodies, like the earth, are composed of about seventy percent water. The prints are both surprising and original, showing a top-down view of the ships. Due to the side-scan sonar, there is a dead zone between the sides, creating a peculiar line in the middle of the print. The outlines of the ships appear to float and radiate like an alien species. Gioti chose to print the images in a golden color to make them attractive and sparkling, enhancing the eerie feeling of the series. Gold, often associated with success, fittingly contrasts with what she views as human failure. She also plays on the phrase "silence is golden," hinting at the collective silence surrounding these stories. Moreover, the glowing golden color could symbolize the multiple chemicals that continuously leak into the sea from these shipwrecks.

Magalie Grondin

Born in La Réunion, France in 1972

In the nineteenth century, a zoologist proposed the existence of a submerged continent beneath the Indian Ocean, which he termed Lemuria. This hypothesis was presented as a means to explain the distribution of lemur fossils found in Madagascar and the Indian subcontinent, yet absent from continental Africa and the Middle East. This landmass was envisioned as encompassing all peoples from Africa to Australia and posited as the ancestral home of humans. For the artist Magalie Grondin, this perspective is profoundly inspiring as it suggests the emergence of a new lineage. "We are the descendants of these primal populations," she says. According to the narrative—subsequently discredited as a scientific statement yet perpetuated as a regional myth—giants from outer space mingled with local lemurs, ultimately giving rise to a mermaid. This figure occupies a central role in Grondin's latest body of work and research, symbolizing the feminine aspect of decolonization and a form of resistance. She envisions this mermaid traversing islands to share stories with women and empower them.

In 2020, a significant oil spill occurred in Mauritius following the grounding of a tanker. Hydrocarbons spread, resulting in the pollution of the Mauritian lagoon. In response, the island's population mobilized, with women cutting their long hair to create a massive sausage and buoys to absorb the oil and mitigate the spill. This event profoundly impacted the artist. She views the solidarity exhibited by these women and their powerful act of cutting their hair to protect their territory as emblematic of the mermaid's embodiment within each woman, instilling them with strength. Hair serves as a potent symbol, representing both identity and connection. Grondin's large installation *Déterrer les sorcières* (Digging up the witches) (2024) features cascading black hair, reminiscent of seaweed, as her work intertwines various elements: human bodies are portrayed as both organic and animalistic, while the cosmos and the ocean coalesce into a unified whole. In her drawings, the mermaid is depicted as bald, perhaps signifying her offering of hair.

Grondin's mermaids appear to float ubiquitously, seemingly taking root in both air and water. She illustrates them within ethereal and abstract landscapes, set against oceanic and cosmic

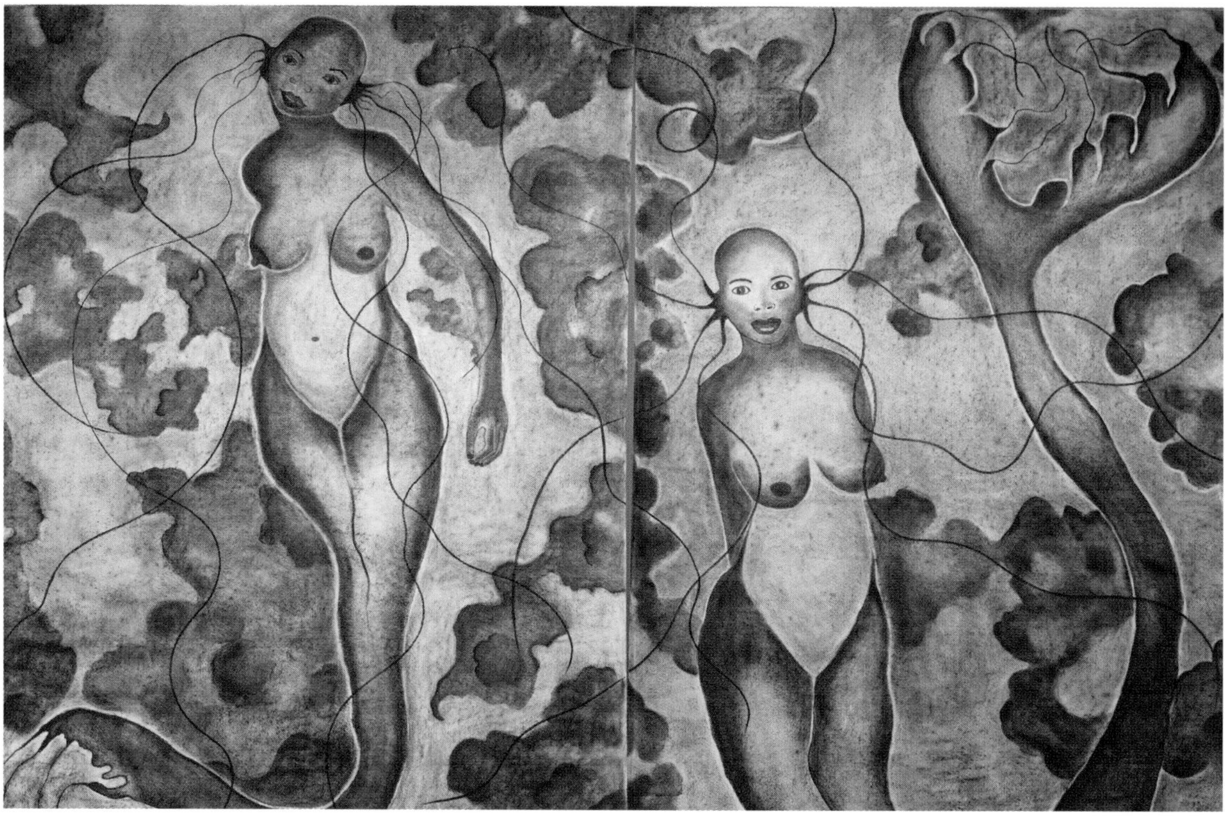

Femmes aux ouïes, zazavavindrano, 2024, Charcoal drawing on paper, 300 × 240 cm

backdrops. "It's a dream," she explains, "I draw inspiration from these beliefs to construct my own universe." Importantly, the mermaid is not merely a myth; in Madagascar, they are believed to exist and are frequently invoked by healers. These benevolent figures provide healing and support during difficult times. In Malagasy oral tradition, these mermaids, known as *zazavavindrano*, possess gills that allow them to inhabit the sea. In the charcoal diptych *Women-with-gills* (2024) these gills extend into long, thin antennae that connect their bodies to the invisible realm of spirits, both celestial and oceanic.

In La Réunion, the ocean is regarded as the domain of the spirits. It is crucial to consider the island's colonial history. Situated in the Indian Ocean, east of Madagascar, the island was uninhabited until the arrival of French immigrants and colonial subjects in the seventeenth century. Subsequently, it was developed as part of the French colonial empire, with slaves from East Africa and later indentured laborers from Asia brought in as agricultural workers. Today's population primarily descends from these diverse groups.

"La Réunion is a small stone lost in the Indian Ocean," the artist confides.

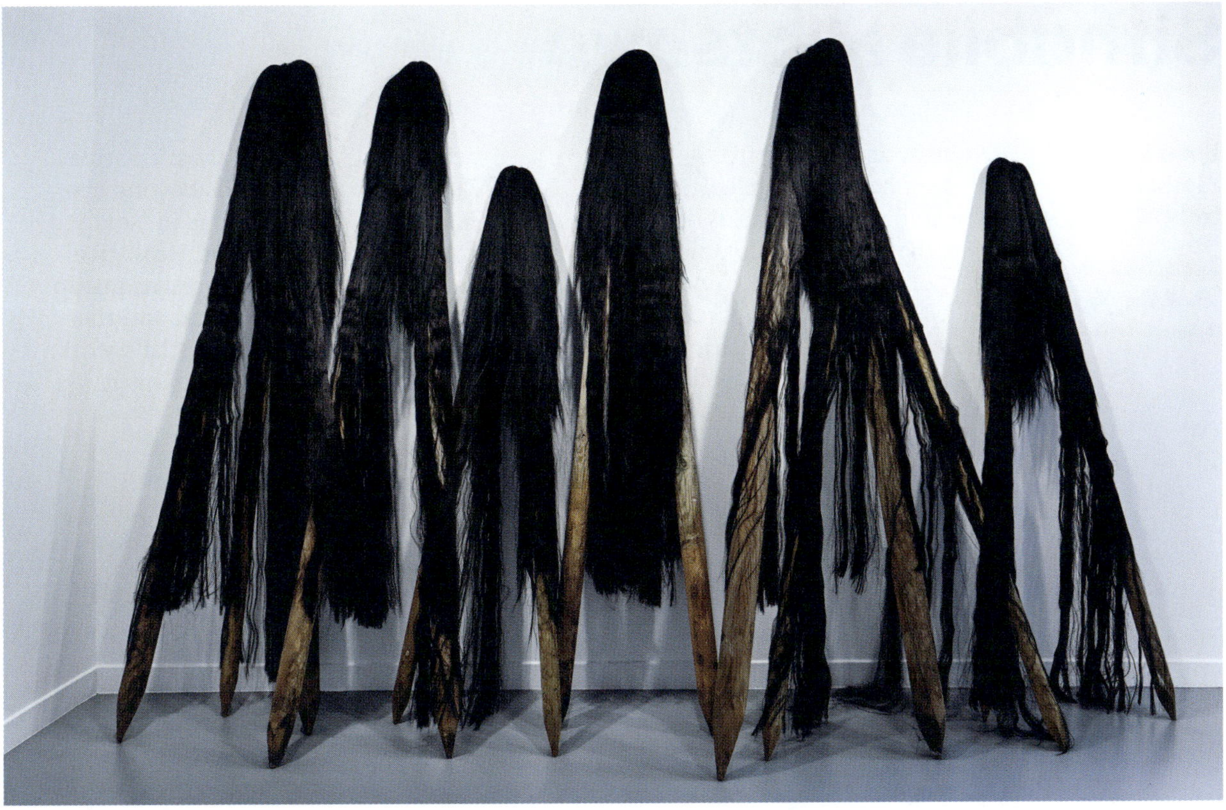

Déterrer les sorcières, 2024, Hair, wood, variable dimensions

"We are a French territory, but we are very far from France. It is difficult to think about our own identity." Recently, she embarked on a journey she describes as "initiatory," traveling from island to island to reconstruct the territory of the lemurs and reconnect with local populations, culture, and the stories of women. "The sea is what connects us. I needed to find other roots, other connections, to seek out my 'sisters.'"

The sea, however, is also treacherous in La Réunion. There are few beaches, and the coastlines can be dangerous. During colonization, escape by sea was impossible. Villages were constructed with their backs to the coast, as all homes faced the mountains. This was the desire of the colonizers, who imposed their vision of the landscape. They perpetuated a fear of the sea, a sentiment that still resonates strongly today. Indeed, the ocean remains associated with the notion of a vast graveyard, particularly concerning the slave trade. The figure of the mermaid allows for the transcendence of these conceptions, facilitating the exploration of new connections with the ocean and neighboring islands. It prompts a reconsideration of the identities of men and women freed from the colonial past and in deeper relation to their environment.

Sithabile Mlotshwa

Born in Zimbabwe in 1975

"As an artist coming from a colonized nation where colonialism utterly eradicated centuries of history, the legacies of colonialism are part of my lived experience. This is why I use my art practice to challenge the whitewashing of history by disrupting settler colonial narratives. As an artist retracing and visually mapping colonial histories, the sea is very important to me because countless stories of Africans, its diasporas and those of my ancestors, are connected to the sea. In fact, you can't tell the African story and its diasporas without it. It was on transatlantic ships that the first African slaves were brought to the colonies. And the sea, where the enslaved people from the Indian Ocean and Indonesian Archipelago were brought to South Africa. The ocean was a stage for horrific cruelty, where Africans were stolen and crammed into cargo holds, barely able to move. Many would not survive the journey–nearly two million of the twelve million people shipped in the transatlantic slave trade died before the ships ever reached North America. There is much that remains unknown about the specific histories of slave vessels, and many of the stories of those transported below their decks have been lost to history. This is why my research specifically focuses on retracing and visualizing this invisible history."

Mlotshwa's long-term research project is titled *In Search of Ubuntu*, which translates to "in search of what makes us human." This concept, originating from the artist's native Zulu language, extends beyond the essence of human beings in isolation from their environment. Ecological concerns, as well as the relationship between humans, the sea, and the land, are central to the artist's work, which values cross-disciplinarity and interconnectedness.

Reflecting on contemporary migration issues, Mlotshwa has recognized that current events are largely a continuation of models imposed by colonization. This realization prompted her to embark on this research project to examine the roots of the neo-colonial system. She began her investigation with the Dutch Empire and the Dutch East India Company (VOC), both of which epitomize a history of trade grounded in organized violence, oppression, exploitation, and environmental destruction.

"I chose the Dutch Golden Age as a point of departure because understanding the current state of our world and finding solutions to the climate crisis, poverty, inequality, racism, and the dismantling of the violent imperial capitalist structures, require understanding how we got here in the first place," states the artist.

Mlotshwa observes that history has a tendency to repeat itself. "Consider the example of fisheries," she notes. "Western trawlers are devastating the seabed along

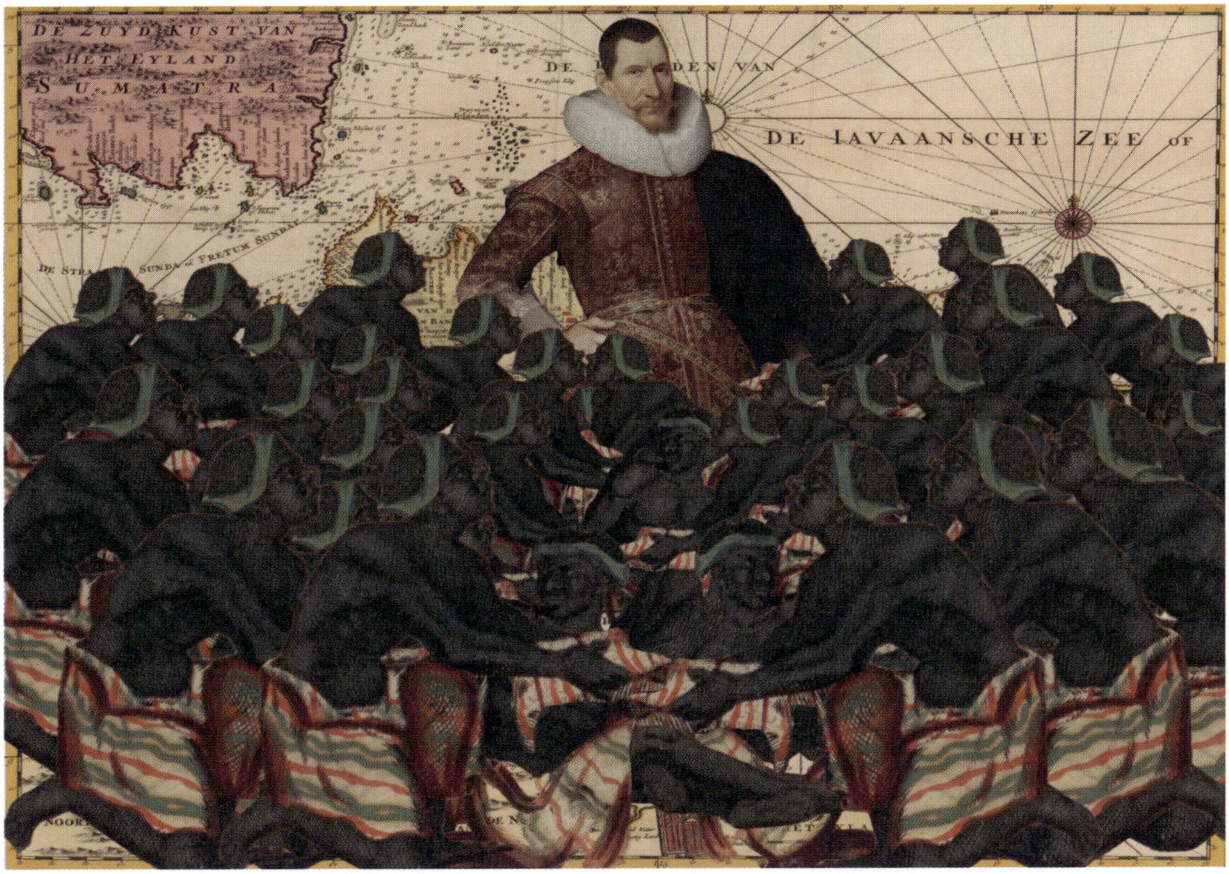

Coen's VOC Blueprint, 2023, Mixed media with gold on paper, 152 × 180 cm

the African coast and depleting fish populations. It is no surprise that local fishermen struggle to survive and are left with little choice but to consider migration. I view my work as a call to a collective awakening."

The colonization and Western conquest of the world were largely executed via maritime routes. In this race for power and wealth, navigation skills and marine cartography were crucial to success, with the Dutch achieving prominence through their mastery of the seas and their powerful ships. Mlotshwa's collage, *Dutch Golden Age* (2024), depicts an admiral and typical figures of seventeenth century Dutch high society standing on a pond filled with gold coins and dead bodies. To the right, Dutch society is depicted as enjoying the fresh goods made possible by the loot and genocide. To the left stands an enslaved man behind the admiral, and on the ground next to the dead bodies, two Indonesians on their knees beg the admiral to stop violating them. In the background money and decapitated heads flowing

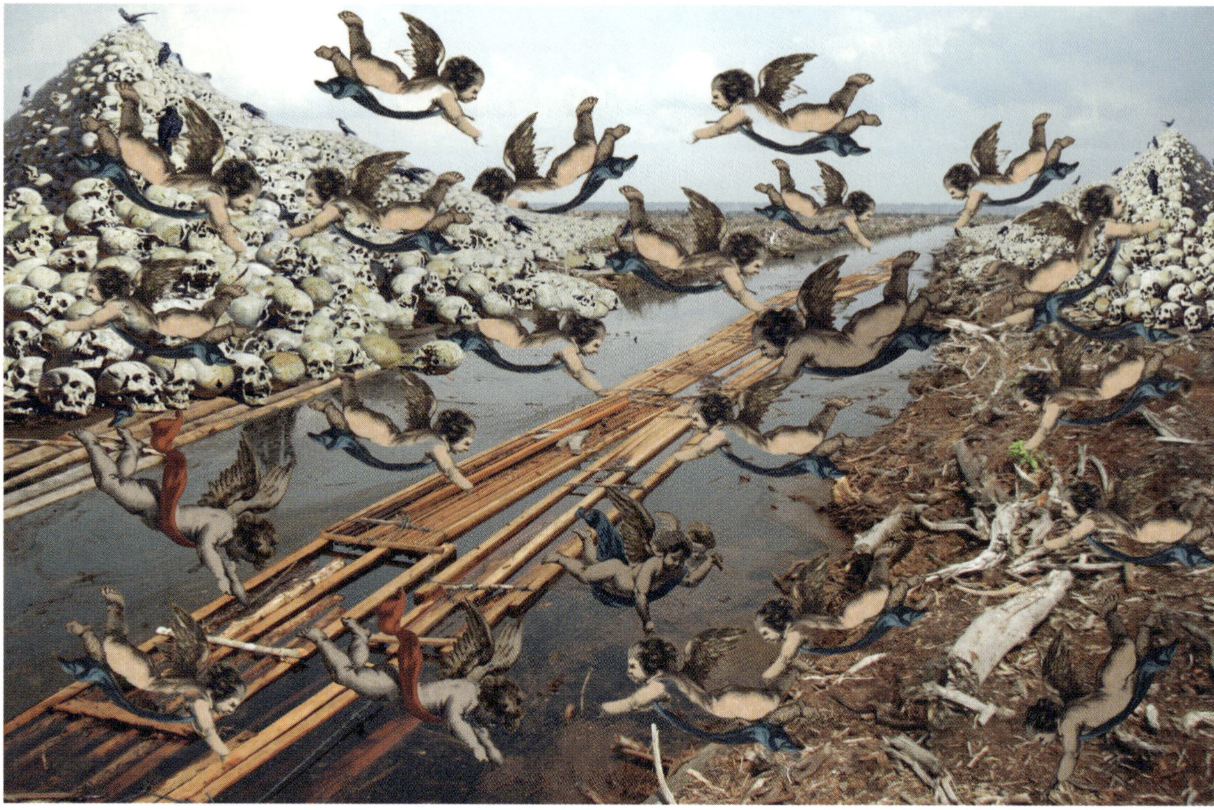

The Birth of Civilisation, 2023, Mixed media with gold on paper, 152 × 180 cm

from the famous ships of that era—used to transport goods and slaves, and as defense against pirates and rival European nations—are visible. Much of the artist's iconographic material is sourced from the archives of the Rijksmuseum in Amsterdam and from historical books.

The artist also favors working with ancient maps, which she transforms and deconstructs, employing collage as a technique to interrogate the writing of history and collective memory. For Mlotshwa, maps have historically held, and continue to hold, significant political power in shaping how people perceive territories and marine spaces. "When you examine old maps, you can already discern how traders connected spaces via the sea and how they disregarded reality to construct their own narratives." Angels and the Virgin Mary were often depicted as if endorsing Western conquests, and Mlotshwa engages with this symbolism. For instance, in *The Birth of Civilisation* (2023), angels are depicted salvaging what they can from the total destruction of Banda Island, which is now part of Indonesia. This massacre was orchestrated by Jan Pieterszoon Coen, the Dutch Governor-General of the Dutch East India

Company, who ordered the complete annihilation of the island's population and environment to establish his own city to prevent the arrival of the British. The collage captures the moment when angels transport the last remaining cut trees along a waterway. "The Dutch did not merely commit genocides; they also destroyed the environment to reappropriate territories and establish their empire," Mlotshwa asserts.

During the seventeenth century, Dutch global dominance was unparalleled, yet the significance of the trade and slavery routes linking South and East Africa to Southeast Asia is often overlooked. In 2022, the Dutch government issued an apology for the atrocities committed during the slave trade, but the apology primarily addressed Latin America, neglecting other regions. "When people think about the slave trade, they typically focus on the transatlantic aspect, while the rest is largely ignored," Mlotshwa observes. She contends that much work remains to be done to uncover the full extent of slavery and maritime capitalism (Mlotshwa refers to a new book, *The Truth About Cape Slavery* by Patric Tariq Mellet, published by Tafelberg in 2024, which is finally opening the path towards such debates and sheds light on these neglected histories). Her artistic practice and research serve as an invitation to delve deeper into these critical aspects of colonial empires and their lasting legacies. At the same time, she acknowledges that because many African countries are still being colonised, it is not yet possible for them to confront their own histories, and that it may take another generation or two before this can happen.

Enrique Ramírez

Born in Chile in 1979

Enrique Ramírez's personal relationship with the sea is rooted in his childhood, when his father worked as a sailmaker. Together, they would navigate the waters to test the sails, instilling in Ramírez a deep connection to the maritime world. His life has been shaped by the sea, and this presence permeates his artistic oeuvre. As Ramírez himself notes, "Chile is like an island." Given the country's geographical configuration, with a coastline extending approximately 6,400 kilometers, it is challenging to escape the omnipresence of the sea.

Ramírez grew up during Augusto Pinochet's dictatorship. According to him, the Chilean people's relationship with the sea is dual in nature: both poetic and political. The poetic dimension could be exemplified by the national Chilean anthem, which celebrates the natural beauty of the country, including the ocean. The lyrics,

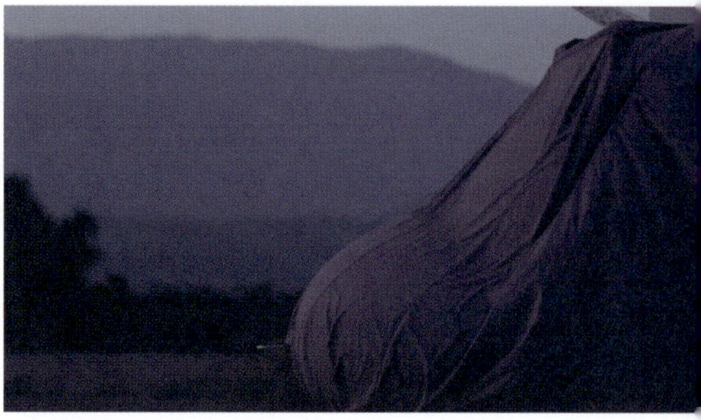

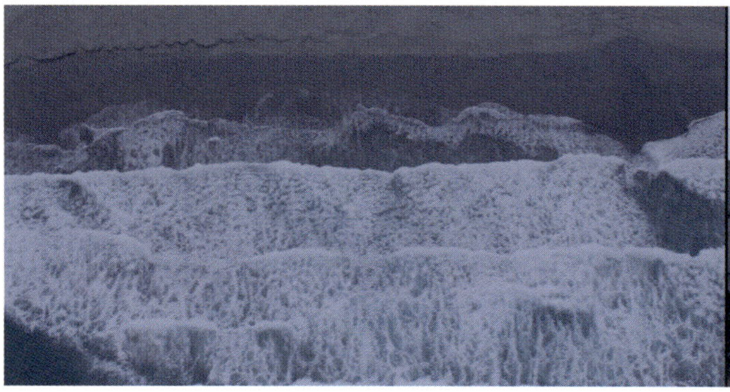

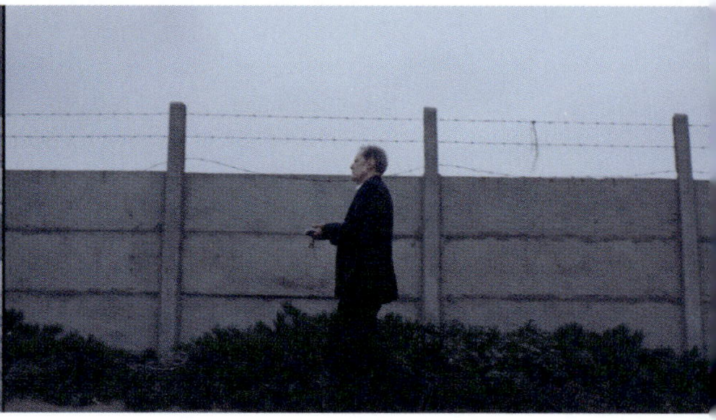

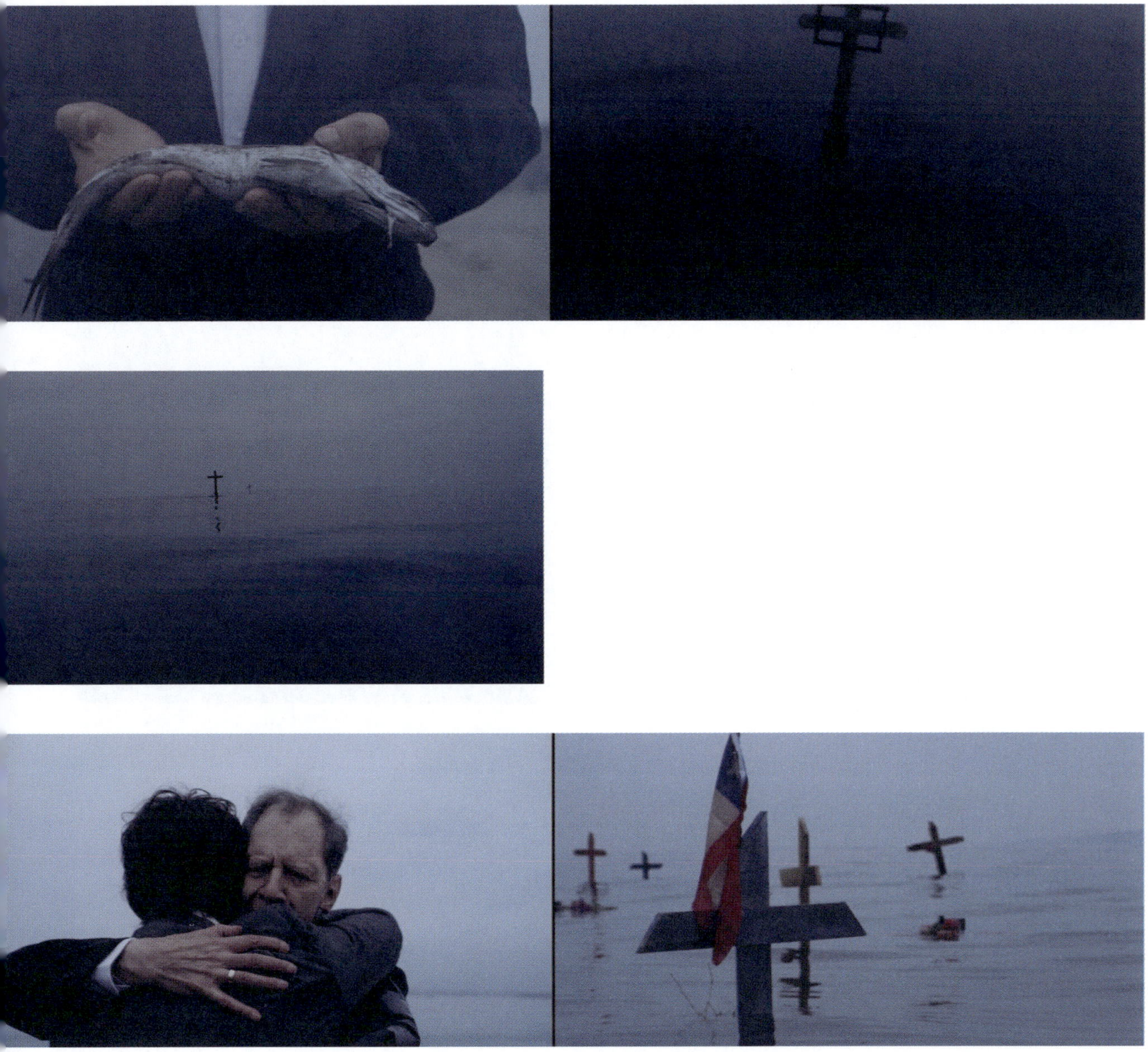

Los durmientes, 2014, Video installation, 15 min

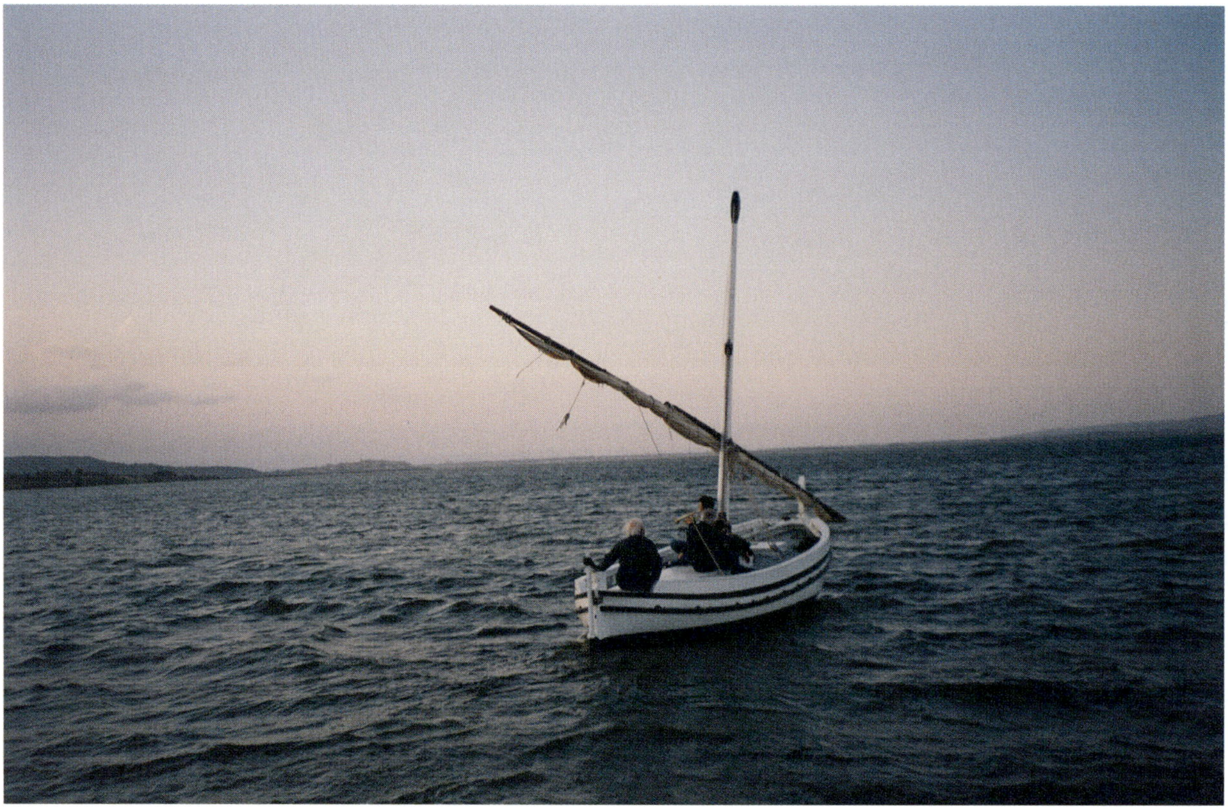

Chant pour la «MAR», 2019, Video, 7 min 44

which the artist incorporated in one of his videos, compare the country with Eden, praising its purity, blue sky, and tranquil sea that bathes its shore. The political dimension, however, is deeply intertwined with the traumatic legacy of the military dictatorship. During what became known as Operation Condor, under Pinochet's regime, thousands of individuals were kidnapped and disappeared. Many of these victims were bound to railway ties and thrown into the sea from helicopters. Consequently, for many Chileans, the sea has come to symbolize a vast cemetery. As Ramírez observes, "For all of us, the sea carries the soul of those who disappeared."

Ramírez has created a film trilogy dedicated to this historical period: *Brisas, Los durmientes*, and *Las tres memorias* (2023), which poetically explores the profound connection between the Chilean people, the sea, and their political history. Notably, the Spanish term *los durmientes* carries a dual meaning, referring both to "sleepers" and "railway ties." In the film, Ramírez collaborates with an actor who slowly walks on the beach, his clothes drenched as if he has just emerged from the sea. A voiceover questions: "Is everything going to the bottom of the sea? What could the depth of this place be? What is the depth of memory?" For Ramírez, "The true site of collective memory in Chile is the sea."

In addition to this series, Ramírez has produced more ambiguous works. *The Song of Mar*, for example, depicts an elderly man, perhaps accompanied by his family, sailing in a small fishing boat on a calm sea at dawn. Although the sail remains furled, the boat moves swiftly, as though propelled by an unseen wind. The folded sail may evoke a sense of nostalgia or a state of waiting. Are they lost? The group sings an ancient-sounding song, with the woman playing a horn and the man, an accordion-like instrument. Their voices carry an uncanny echo, and at one point, they collectively call out "Mar! Mar! Mar!", invoking the sea to return the bodies of the disappeared. This time, their calls are not confined to the Chilean context but resonate with the contemporary migrant crises in the Mediterranean Sea. However, no explicit reference to these events is made, allowing the audience to interpret the haunting and poetic imagery independently. In addition to video works, Ramírez has written poetry, composed music, and created performances. When the word MAR is inscribed thousands of times on a blackboard, for instance, it metamorphoses into the crests of waves, evoking the imagery of a restless sea.

Lisa Reihana

Born in New Zealand in 1964

The Māori were the first inhabitants to reach the shores of New Zealand approximately 1,200 years ago. They navigated aboard canoes, journeying from distant Polynesian islands.

The ocean has thus always been central to the development and identity of Māori culture. It provided the people with food and resources and embodied their history, enabling them to connect with their ancestors. Later, with the arrival of Captain Cook in 1769, the ocean also marks the onset of colonization in the country.

Lisa Reihana's art practice explores the history of New Zealand through the lenses of colonization, indigenous knowledge, gender issues, and identity. Her large-scale installation exhibited at the 2017 Venice Biennale, *In Pursuit of Venus*, interrogates the historiography of the Pacific, as it has been predominantly written from the perspective of the colonizers. In *Nomad of the Sea*, she imagines the story of the first child born to a European woman and a Māori chief, a reflection linked to her own biography. More broadly, her multidisciplinary artworks are

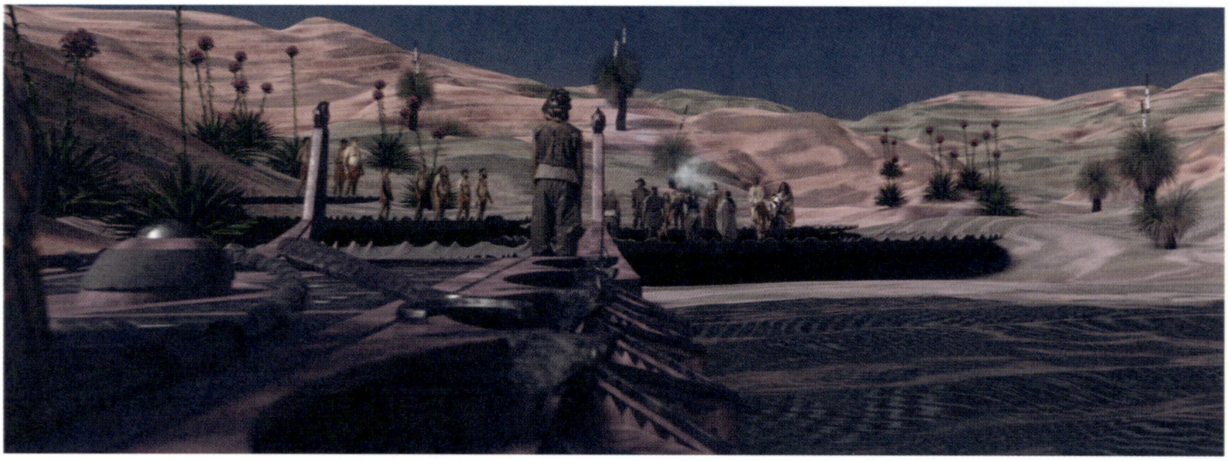

Groundloop, 2022, Single channel digital video with multi-layered audio, 22 min

deeply rooted in her Māori heritage and emphasize the profound connection that indigenous people have with their environment, conceived as an integrated whole. A typical example she frequently references is the way Māori present themselves: not by their names, but by the place from which they originate.

Above all, the artist highlights the extensive knowledge of the sea shared by Māori people. In the video *Groundloop* (2022), the interconnectedness of humans and the natural world is palpably conveyed. The film begins with a shamanic ceremony. A dancing woman, who represents the Aboriginal practice of fire-stick farming, holds a fire-stick made of tree bark, recalling traditional practices that care for the land. The music is loud and dramatic, contributing to a sense of temporal dislocation or entry into a dreamlike universe. The camera transitions to the shore. Magical creatures merge with seabirds, and viewers seamlessly transition from the seabed to aerial vistas.

Reihana often draws parallels between the ocean and the sky as expansive, open spaces. She suggests that the contemporary endeavor of space exploration parallels the experiences of the first navigators as they embarked on their maritime journeys. She is interested in the encounters that happen on the way.

Groundloop presents a fictional narrative about the interactions among different indigenous groups from the Pacific. The point of departure is an imaginative scenario: what would have transpired if James Cook had never reached Australia? In this alternative history, she envisions that Cook would not have sailed to Botany Bay in Australia, and questions whether such groups of indigenous travelers would have been welcomed. The film depicts their meetings and journeys, emphasizing the shared philosophies of life among these diverse communities and tribes, philosophies deeply rooted in animism, and a profound connection with the natural world. Although they hold different beliefs, their operational principles in the world are similar, centered on respect for the natural world.

In reality, when examining the seabed, it becomes clear that New Zealand and Australia are part of the same landmass. In the film, the navigators follow this route, guided by the fish swimming beneath their keel and the stars above. "For us, the Pacific is a superhighway," she says. "The ocean is what connects us all." For Reihana, it was intriguing to imagine this meeting between an ancient race of First Australians and a relatively young one–the Māori, whom she refers to as the "teenagers of the world" because New Zealand was the last land mass to be populated.

The film concludes with the female navigator and her fellow travelers arriving in Australia. A welcome ceremony is held during which two Tongan women perform, one dressed in a silky blue garment dances with the other wearing red and black outfit. This symbolizes the devastating encounter between the ocean and the Hunga Tonga-Hunga Ha'apai volcano, as well as between tradition and contemporaneity.

Despite the sea being considered sacred and endowed with a spirit, New Zealand faces significant pollution issues, particularly due to intensive farming activities. Current debates raise the question of granting legal status to the marine coastal area to protect it. The artist expresses her anxiety about the global situation and the state of Mother Earth, Papatūānuku, who is considered the origin of all life.

Bao Vuong

Born in Vietnam in 1978

"I would like to say that in the face of the vastness of the waters, like in the face of any difficulty in existence, everyone can find the courage necessary to move forward despite the waves, the storms, and overcome the distance to be crossed that still separates them from the shore, from rest, from a new country."

(Text written by the artist for his exhibition at the Toulon Asian Arts Museum, 2024).

At first, painting the sea, and especially a dark, black sea, was for Bao Vuong a way to tell his own story, the story of his family. It was also a way to pay tribute to all the refugees who did not arrive, and perished at sea.

His parents fled Vietnam in 1979, when he was one year old. It's a story he was too young to remember, but that remains very present through his parents' memories. Their crossing lasted eleven months. They left the

Mekong Delta and first arrived in Malaysia, where they were quickly pushed back. Their boat had run out of fuel and drifted until a fishing vessel rescued them.

Today, this story is repeating itself, and it is likely to repeat itself even more in the future as climate change pushes increasing populations to flee their countries and risk their lives by taking to the sea. The black sea is also a more general metaphor for the dark moments of our lives, the moments of solitude, or of resilience when great waves overwhelm us. The artist asks how to move forward in these cases, how to find the strength and resilience to overcome the waves and find one's luminous self. For him, art allows doors to be opened as everyone can project their own story within the work. A painting is silent, but from this silence a multitude of stories and emotions can be born.

After studying in France, Vuong returned to Vietnam to pursue his art practice. At that time, he was painting with tar and doing portraits of refugees, portraits of women, of mothers, who carried a lot of suffering. This material allowed him to embody his story. Tar is what was missing during their boat journey, and it is also what sticks. In Vietnam, where the story of the boat people is still taboo and officially did not take place, the exhibition was censored on the opening day.

This moment marks a real break in his practice. He decided not to be carried away by this wave of censorship, and kept only the sea and the sky from these previous paintings. This was the beginning of his series *The Crossing*. Without its figurative dimension, the work actually became stronger: instead of contemplating an otherness and looking at someone's suffering, it is the viewer who makes the crossing. In each painting, the horizon line is placed differently, and the point of view changes: one is either in a boat, or immersed, or perhaps, even, on a beach.

The Crossing series has no end. The artist says he will continue to paint these landscapes of the black sea, as it may be a lifetime's work. "I never tire of it," he says. "My parents spent eleven months on the boat. Maybe I should do a painting for each night. In reality, as long as there are people risking their lives at sea and living through the same nights, this series cannot end."

"Today I no longer paint with tar, but use oil paint for the sea. However, the sky is done in pencil with a mixture of incense ash. Often focused on the clouds, this dust is like a prayer I am addressing."

The artist's family is Christian and also Confucian, and all religions are syncretized. "What's important to me is to connect to a form of spirituality, and art is one of them. Painting waves can take a long time, sometimes I repeat the word 'wave' like a mantra, and my practice approaches a form of meditation. I like to believe that my hand is sometimes guided by those who never returned from their crossing and who whisper to me 'Bao, tell our story.'" →

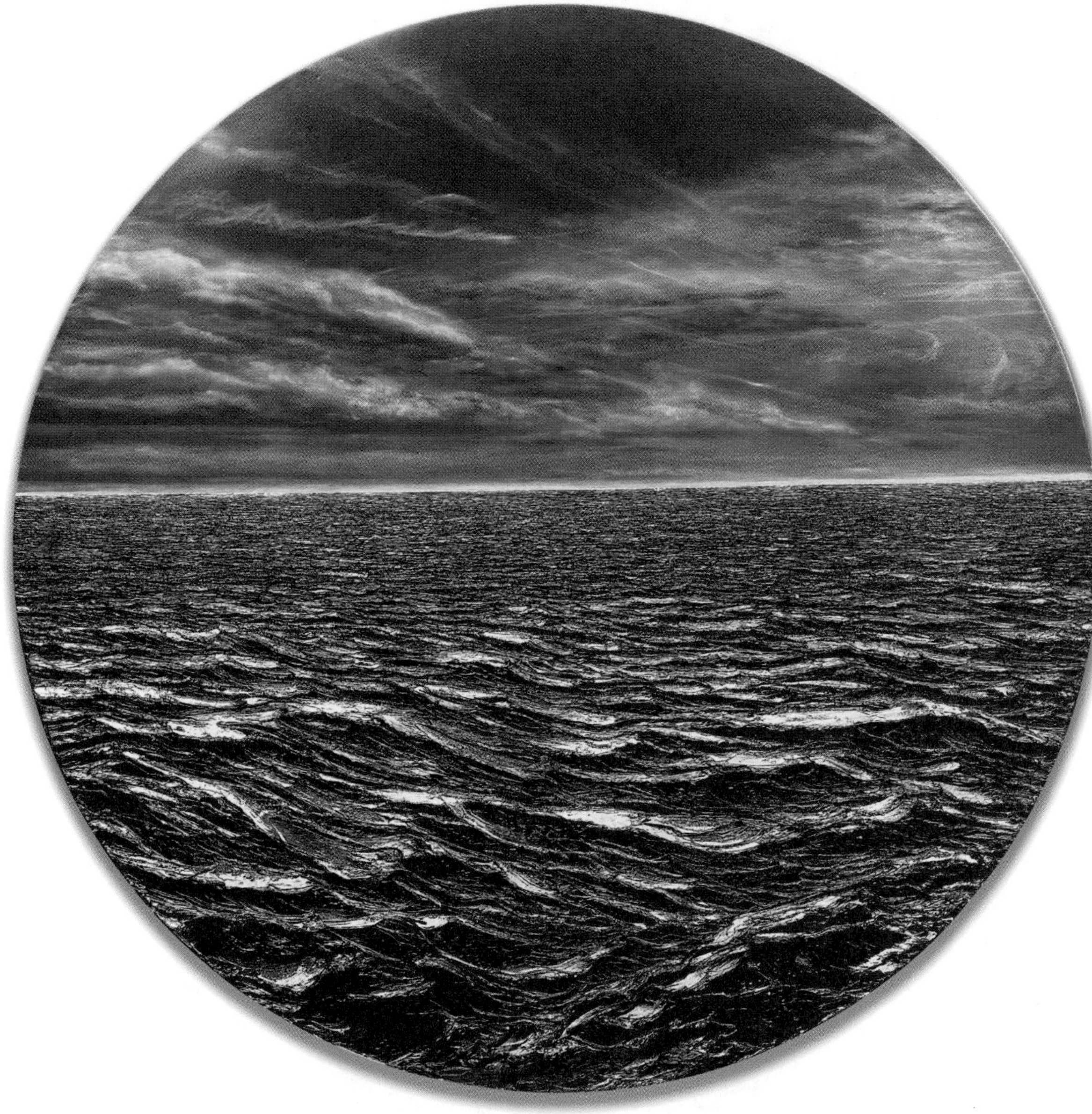

The Crossing 44, 2021, Black oil painting, black acrylic graphite powder and golden leafs on canvas, 140 × 140 cm

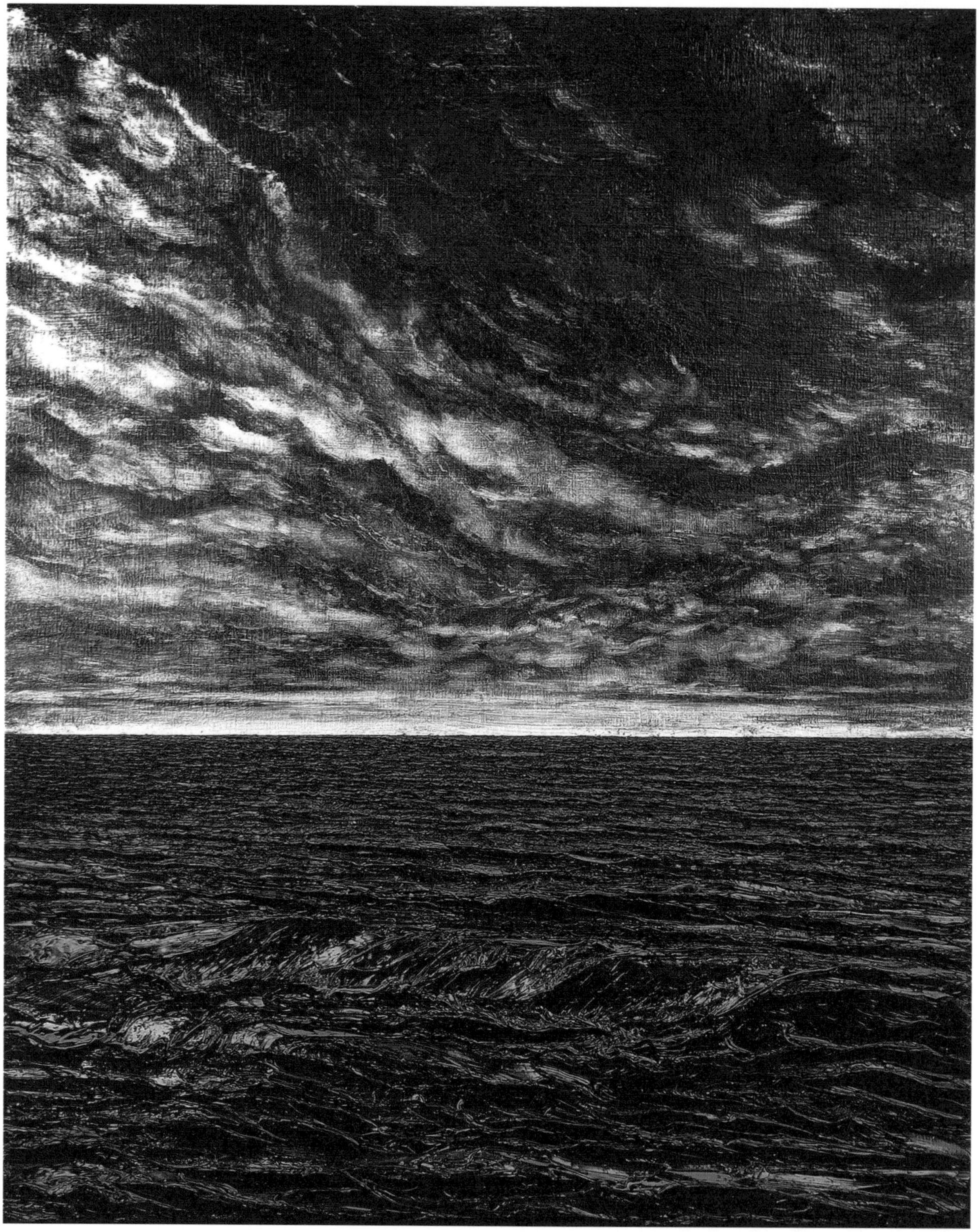

The Crossing 237, 2024, Black oil painting and golden leafs on canvas, 50 × 140 cm

He likes to let himself be carried away by the painting and the sea he represents, both literally and figuratively. "I start with a sketch and let myself drift, carried by the waves. It's also my philosophy of life, and it's full of surprises. I like to be where the current takes me."

At first, he used photos to represent the sea, but he no longer really needs them, as he can now imagine and appropriate the currents. "I am not looking for hyper-realism but it is important that my representation is accurate, so that a navigator can find what he feels at sea." Over time, the oil dries and shrinks slightly. "The paint lives, like the sea, and I love this idea. Often people want to touch my paintings, scratch, caress the material of the sea," he says. "It is a material that attracts and repels at the same time."

Painting waves is an attempt to capture the uncapturable: "As soon as you see or name the wave, it is no longer. We don't know where it begins or where it ends. It is almost abstract, and yet very concrete, of course, as all those who have sailed know. Moreover, no two waves are alike." This idea is very conducive to meditation on the human condition. Vuong sees life as a passage, like a wave. According to him, each human being is like a drop of water that joins the sea once their life is over.

The oceans, for him, are not spaces of separation. On the contrary, they are what connect us to one another. "If the sea seems to divide the world into several parts, it is also what links people, particularly in their resemblance to the difficulties of existence and their willingness to overcome them."

Ambivalences & Liminality

Then, as if breathing, the sea swelled beneath us. If you must know anything, know that the hardest task is to live only once. That a woman on a sinking ship becomes a life raft—no matter how soft her skin. While I slept, he burned his last violin to keep my feet warm. He lay beside me and placed a word on the nape of my neck, where it melted into a bead of whiskey. Gold rust down my back. We had been sailing for months. Salt in our sentences. We had been sailing—but the edge of the world was nowhere in sight.

Ocean Vuong

1. Yoan Capote
Born in Cuba in 1977

2. Maureen Gruben
Born in Canada (Tuktoyaktuk) in 1963

3. Deborah Drisana Jack
Born in the Netherlands in 1970

4. Ange Arthur Koua
Born in Côte d'Ivoire in 1989

5. Louisa Marajo
Born in Martinique, France in 1987

6. Barthélémy Toguo
Born in Cameroon in 1967

7. Zai Kuning
Born in Singapore in 1964

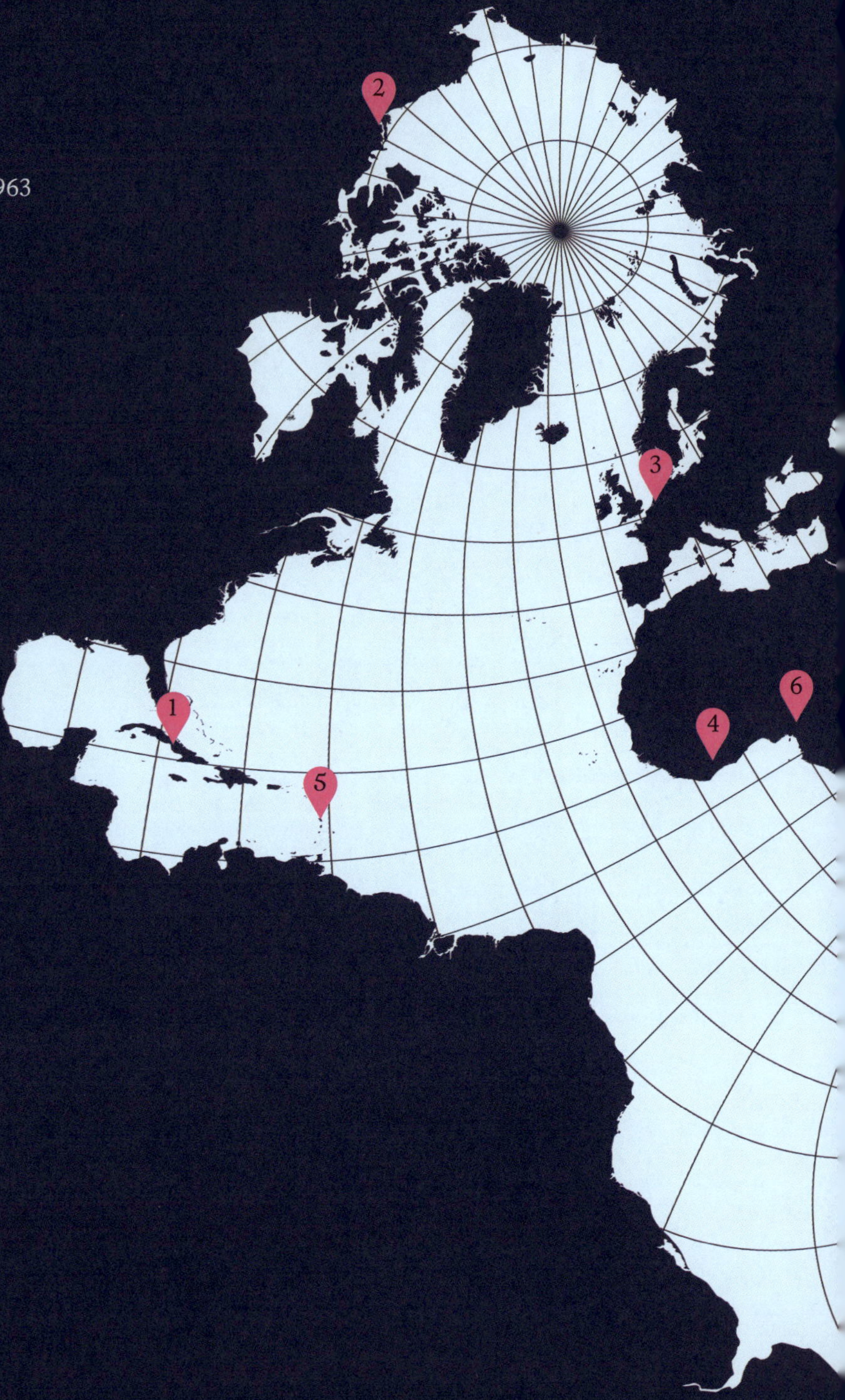

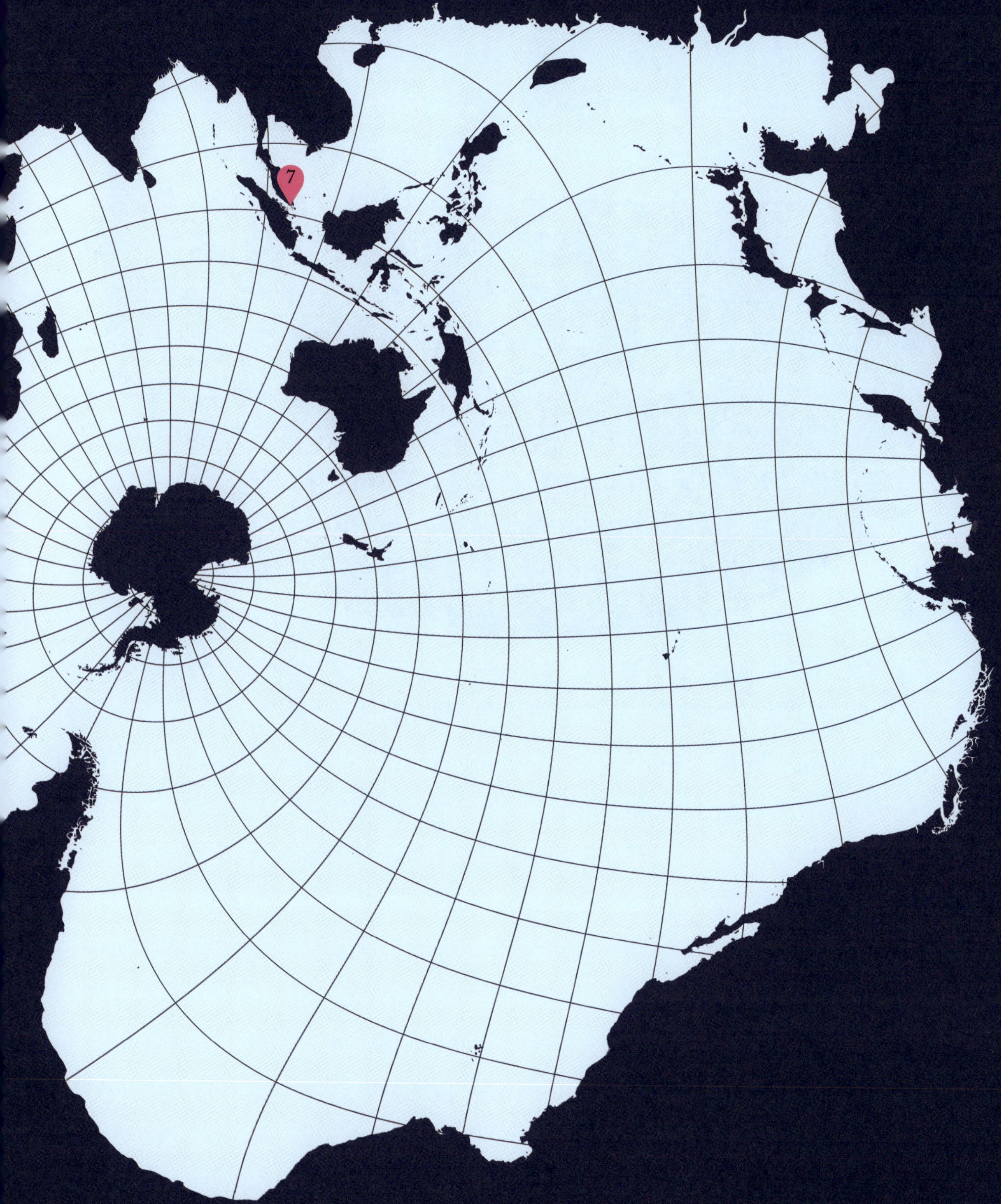
7

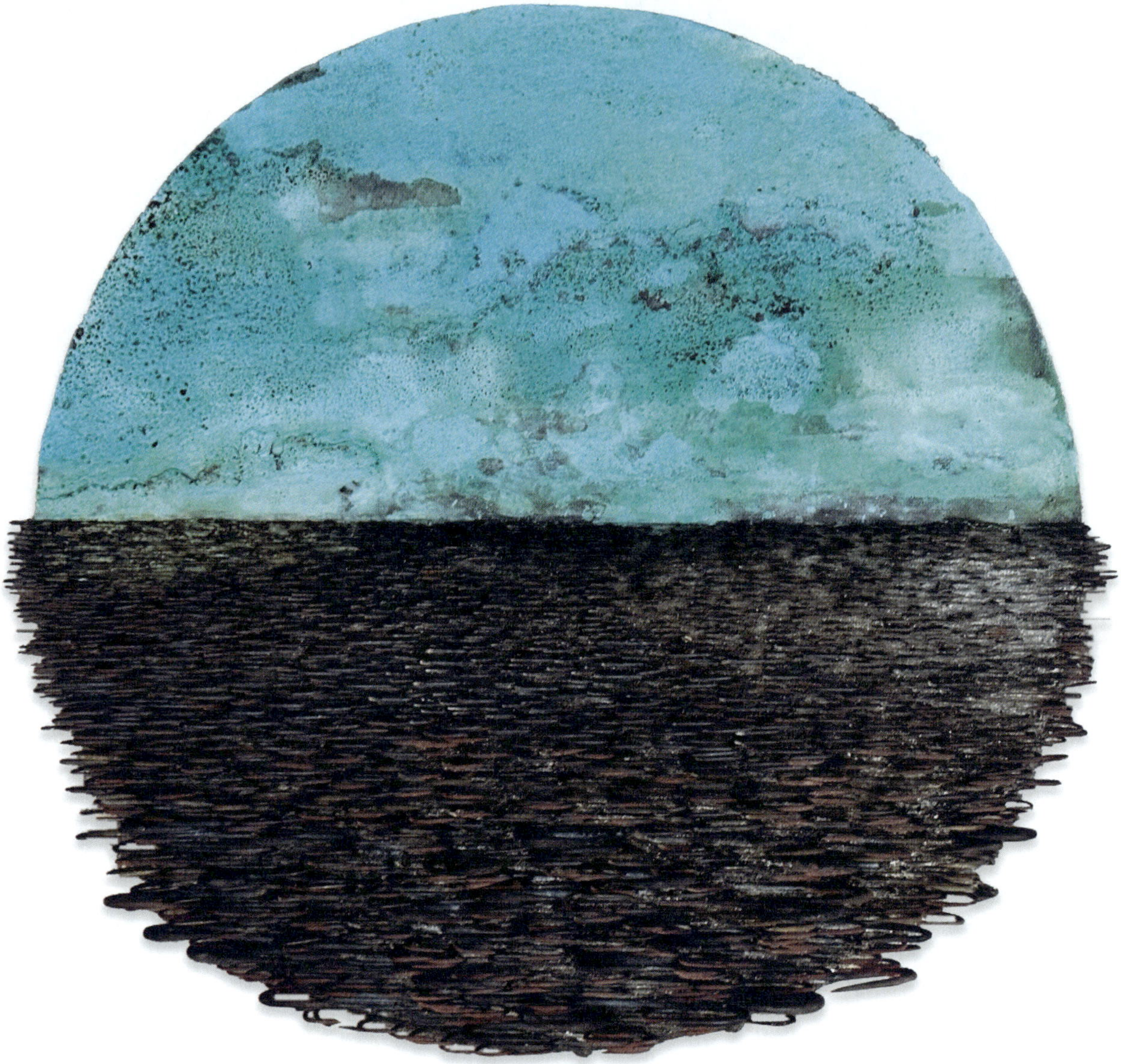

Palangre (Tondo), 2024, Steel fish-hooks and cast bronze plate with patina, 42 × 42 cm

Yoan Capote

Born in Cuba in 1977

"For Cubans the sea represents a double isolation, geographic as well as political. To this day people are not free to leave the island country, which has been embroiled in an ideological struggle with its powerful neighbor, the U.S., for more than sixty years. Thousands of Cubans have died trying to cross the relatively short stretch of sea that separates the island from the coast of Florida in precarious, overcrowded vessels. Hence, a seascape can be also understood as the barrier that keeps people imprisoned against their will, choppy waters becoming the ominous image of a graveyard" (Artist's statement for the exhibition *See scapes*, Be Brown Fine Arts, London, 2024).

Yoan Capote's practice is deeply rooted in the history of Cuba, a country that has experienced prolonged isolation, dominated by a single political ideology and a lack of freedom. Throughout the twentieth century and continuing to the present day, many waves of Cuban migrants have fled to the United States. In the 1990s, driven by economic and political hardships, many Cubans attempted to escape by sea, leading to what is known as the *balsero* crisis, during which approximately 36,000 Cuban rafters were intercepted at sea in 1994. During this period, the American government chose not to welcome these undocumented migrants, opting instead to return them to Cuba. These years also mark the beginning of the artist's practice as he moved to Havana to study art.

Capote's seascapes are constructed from heavy materials that symbolize this painful history. His use of fishhooks, in particular, reflects the harshness of economic life, the struggle for survival, and the sea as a dual symbol of nourishment and danger. Upon closer inspection of his compositions, where the surface of the sea may initially appear voluptuous, one realizes it stands as an insurmountable barricade, blocking a forbidden horizon. Similarly, the multitude of saw blades that constitute the expanse of water in certain compositions prevent any possibility of escape. The waves themselves, formed by a juxtaposition of blades, appear to be as guillotines, as seen in the work *Aguas Territoriales (Procusto)*, 2024. Capote has even painted a seascape using human blood.

Simultaneously, the process of collecting hooks, barbed wires, and fences invites a reflection on art as a space where collective suffering can be expressed and condensed, thereby opening up the possibility of healing in real life. The artist often patiently flattens these nails, as if to neutralize their dangerous effects. His skies, frequently tormented, are also pierced by luminous rays, suggesting dawn and the potential for a new beginning.

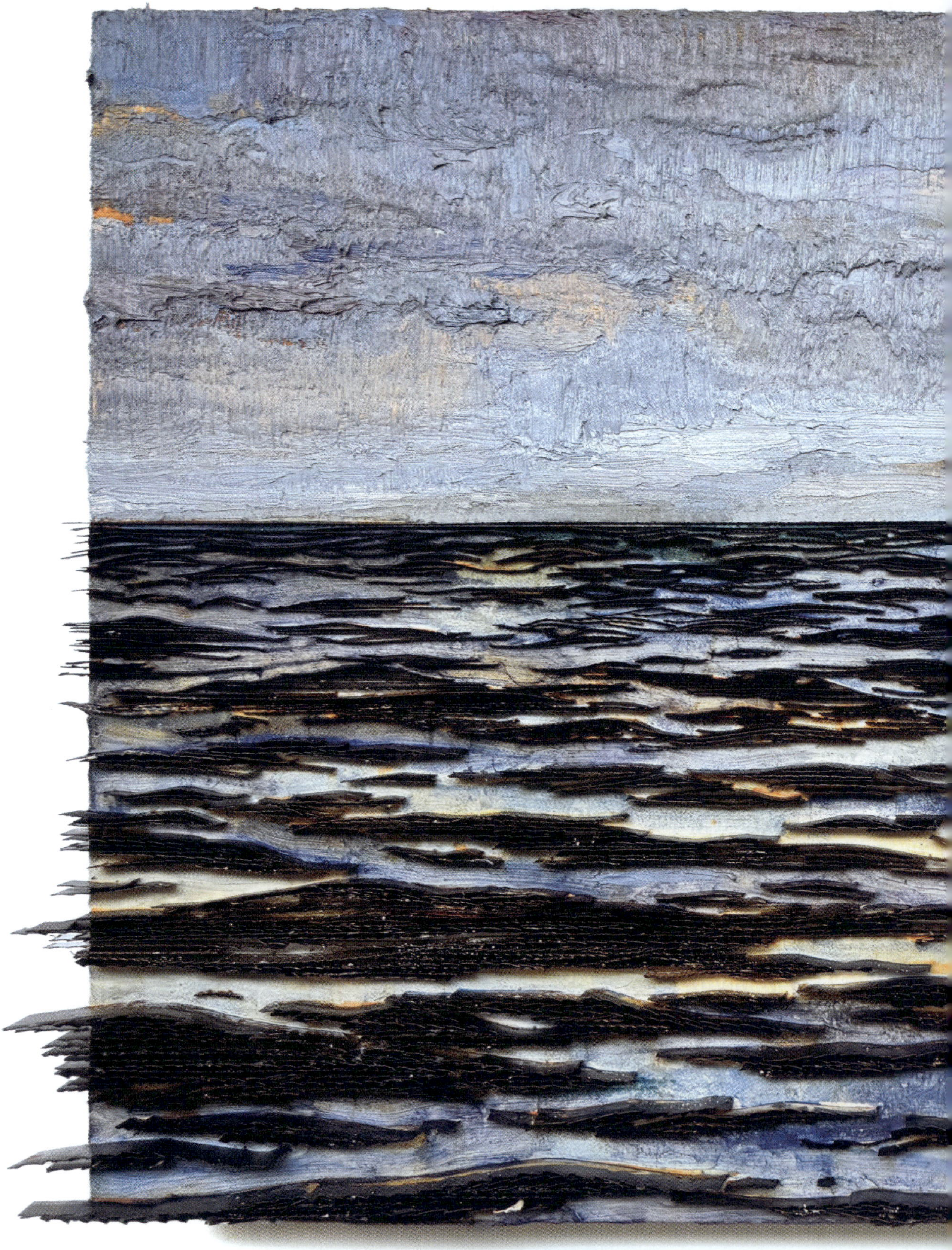

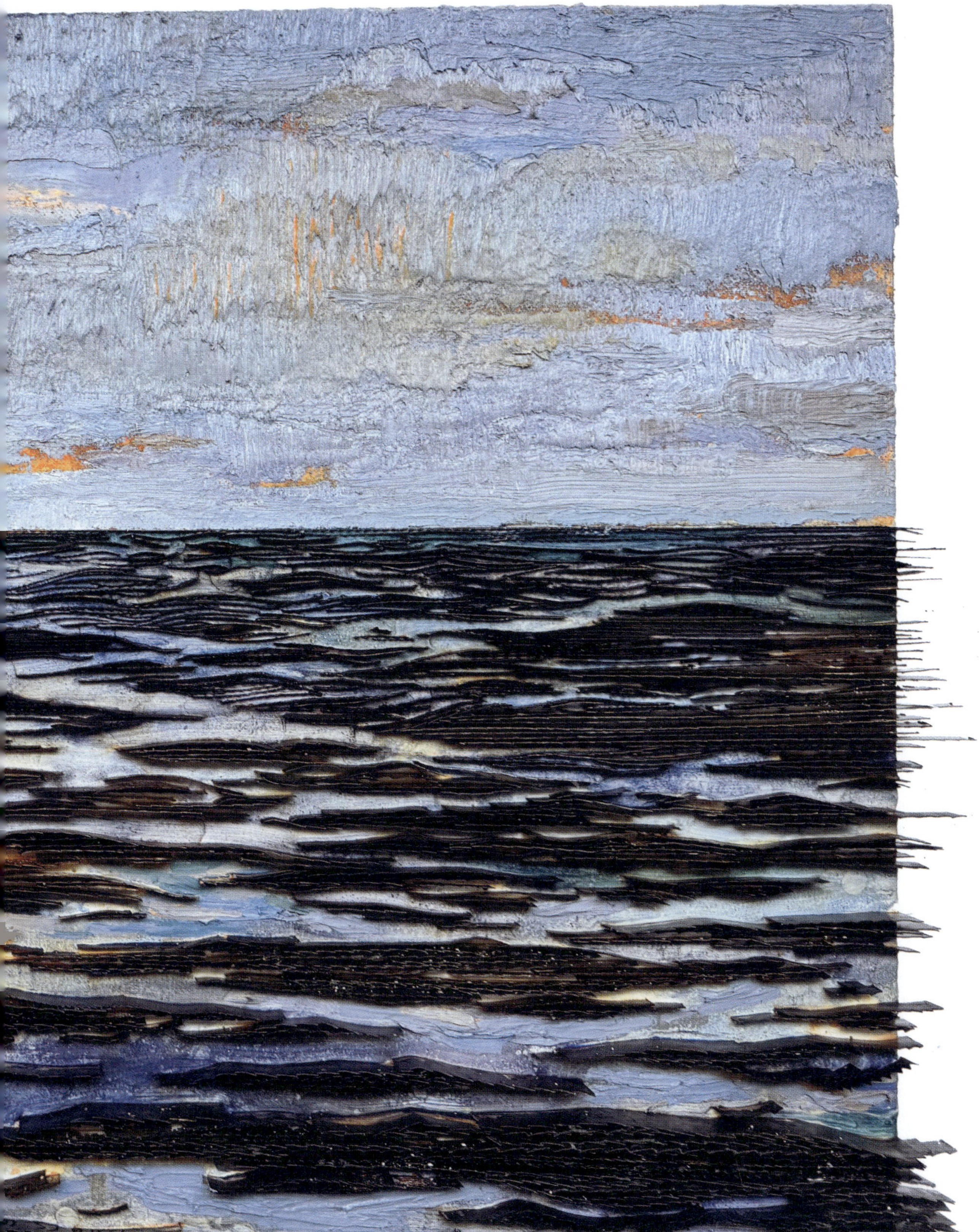

Aguas Territoriales (Procusto), 2024, Saw blades, oil, plaster on jute mounted on wooden panels, 100 × 150 cm

Maureen Gruben

Born in Canada (Tuktoyaktuk) in 1963

Maureen Gruben's multi-media practice incorporates diverse organic and industrial materials that are often salvaged from her local environment. She was born, raised, and now lives in Tuktoyaktuk, a hamlet in the Inuvialuit Settlement Region of the Northwest Territories, Canada, where her parents were traditional Inuvialuit knowledge keepers. Her position on the shore of the Beaufort Sea allows her to witness the effects of global warming first-hand, with her home losing approximately one meter of coastline each year due to accelerated coastal erosion. For her community, the sea, which has always been seen as a precious resource, has also become a threat.

Stitching My Landscape (2017) is both land art and performance. Installed on the sea ice outside Gruben's home in Tuktoyaktuk, the work consists of 111 ice-fishing holes "stitched" with 300 meters of broadcloth. Her process of rolling fabric from hole to hole is an act of endurance that generates a familiar pattern of both raw stitching and beautifully worked trim that adorns Inuvialuit drum dancing parkas. Red cloth on snow-covered ice also recalls a memory of her brother harvesting seal and stretching out fresh gut. The piece evokes the strength of family and community and the potential for healing–and being healed–by nature.

Nuna (2023) (meaning "land" in Inuvialuktun), repurposes the textile from *Stitching My Landscape*. For this piece, Gruben sewed the long strips of fabric into two 17 × 8,5 meters panels. Working with Tuktoyaktuk community youth, she brought them back out to the sea ice in early spring and laid them across one another. Over a two-month period meltwater formed beneath the installation, creating a large, cross-shaped pool. This appearance, a striking collaboration between human actions and the shifting season, emphasizes that we are embedded within and constantly affecting complex systems much larger than ourselves. Drone footage captured the initial placement of the work; an intimate first-person perspective of visiting the site weeks later, and pulling the panels from the meltwater, was filmed on the artist's phone. The red cross, a universal symbol of aid, draws attention to the urgent need to address climate change and mitigate its impacts on the Arctic landscape.

For *Nakataq I* (2022), "A marker to help in finding something again," Gruben etched patterns onto photographic aerial survey prints that were abandoned in Tuktoyaktuk by oil companies after briefly operating there in the 1980s. These prints chart ice coverage in relation to Beaufort Sea oil wells. By incorporating designs based on the fox stretchers and traps of her father Eddie Gruben, Gruben honors his legacy and underscores tensions between land as home and land as finite industrial resource. This theme is also present in *Aidainnaqduanni, Aurora* (2020), which translates to "we are finally home" in Inuvialuktun.

This installation features vintage polar bear rugs paired with found survey tripods. The tripods are also remnants from past oil extraction work. By returning these museum-donated rugs to sea ice the piece vividly portrays the fragility and resilience of the Arctic environment.

Finally, Gruben's 2024 exhibition *The Land that used to be* features *Qikuryuaq* (Clay Hills), an installation comprising approximately 1,000 beads representing the population of Tuktoyaktuk. Hand-made with local clay sourced from slumping permafrost, these beads highlight the severe impacts of climate change facing Tuktoyaktuk. The community will need to be relocated inland in the next thirty to fifty years. Handmade sleds are also featured in the exhibition. A salvaged child's sled forms the basis of the small sculptural work *untitled (sled)*, 2023, while the large format photographic print *Moving with joy across the ice while my face turns brown from the sun* (2019) depicts an installation of fourteen sleds borrowed from community members. These sleds evoke movement across changing territory and echo the aspiration of *Qikuryuaq* to, in the artist's words, "hold the land a little bit longer before it is gone."

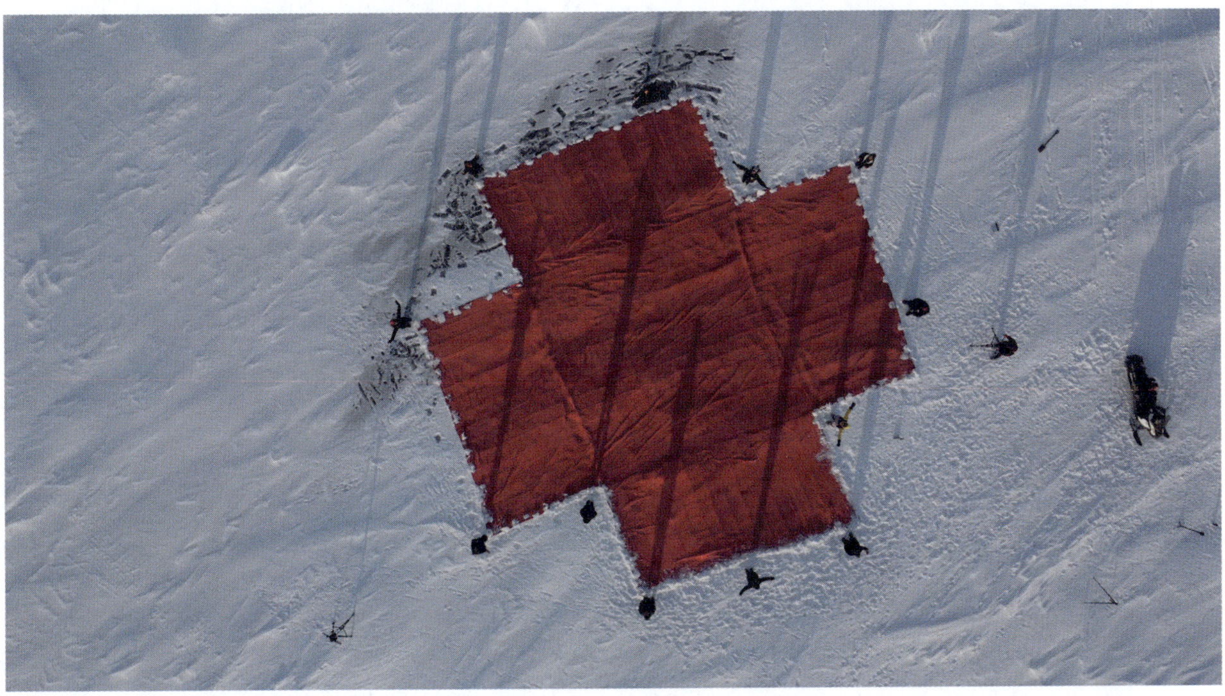

Nuna, 2023, Video, 9 min 12

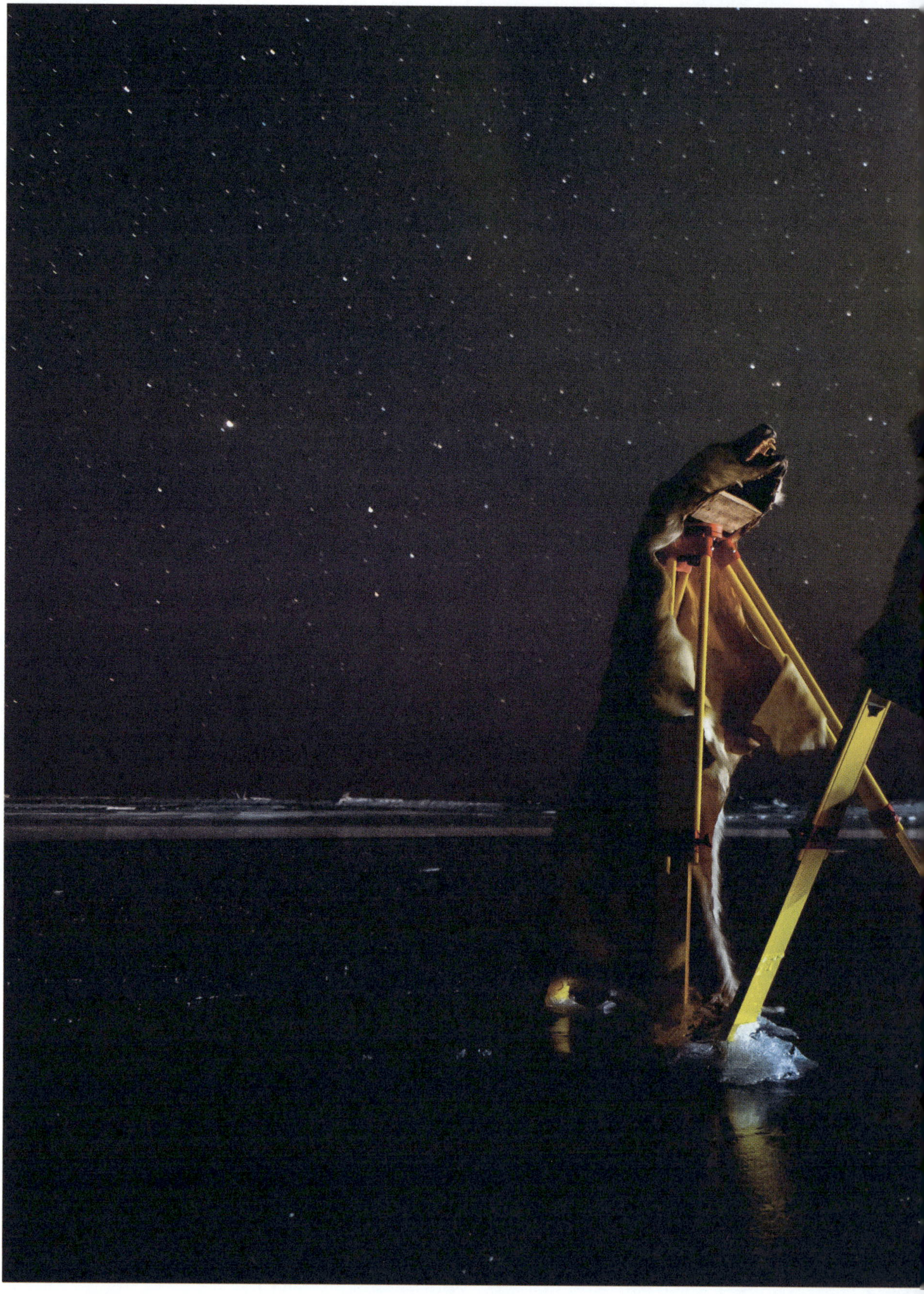

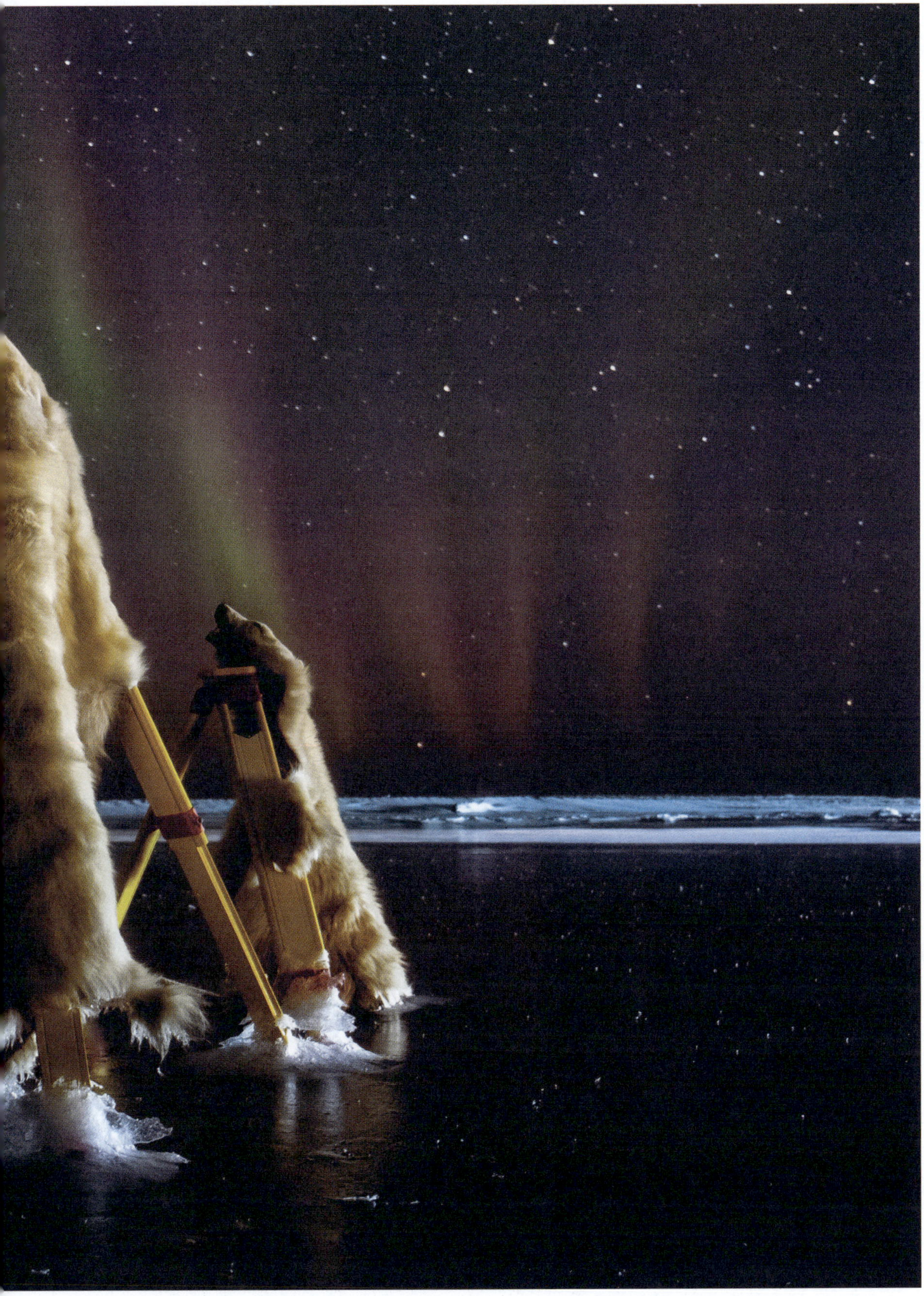

Aidainnaqduanni, Aurora, 2020, Archival inkjet on Hahnemühle Photo Rag Baryta, 80 × 120 cm

Deborah Drisana Jack

Born in the Netherlands in 1970

An artist and poet, Deborah Jack grew up on the southern side of the island of Saint Martin in the Caribbean Sea. In the seventeenth century, when Western colonizers "discovered" the island, they divided it into two separate countries: its northern French side, called Saint-Martin, and its southern Dutch side. Since then, the history of this island has been deeply tied to human and natural exploitation, especially with the activity of salt extraction, a flourishing colonial industry involved in the slave trade. Through her multimedia artworks, Jack explores this cultural heritage together with today's challenges faced by the local population in the context of climate change and land transformations, and overtourism. With a metaphorical and often poetic approach, she points to the hidden violence and complexities that constitute the Caribbean identity. Rather than trying to pin it down in clear definitions, she explores its multifaced dimensions, in line with the concept of the "Black Atlantic" coined by sociologist Paul Gilroy to reflect on diasporic communities. The sea, and its ever-changing shores, becomes a rich site to question these ambiguities and entanglements.

"My work engages with an oceanic and archipelagic conceptual and visual discourse through a Black feminist lens. It is the result of my investigation of the tension that exists in spaces that are both/and: sites of trauma and sites of healing, of arrival and departure, of corrosion and preservation."

Jack sees the shoreline as a liminal space. She observes the fluidity of the water as it interacts with the shore, the lines that are created by the encounter and the temporary quality of those lines. In her work, this perpetual motion embodies a form of memory and abstraction. "I am interested in how the collision of the sea with the shore causes subtle erosions that can serve as a metaphor for the erosions of memory." She readily quotes Saint Lucian poet Dereck Walcotts' poem "The Schooner Flight," which refers to sailors who love the sea that drowns them. This duality is constantly at play in her work: it also alludes to the enslaved people who found freedom in jumping from ships rather than submit to enslavement, and to the quality of salt, a medium she often uses, which preserves and corrodes at the same time.

For her, climate change and its effects on rising sea levels and coastal erosion make this interplay even more important. It calls on us to examine our relationship to the sea and our increasingly shared vulnerability.

"Does water have memory?" she asks. "What is the tension when the water and the land connect, when different bodies of water connect and when bodies and water connect? What are the shared vulnerabilities of colonizer and colonized when rising sea levels threaten the existence of both? The sea absorbs and expels. The currents and water cycles mean that the Black Atlantic is everywhere and on every coast." →

1. 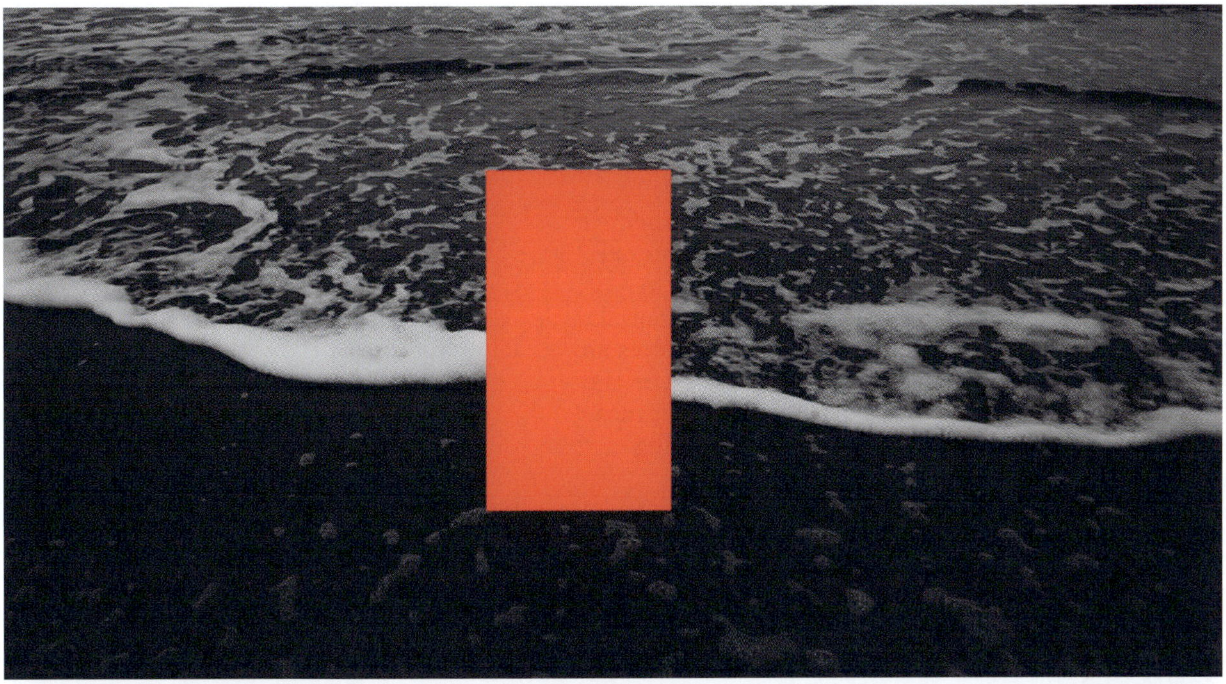

2. 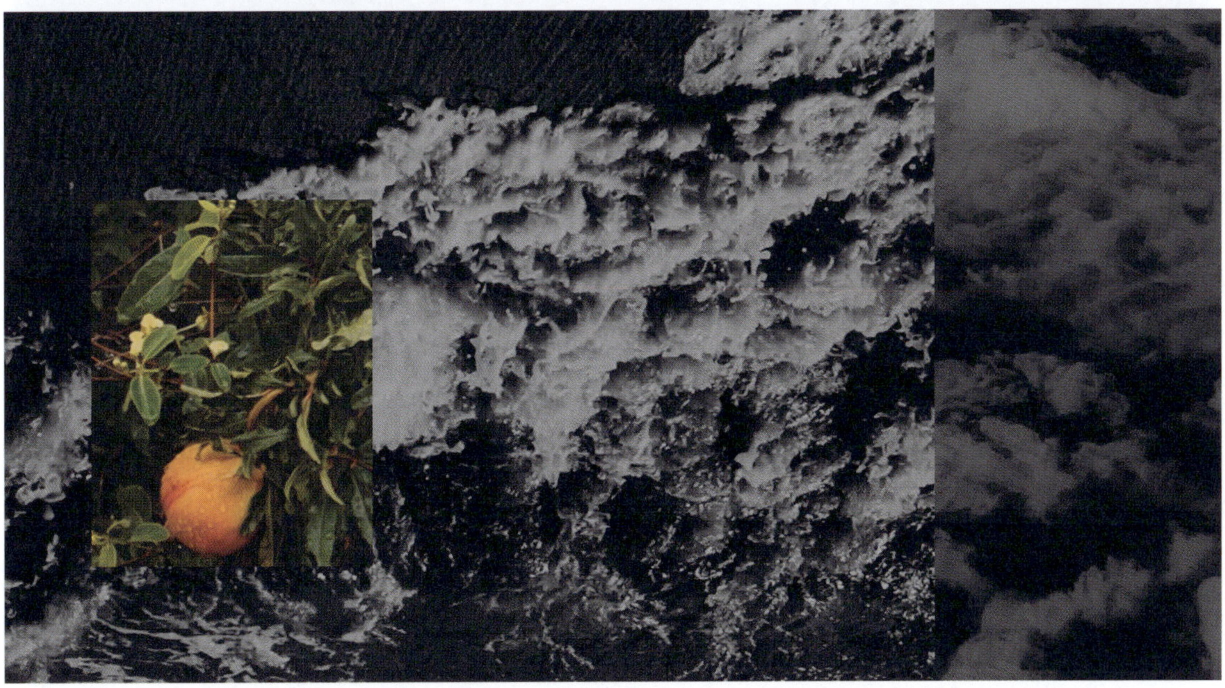

1. *Drawn by water: Sea Drawings in three acts, Act One: Weight/Wait on the Water*, 2018, single-channel video, 6 min
2. *The fecund, the lush and the salted land waits for a harvest...her people... ripe with promise, wait until the next blowing season*, 2022, 7-channel video, 8 min

The *Drawn by Water* series is a continuation of a central concept in her practice, that Atlantic hurricanes are the ocean's way of commemorating the Middle Passage and those who died on the voyages of slave ships, in a myriad of ways. For her, the hurricane represents the over one million perished souls swirling in what she sees as a seasonal memorial. "It reminds us that Nature requires an outlet for this trauma, imposed on the land and seascape. Today, hurricanes have intensified due to a warming planet, an effect of the extractive capitalism that coincides with the extraction of human bodies and fossil fuels. The series also addresses my belief that the shore and the coast are sites that resist mapping, that those grids and lines imposed on the landscape and sea are there for division, ownership, and control. However, I firmly believe that the micro erosions that occur at the edges, at the site of interaction with the sea, cannot be mapped."

Ange Arthur Koua

Born in Côte d'Ivoire in 1989

"Côte d'Ivoire, my country, has a border with the Atlantic Ocean. Along this border, which stretches from east to west, live the peoples known as 'lagoon peoples,' who are actually a subgroup of the larger Akan ethnic group from which I come. The entire societal life of these peoples is based on the ocean and the Atlantic Ocean is their 'mother.' Although this part of the ocean is not used by most migrants to reach the West, it remains the only opening to the sea for my country—an opening that allows one to dream of distant places, where the emptiness on the horizon that the ocean presents nourishes the desire to travel into the unknown. The Atlantic Ocean is the same everywhere on the globe. For me, as an admirer of Akan beliefs, the Atlantic Ocean and all other oceans are 'continents' of the soul, as they have welcomed many people in pursuit of their dreams." (Conversation with the artist, October 2024)

Ange Arthur Koua's compositions often combine painting and fabrics. Just like the Akan people, the artist believes in the existence and immortality of the *wawè* (soul), which can be found in the deceased's clothing. Koua collects old fabrics and cloths, imbued with these souls, and uses them as his medium to create large-scale paintings. *Waves* (2021), for instance, features recycled burlap cloth and jeans, which stand for those who tried to leave the country by sea, and who perished along their journey. For the artist, this is a way to give tangibility to what he calls the invisible continent of the ocean, constituted by these lost souls. His tapestry holds their memory and stories, as well as the hopes for the younger generation to reach, one day, what they see as an Eldorado. Clothing and human silhouettes in juxtaposition seem to form a boat, yet this time a solid one that would resist time and erasure. Burlap, for the artist, refers also to the bags that are used to transport raw goods from Côte d'Ivoire to Western countries. "How come goods can travel over the ocean, while our people cannot?" asks the artist.

In 2021, Koua created an ephemeral artwork on the ground titled *Those Who Leave*. He worked on the shore of the Bandama River, one of the largest rivers of the country, which flows into the sea. This charcoal and chalk mural, which was accompanied by a poem, conveys the same idea of migrating souls while exploring the relationship between those who stay and those who leave, in a literal and metaphorical sense. The sea is here perceived as this unique site where essential transitions take place, on the edge of which one can contemplate either a better future or death.

"Over our oceans, the cries of souls linger. On our seas, breaths can still be heard amidst the sound of the waves. I hear regrets, I hear torments . . . boarding the sea on a makeshift raft, with a suitcase full of dreams and a passport to death." (Arthur Koua, quoted by curator Armelle Voisin in the catalogue of the exhibition *Where the Sea Is*, Brest, France, 2021)

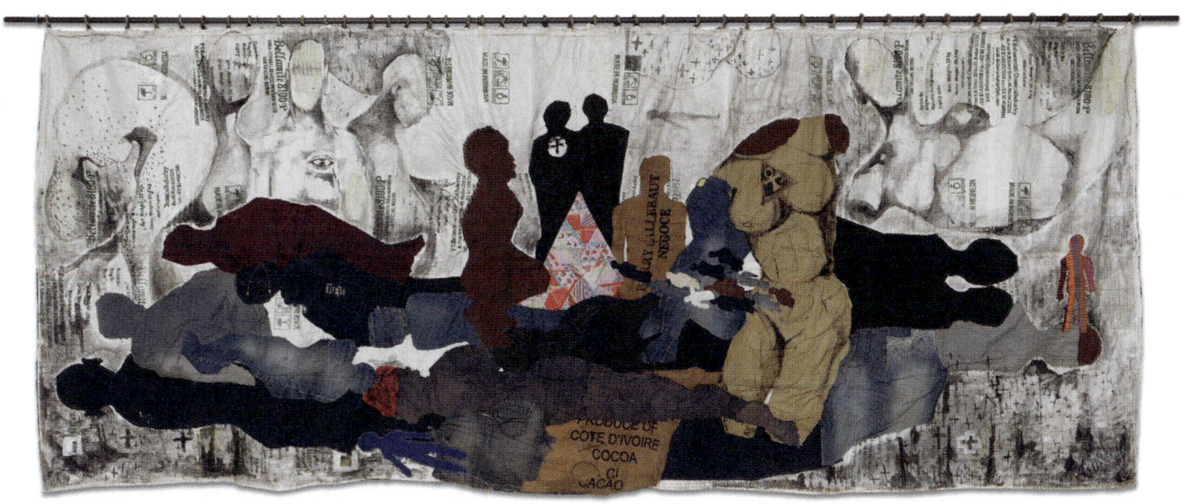

Vagues, 2021, Fabric and acrylic on plastic sheets, 200 × 470 cm

Louisa Marajo

Born in Martinique, France in 1987

"In Martinique, the sea is everywhere, it is within us."

Louisa Marajo lived in Martinique until the age of eighteen. She moved to France to pursue art studies and now lives between Paris and the Antilles. Her works–sculptures, paintings, and installations–are deeply rooted in the history of Martinique, where the artist grew up. They question the colonial history of the island, its fragmented identity, and the diaspora of its populations while highlighting the current ecological and socio-economic problems of the archipelago. Everything in her work reflects a strong sense of resistance despite the chaos of the compositions. For the artist, the sea perfectly embodies this tension, serving as a space of gathering and mingling, a bridge between nations, and a source of shipwrecks and dismemberment.

Marajo was told that the first book she ever held in her hands was a book about Hurricane Hugo, a tornado that devastated Guadeloupe in 1989. This, she explains, is the root of her obsession with chaos, wreckage, and fields of ruin. She speaks of a "chaos factory" when she juxtaposes remnants, scars, and debris as a way to embody complexity and resilience.

Marajo enjoys remixing old works, reusing images from her own creations in order to provoke accidents and to think anew about what is taken as given. She roughs up her own work to push it further and explore new territories. This approach recalls the definition of painting given by French philosopher Gilles Deleuze, for whom painting always arises from a catastrophe.

This catastrophe is clearly at work in Marajo's creations: for example, planks are stacked, sometimes broken or burned, architectures seem on the verge of collapse, images ignite. They evoke the aftermath of a typhoon, when nature has swept away everything in its path, haphazardly destroying and recomposing it into an incongruous and spectacular whole. Sometimes, it feels as if the artist was challenging the sea itself, especially with installations bearing such titles as *Awakening of the wave* (2019). Her architectural artworks resemble abandoned sets, and it is unclear whether something is being installed or, on the contrary, being uninstalled. This ambiguity evokes waves that rise and ebb away.

The issue is to recover some sort of unity in a chaotic, fragmented assemblage. The idea of fragmentation is often associated with Caribbean culture, notably under the influence of Edouard Glissant's writings. The poet and philosopher, also born in Martinique, coined the concept of "archipelagic thinking" to reflect on the diversity of the islands, which are nonetheless connected within an archipelago. "People always cite Glissant, but we could also quote Edgar Morin," the artist points out. →

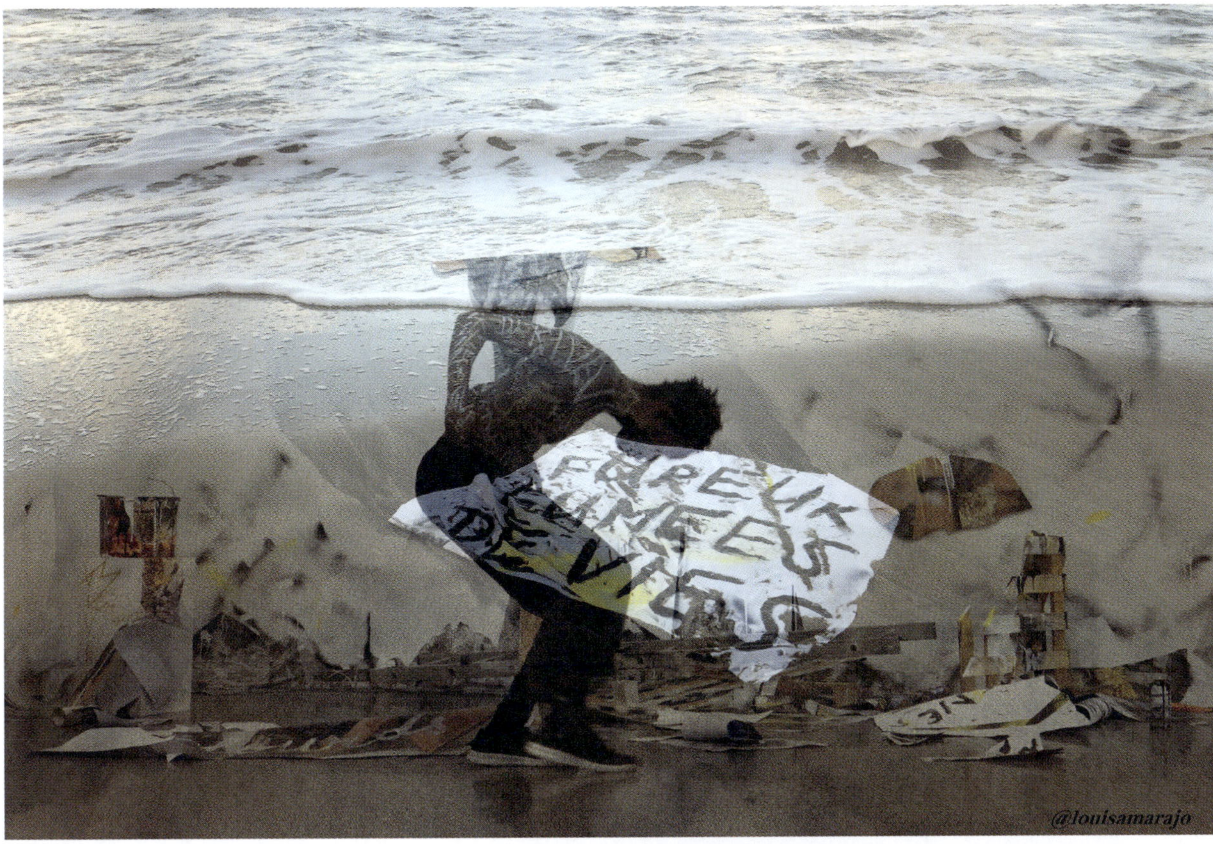

1.

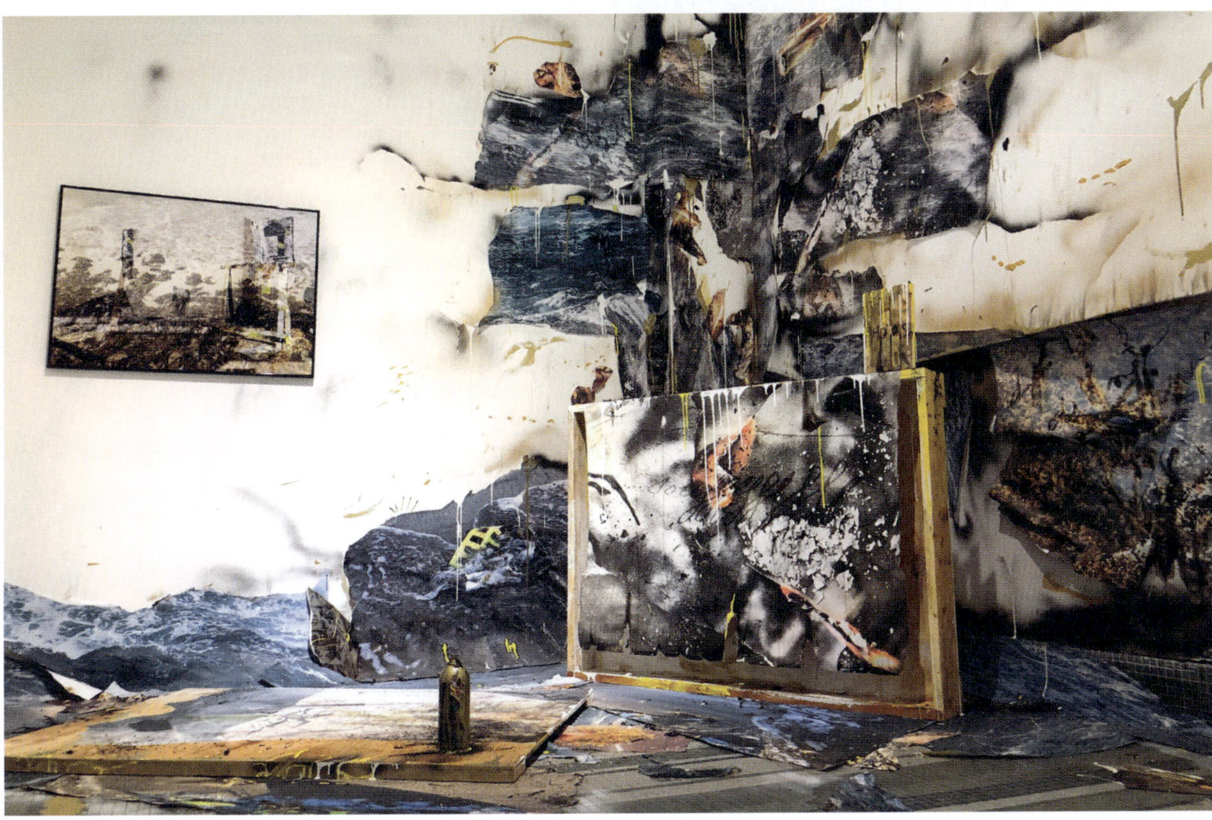

2.

1. *Fureurs*, 2021, Inkjet print on bamboo hahnemuhle paper, 80 × 55 cm
2. *The Burning Wave*, 2022, Site-specific installation

The French philosopher is the philosopher who developed The Theory of Complex Thinking, which he views as a source of richness.

This complexity is at work in each of Marajo's compositions. In *Fureur* (2021), for instance, the silhouette of a man stands out on the beach. His body is traversed by other images, by the swelling sea, and by a play of reflections and lights. He dances, tramples on pieces of paper and wood, on ruins, and on previous works of the artist, and it feels like a ritual. We wonder if he is praying to the sea, suffering, or beseeching fate to bring forth something new.

From 2019 onwards, Marajo began working with sargassum, the seaweed that increasingly washes up on the Martinican coasts. These algae are beautiful and golden, yet dangerous. Gradually, she introduced her photographs of sargassum into her architectural compositions to create ambiguity. Ultimately, it remains an element, like an additional residue, churned by the sea. "This seaweed is very problematic," she says, "It affects the health of people and the economy." Furthermore, the artist notes the similarity in media discourse when it comes to talking about waves of immigration and waves of harmful seaweed. People become like washed-up algae. The local government installs nets and tries to take other measures, but in France, everyone has lost interest because the problem is so distant. Above all, according to Marajo, only the symptoms are treated, not the causes, which are complex, numerous, and yet well identified.

The Antilles are also affected by dust haze, fine particles that arrive from the Sahara and are toxic to humans. These phenomena add to previous scandals like chlordecone, a pesticide that affects humans directly. Until 1993, it was used on banana plantations, and its consequencess are still felt. The artist speaks of a system that destroys everything and that violates the planet.

True ecology, for her, would mean deeply questioning ourselves and rethinking everything radically.

I have gardened the sea
With my atmospheric flowers
Standing totems
Between sky and Earth
I have gardened the sea
Setting our shores ablaze
Syncretic
Unbolted chaos
My flowers have grown
And transformed
The boundaries

Louisa Marajo, 2022

Barthélémy Toguo

Born in Cameroon in 1967

"The sea voyage is a journey not only of pleasure but also of death. Today, the ocean is a space of tragedy, but that's not what I want to represent. I imagine fish that live in the deep waters, where I have never been. It is a celebration of life." (Conversation with the artist, September 2024. All quotes from this conversation.)

Toguo conceives the sea as a dreamlike space in which he can dive. He has never gone scuba diving; he prefers to dream of these aquatic landscapes. He invents surreal fish that do not exist, beings that are distant. "I love the shape of fish and invent other imaginary ones. With the sea in mind, I often draw the blue dotted lines in my work. I imagine rain and oceans transforming a scene or landscape into a marine universe. I marvel at the fluidity of the water, which represents a gentle, non-violent world; its colour is undeniably beautiful. These depths are also dangerous, but it's clear that this substance offers more visual softness

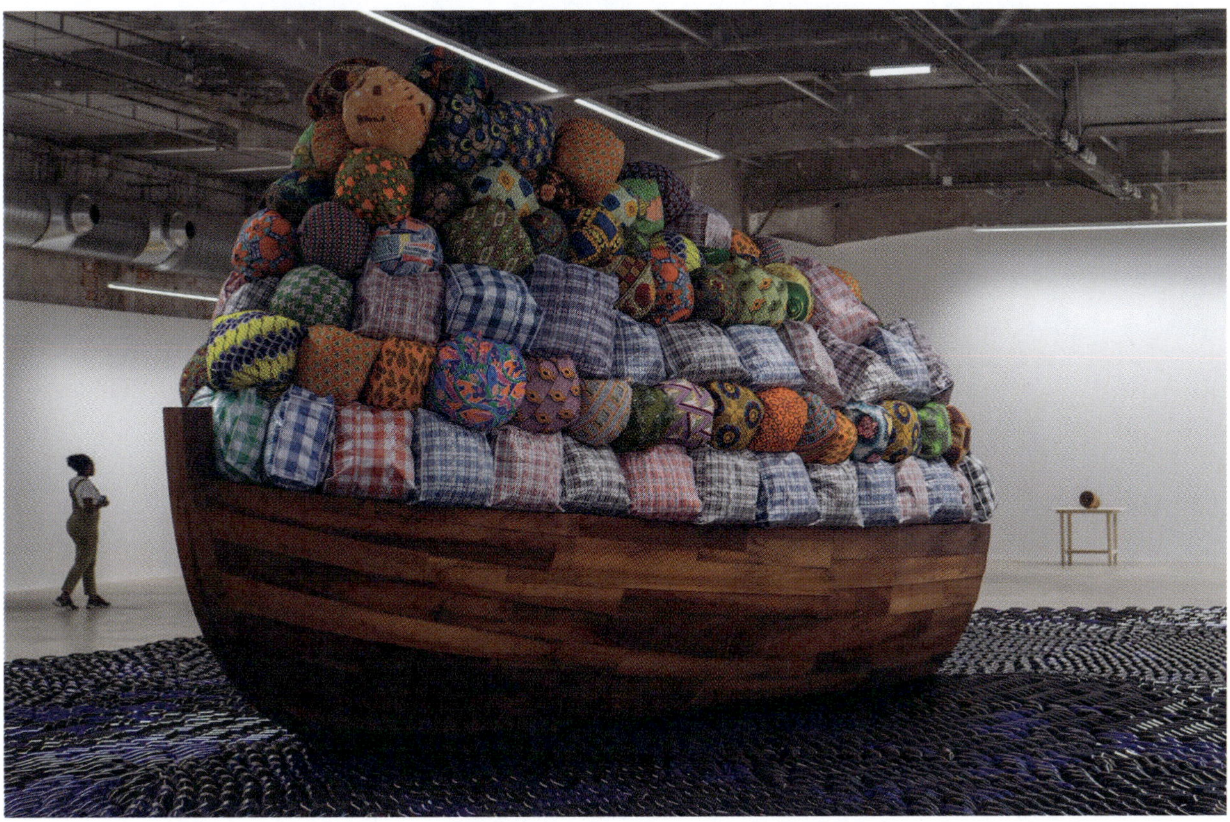

Road to Exile, 2008, Installation, 220 × 260 × 135 cm

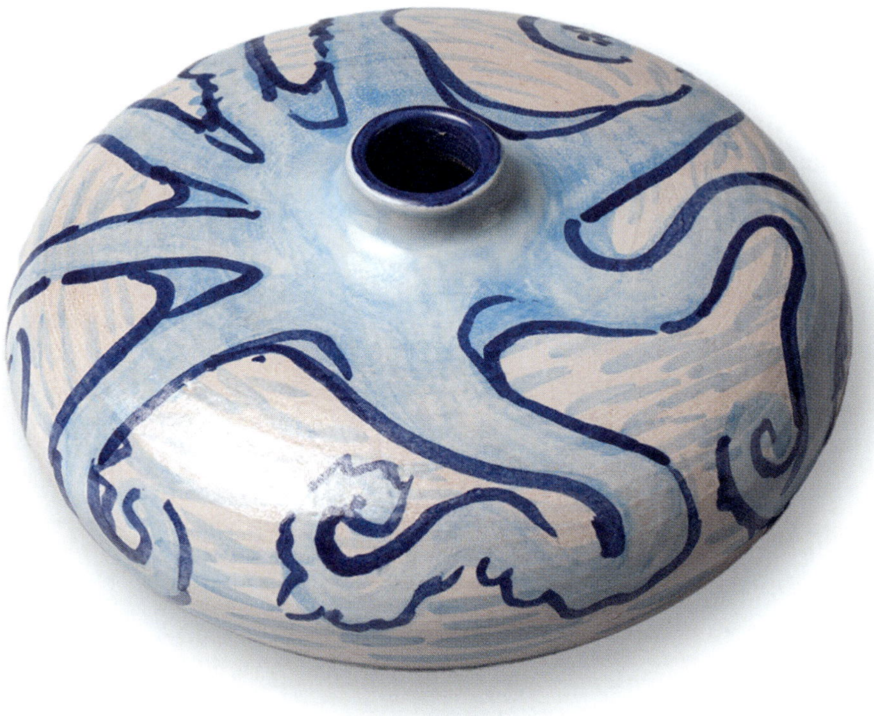

Balade aquatique XV, 2022, Enameled porcelain, 8 × 15 cm

than a rock, and the same applies to touch. The dream of falling into water rather than onto a rock creates greater serenity."

After Toguo's 2016 exhibition *The Deluge*, where he explored the excesses of water, he wanted to go "beneath the deluge" to see what was happening in the ocean depths. For him, water always represents both joy and disaster at the same time.

It is a theme that lies at the heart of his work. *The Deluge* was linked to a cascade of violence, and an installation of coffins occupied the central space of the exhibition. "Everyone leaves with these ideas of catastrophe," he confides, "but there is always hope. The sea brings this hope."

In his works, blue human bodies extend into jets of water, their limbs are fluid, jars of water overflow; they are

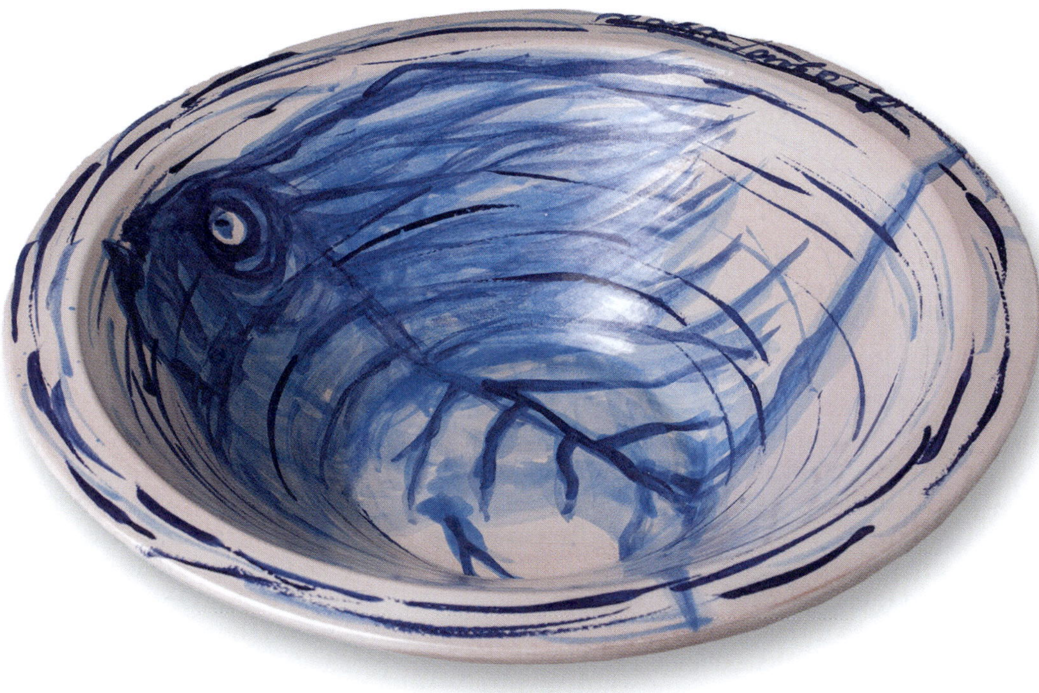

Balade aquatique XXVI, 2022, Enameled porcelain, 9 × 35 cm

shared and give birth to life. For Toguo, water is the essential element of life. Several of his exhibitions directly reference this issue, including the problem of the lack of drinking water in many parts of the world and the right to water. *Water is a Right* was the title of one of his recent exhibitions, as was *Water Matters*, a series of bottles from all over the world with painted scenes of people giving and receiving water. While he speaks here of fresh water, the sea is never far away, because social and ecological issues intertwine when it comes to water and the respect for natural resources.

Working in the realms of painting, drawing, sculpture, photography, performance, and installation, Toguo addresses enduring and contemporary issues of borders, refugees, and migrations as well as ecological issues, with the idea that sociopolitical and environmental concerns are deeply linked. Some of his large-scale installations feature colorful bundles made with African fabrics, sometimes piled atop overloaded boats. In 2013, he created Bandjoun Station, a foundation in Cameroon that hosts artists and academics from around the world in residential workshops, and develops propositions in harmony with the local population. He was named UNESCO Artist for Peace in October 2021.

Zai Kuning

Born in Singapore in 1964

Zai Kuning has spent extensive periods of his life navigating between the geographical and cultural milieus of Singapore, Indonesia, and Malaysia. His research-driven artistic practice reflects on the fluidity and interconnectedness between these worlds, a fluidity that is particularly embodied in the mode of living of the Orang Laut, a community of sea-faring people from the Riau Archipelago with whom he has immersed himself and collaborated. This experience has been an integral part of his broader research on the history, maritime trade, and seafaring traditions that have shaped Malay people and culture over centuries.

Zai's video work *Riau* (2005) relates his own personal narrative as he searched for and gradually became accepted by the Orang Laut community. The video follows his investigative process and documents the seafarers' way of life and traditions, including their musical and performative cultural practices.

"The Orang Laut were one of the first people to inhabit this region. They are still around but as nomads

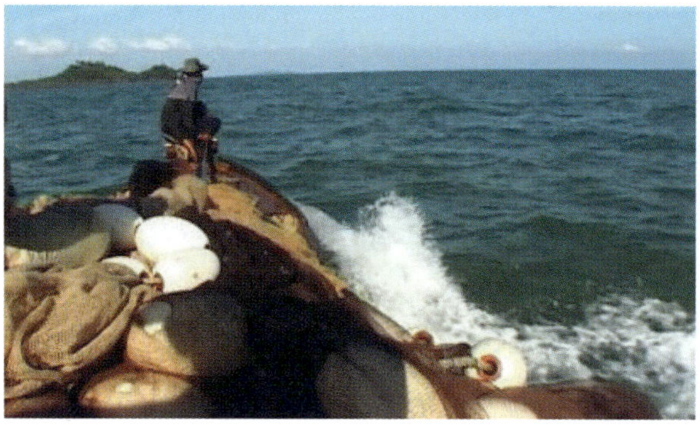

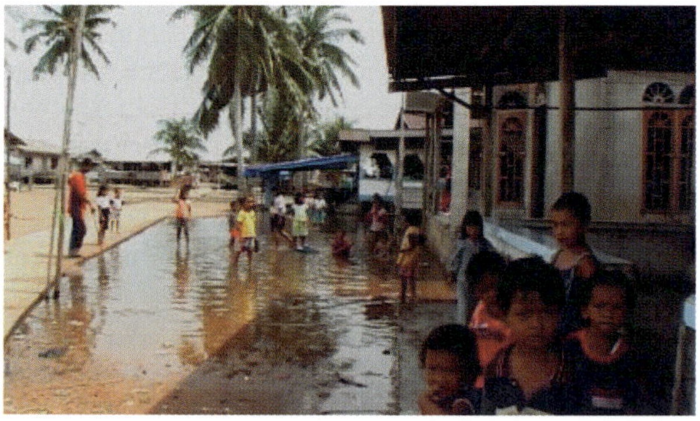

Riau, 2003, Video, 29 min 45

and animists, they are victims of discrimination. The nature which they live in is all gone because of commercialism and 'development.' When you see beach resorts and mines, it means progress, but it also means that the Orang Laut are dying slowly. There is no place they can go without having to justify their presence. They have not only lost their natural habitat, but are also discriminated against by migrants who have now become settlers and are the majority. Like many other indigenous peoples in the world, they have become pariahs in their own domain; their gentle way of living based on interconnectedness has become impossible due to capitalism." (Catalogue of the exhibition *Dapunta Hyang: Transmission of Knowledge* – Zai Kuning, National Arts Council, Singapore, 2017).

For seafaring communities, national boundaries hold little significance, as they have traditionally navigated freely between diverse geographical regions. Ancient cartographic records reveal that the Riau archipelago comprised parts of present-day Malaysia (Johor-Melaka), Singapore, and Indonesia (Bintan, Batam, Daik, and Lingga). Moreover, the maritime domain of the Orang Laut extended even further, encompassing territories as distant as Taiwan, the Philippines, and Thailand. This conception of space, with the sea as the primary frame of reference, lies at the core of Zai Kuning's installation work Dapunta Hyang: Transmission of Knowledge, exhibited at the Venice Biennale in 2017. For this project, the artist constructed a full-scale boat using the techniques of the Srivijaya maritime kingdom, a thalassocratic empire that dominated the broader Malay world between the eighth and twelfth centuries CE. The imposing presence of this seafaring vessel serves as a symbolic reminder of the fluidity and interconnectedness that have characterized the region's cultural and commercial exchanges over centuries. Through this work, Kuning aims to heighten awareness about multifaceted cultural origins, especially in a cosmopolitan port city like Singapore, where diverse ethnic groups have engaged in maritime trade and navigation for generations.

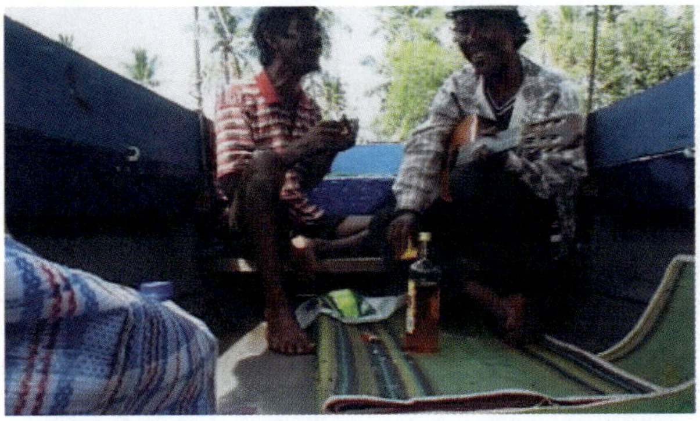

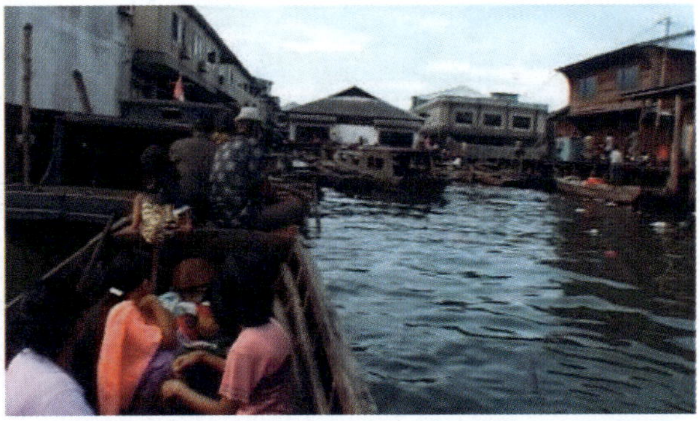

Riau, 2003, Video, 29 min 45

Rémy Rioux

CEO of the French Development Agency and Chairman of Finance in Common (Paris)

At the beginning of the sixteenth century, Joachim Patinir painted *Landscape with Charon Crossing the Styx*, a work that embodies humanity's collective imagination regarding water and the ocean. This painting remains strikingly modern, especially as climate change now threatens both human and non-human life. In Greek mythology, Charon transports the souls across the river Styx. Here, he ferries one of the Oceanids, the many daughters of the Titan Oceanus. The river Styx divides the painting down the centre with Charon in the middle hesitating between Heaven, the land of plenty, Hell, where resources are scarce, and Limbo, the blurred in-between state.

This Limbo could well symbolize the international community's weak commitment to the oceans so far, despite the 2030 Agenda including ocean conservation as its Fourteenth Sustainable Development Goal (SDG 14). The ocean is not a "Terra Incognita" anymore but this SDG remains the least funded.

The ocean's unusual status–mostly outside national jurisdiction–makes it a challenge for financial actors, particularly Public Development Banks (PDBs) which act on behalf of governments and primarily focus on their land-based territorial issues.

But oceans know no borders or mandates. The 536 PDBs worldwide, all gathered in the Finance in Common (FiCS) movement, are now coordinating to finance concrete actions to protect the oceans and support a sustainable blue economy. Examples include the Chinese and French PDBs partnering to clean up Hahn Bay in Senegal, and the Kiwa Initiative led by the European Union, Australia, and Canada for investing in the sustainable management of Pacific Island coastal ecosystems. I could also highlight the mobilization of public finance for countries that rely significantly on maritime trade for growth and choose to pursue sustainability. Indonesia, for example, is developing eco-fishing port projects to enhance the added value of fishery products and mitigate the sector's environmental impacts. French overseas territories are also at the core of AFD strategy.

Public financial actors are essential for instigating quality projects, securing funding, catalyzing private investment, and measuring impact. These are the goals of the FiCS Ocean Coalition, aiming for a systemic transformation towards a sustainable ocean-based economy.

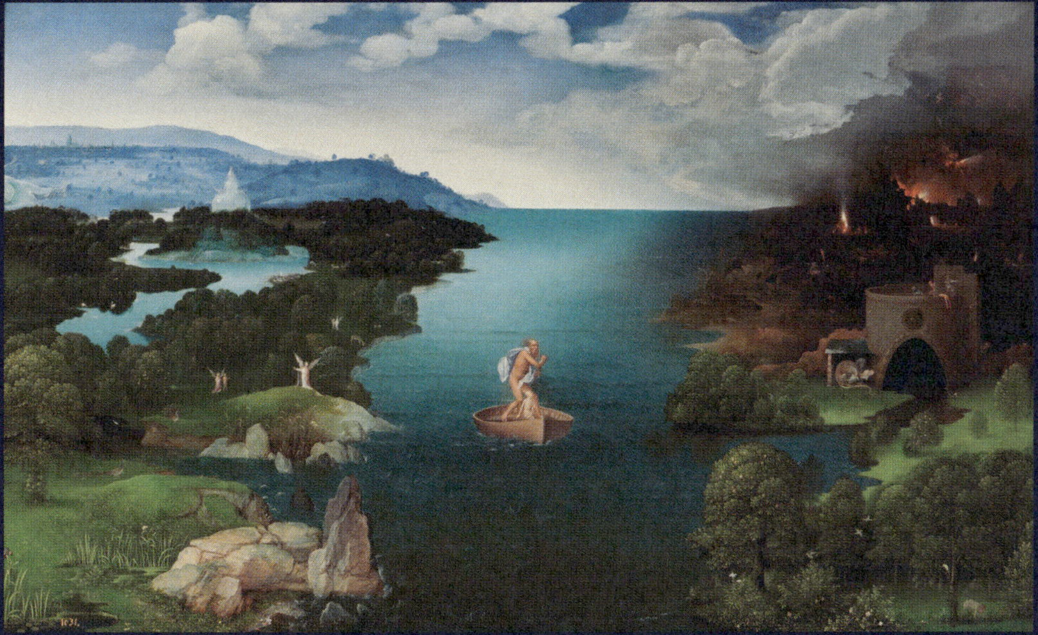

Joachim Patinir, *Charon crossing the Styx*, 1520-1524, Oil on panel, 64 × 103 cm

Just as in the allegory of the Styx, the ocean separates and connects us, symbolizing human fear and the need for cooperation in crossing unknown territories. From Renaissance transatlantic journeys to modern and all-too-often tragic Mediterranean crossings, the ocean is inherently an issue of development inciting us to understand one other and act for the common good.

And just as the *Ocean of the Streams of Stories* written by Somadeva Bhatta in the eleventh century contains multiple layers of stories within stories, I believe this "Ocean Manifesto" illustrates the cosmogony of the ocean. Similarly, the Metis Art and Development Fund offers development projects in the diverse contexts in which they are implemented, as artistic stages. Because our role as PDBs is to take into account the entire cosmogony to better address the ocean's multiple challenges. Like Charon steering his rudder, let us ensure that the container ship representing the international financial system is usefully guided by the "tuggy boat" of public finance, to steer ocean finance in a positive direction.

A Living Entity to Deal With

The sea has a vast expanse, its calm swish and its wild laughter, its mirages, its leaping whales and roaring dragons. Its surf rises like hill crests and its ebb resembles mountain ranges.
One might think of the sea only as the sea and of the mountains strictly as mountains only, which would be a mistake. To me mountains are seas and seas are mountains, and the seas and mountains know that I know.

Shitao, Sayings on Painting from Monk Bitter Gourd

Ocean! You permit, you are the accomplice, maker of stars

Edouard Glissant

1. Achille Adonon
Born in Benin in 1987

2. Felipe Fittipaldi
Born in Brazil in 1982

3. Naiza Khan
Born in Pakistan in 1968

4. Gideon Mendel
Born in South Africa in 1959

5. Paribartana Mohanty
Born in India in 1982

6. Shneider Léon Hilaire
Born in Haiti in 1990

7. Judy Watson
Born in Australia (Waanyi) in 1959

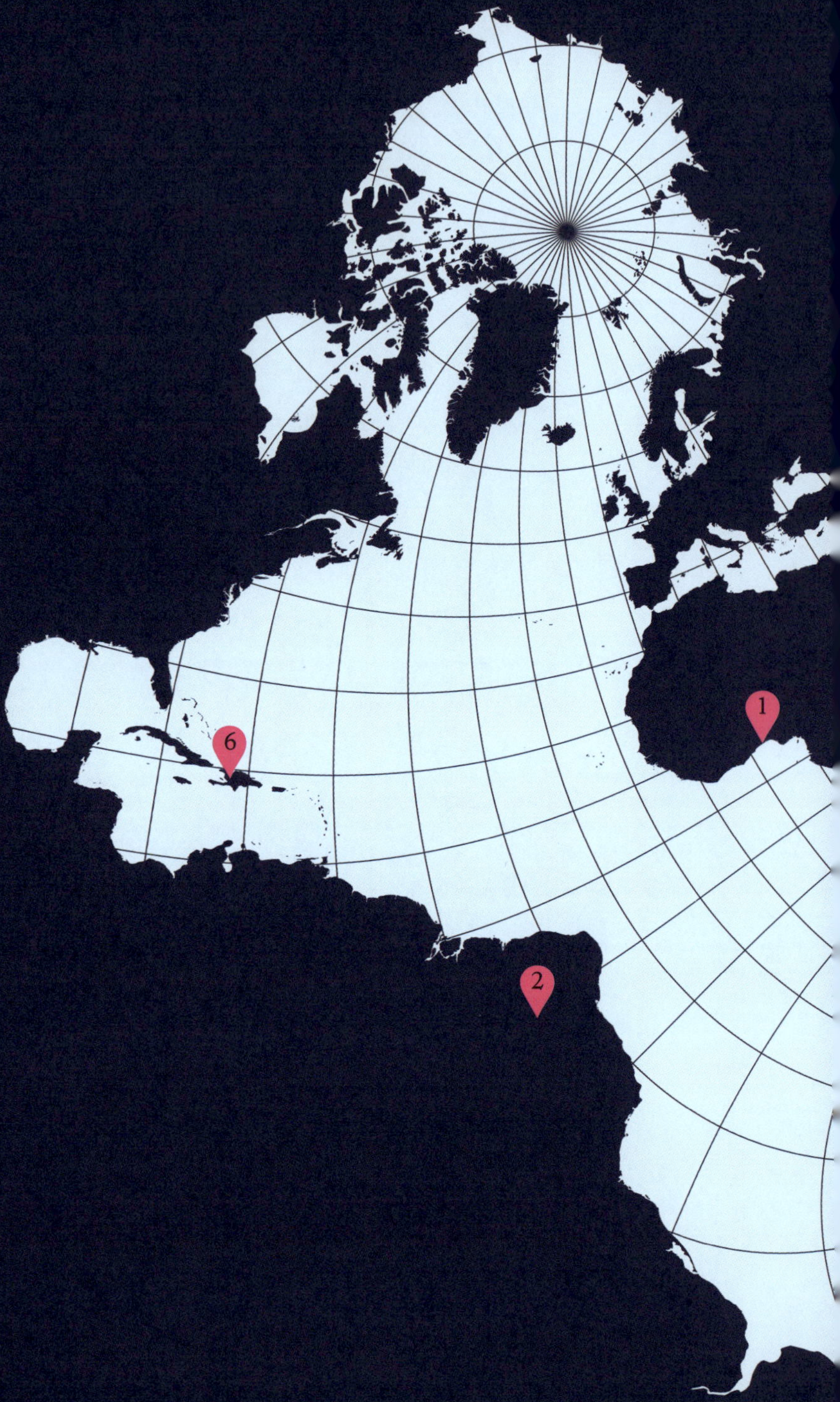

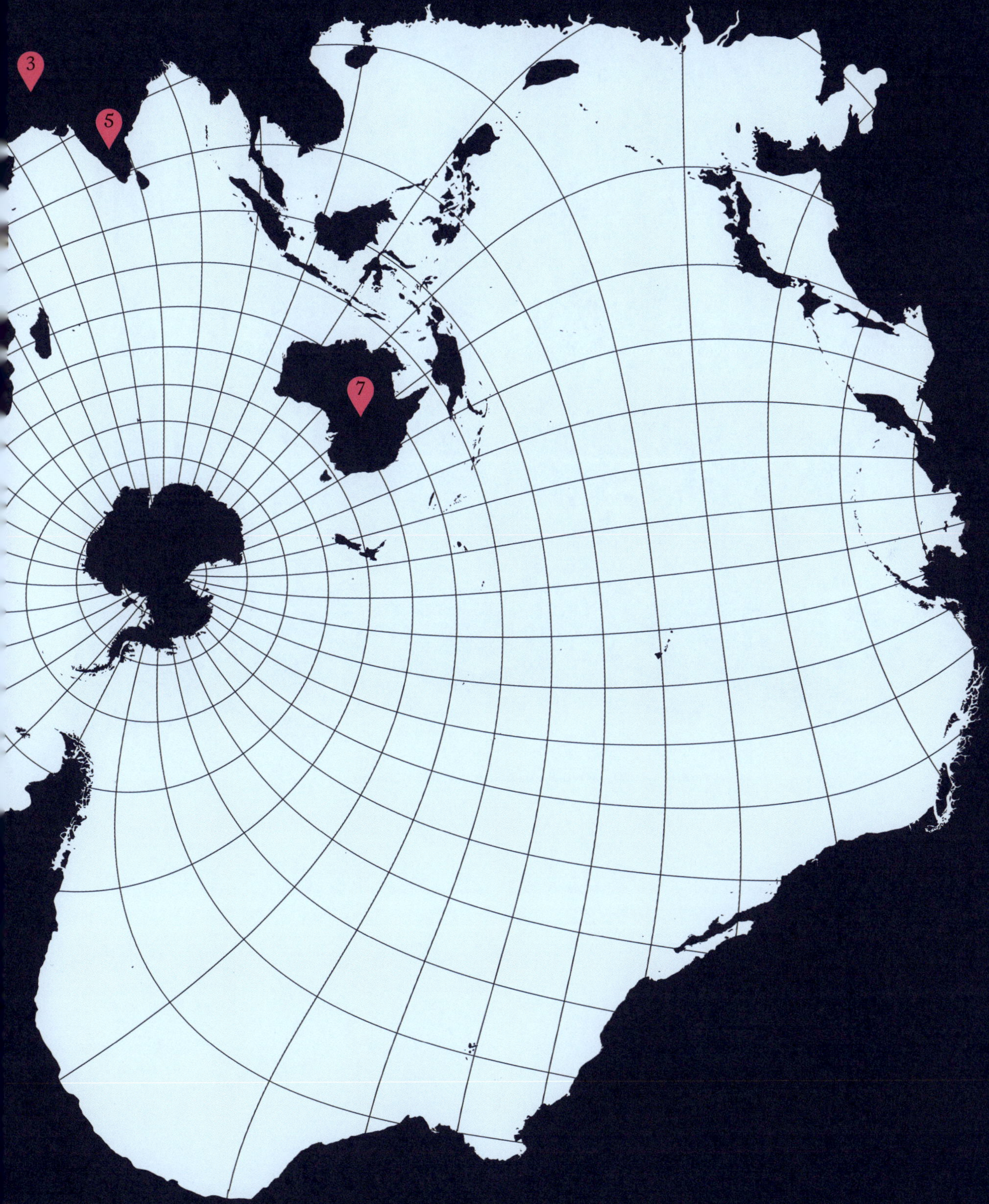
3
5
7

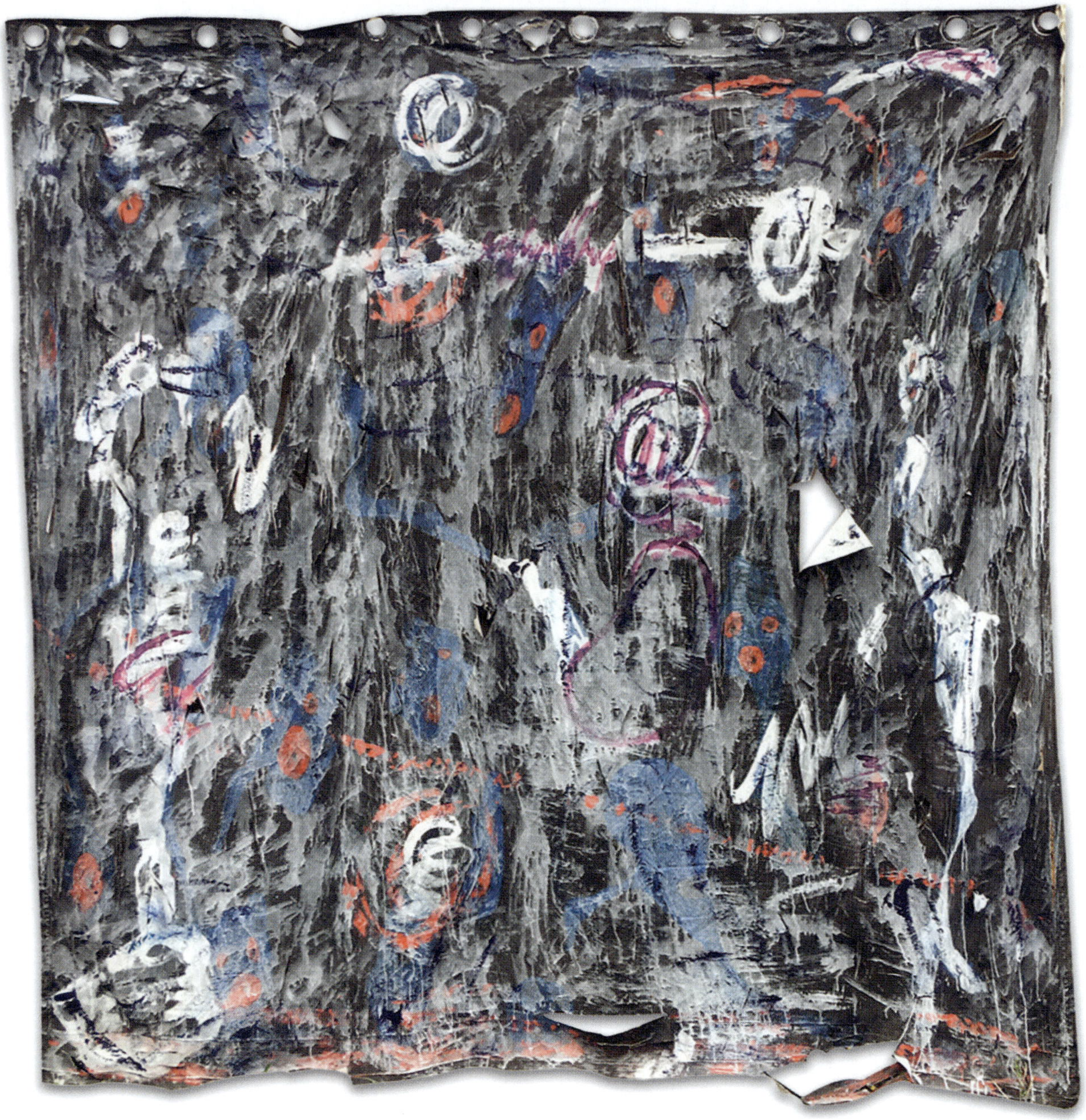

Accalmie I, 2020, Mixed media on canvas, 367 × 256 cm

Achille Adonon

Born in Benin in 1987

Les Fantômes de la mer (Ghosts of the Sea) (2018) is a series of abstract paintings that explore the enigmatic manifestations of Mami Wata, the water spirit omnipresent in Benin and neighboring regions. In his practice, Achille Adonon likes to reflect on the influence and spiritual significance of the ocean in the local culture: "I grew up with these beliefs. My work on the ocean is intimately connected with the myths surrounding Mami Wata. Today, due to the impact of human activities on the oceans, these spirits are endangered. While I cannot speak on their behalf, I feel their suffering as a result of our behaviors and actions."

Adonon's daily life has been shaped by witnessing the followers of Mami Wata perform rituals along the beach, an experience that profoundly influences both his personal perspective and artistic expression. In his view, recognition of her existence, accompanied by respectful engagement, is believed to offer protection and prosperity. Conversely, Mami Wata can be destructive when her presence is ignored or disrespected. In such instances, the spirit is manifested as a seductive entity, embodying themes of lust and perversion.

Within this series, the metaphor of Mami Wata functions as a narrative tool to address contemporary ecological crises. The marine spirits are portrayed as benevolent forces whose relationship with the world has been disrupted by the pollution and degradation of the oceans. Here, the concept of lust extends beyond its traditional associations, symbolizing the destructive human actions that are transforming the seabed into a metaphorical battlefield.

In contrast, in *Accalmies* (Respite) (2020), Adonon expresses a sense of serenity as he surrenders to the energy of the spirit. "Everything we do is connected to the spirit, and we are guided by it. In my artistic practice, there are moments when I relinquish control, becoming a mediator between the material world and the entity that speaks through me." For him, this spiritual connection serves as a source of healing and renewal. In both series, though, the artist's brushstrokes are very vivid, communicating a strong sense of dynamic and energy that characterizes the aquatic life with its currents, flows, and inner life.

Adonon emphasizes the importance of re-establishing a relationship with the spiritual beings that bridge the visible and invisible worlds, arguing that only by listening to them can humanity learn to respect and preserve the environment. To him, the ocean represents not only a source of material wealth and spiritual escape but also a fragile world that urgently requires protection.

Felipe Fittipaldi

Born in Brazil in 1982

"I was born and raised in Rio de Janeiro, a city whose atmosphere is defined by the relationship between its inhabitants and the ocean as an integral part of the people's identity, shaping their communal and individual experiences. Growing up amidst the city's iconic beaches and marine landscapes, I experienced firsthand the beauty and importance of this natural environment. This direct connection often heightens awareness of environmental issues, as you witness the impacts of these problems on familiar and cherished spaces. The emphasis on environmental issues manifested in my visual projects arises from the profound connection I've had with the ocean since childhood."

With the *Eustasy* project (2014-23), Brazilian photographer and video maker Felipe Fittipaldi documents the impact of climate change on Atafona, a small town located in the delta of the Paraíba do Sul River in Southeast Brazil.

"Over the past few decades, the sea has been gradually submerging the town, forcing hundreds of people to abandon their homes. Its dunes now conceal about 400 buildings, including public spaces, residential blocks, a hotel, a gas station, and a church. A combination of factors, including rising sea levels and disastrous human interventions along the river, has made Atafona one of the most significant cases of coastal erosion in Brazil."

The Paraíba do Sul River is a vital resource for Brazil's largest cities, supporting approximately fourteen million people. The burgeoning hydric deficit at the estuary, a result of human exploitation, is one of the primary factors propelling its rapid erosion. The weakened water flow can no longer maintain a balance with the ocean in the delta, leading to insufficient sediment replenishment and an inability to counteract the advancing sea.

Eustasy is the outcome of a long-term visual exploration of the complex relationship between a community and its environment, characterized by dependency, melancholy, and individuals confronting the relentless degradation of their physical world. According to the United Nations Development Program, by 2050, around 200 million people worldwide could be displaced due to shoreline erosion, coastal flooding, and agricultural disruption.

Viewing the ocean from this perspective, as a living entity with its own life and cycles, allows us to better grasp its movements and prepare for its impending expansion.

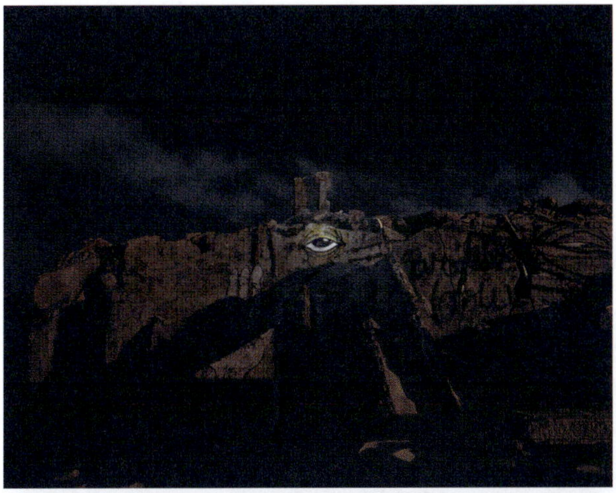
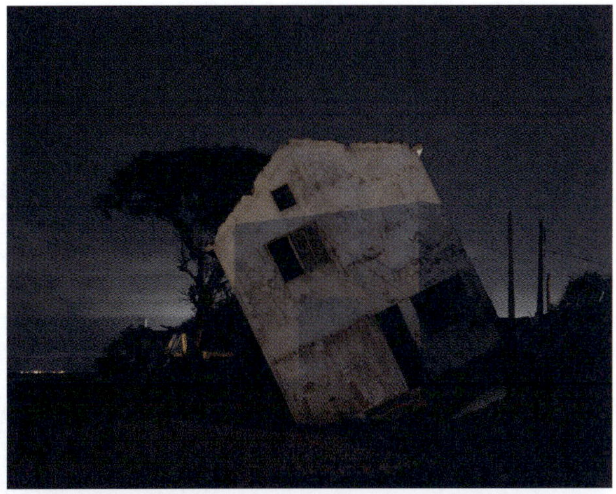
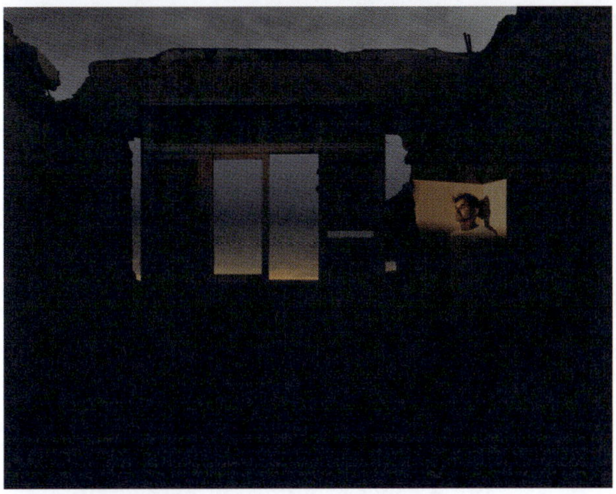
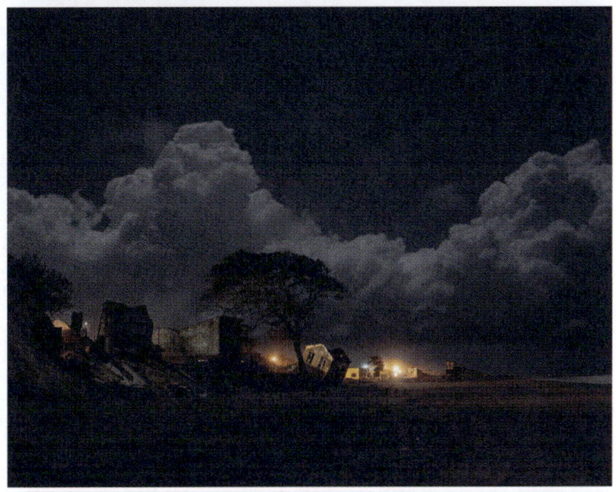
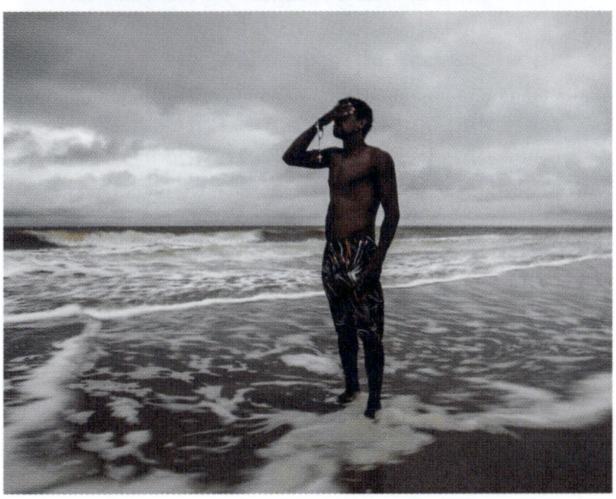
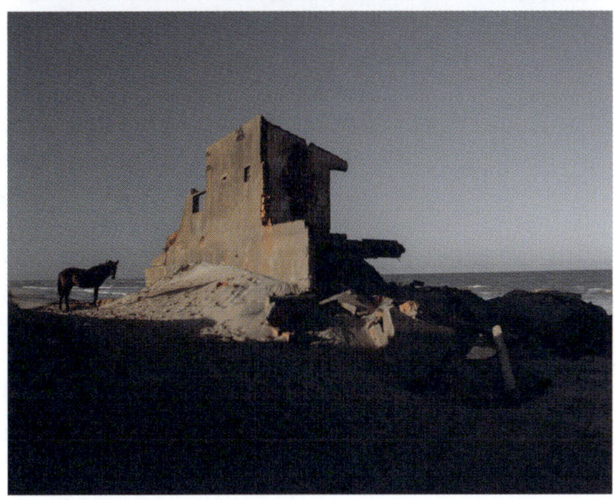

Eustasy, 2014-on going, printed in Hahnemühle Photo Rag, 48 × 60 cm

Naiza Khan

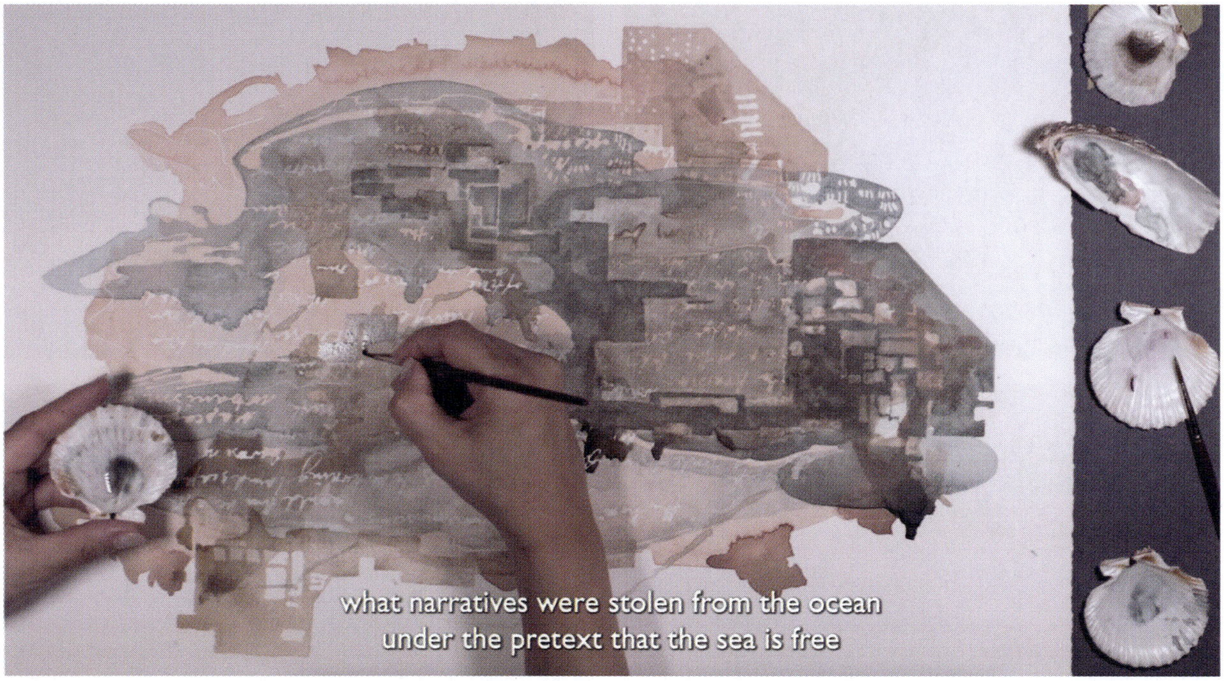

Mapping Water, 2023, 4K video, 20 min

Born in Pakistan in 1968

Living in the port city of Karachi, Naiza Khan has developed a strong and complex relationship with the sea over time. The history of the metropolis is deeply connected to the ocean, which has brought the city both prosperity and sources of conflict. Under the British, the port was considerably developed and it is through the sea that Karachi has been and continues to be connected to Africa and Asia. Today, this network is strengthened with the emergence of the new Silk Road.

Naiza Khan's study of the Manora peninsula, which began in 2008 and lasted over ten years, resulted in a series of artworks titled *Manora Field Notes*. This long-term research project was grounded in regular site visits, discussions with diverse local stakeholders, including fishermen, boat pilots, temple keepers, and local cafés, and parallel dialogues with scholars. This living, situated knowledge was essential in informing the artist's own narratives, which were also shaped by her study of archival materials, particularly historical accounts of storms and cyclones. For Khan, the proximity of the island, located just twenty minutes by boat from the mainland pier, allowed her to move between different temporalities, as "the water body pulled [her] in." Manora's history as a multi-religious space with numerous sites of worship, as well as its role as a defense outpost from the eighteenth century onwards, further underscored the island's significance as a microcosm of Karachi's evolution, reflecting the city's multiethnic character shaped by its position as a hub on maritime trade routes.

This long-term research continues to inform Khan's work, serving as a foundation for her broader explorations of port infrastructures, climate change, and our connection to the ocean. Her challenge is to develop a visual language that can embody the complexities of the tangible and intangible layers of history, coastline changes, and power structures that characterize the dynamic relationship between Karachi and the sea.

Increasingly, the artist is also considering the monsoon as another key factor influencing the region's development and how climate conditions are affecting the city. During the monsoon season, life in Karachi often grinds to a halt, with the city recently experiencing massive flooding events. The local landscape is in a constant state of change, caught within a cycle

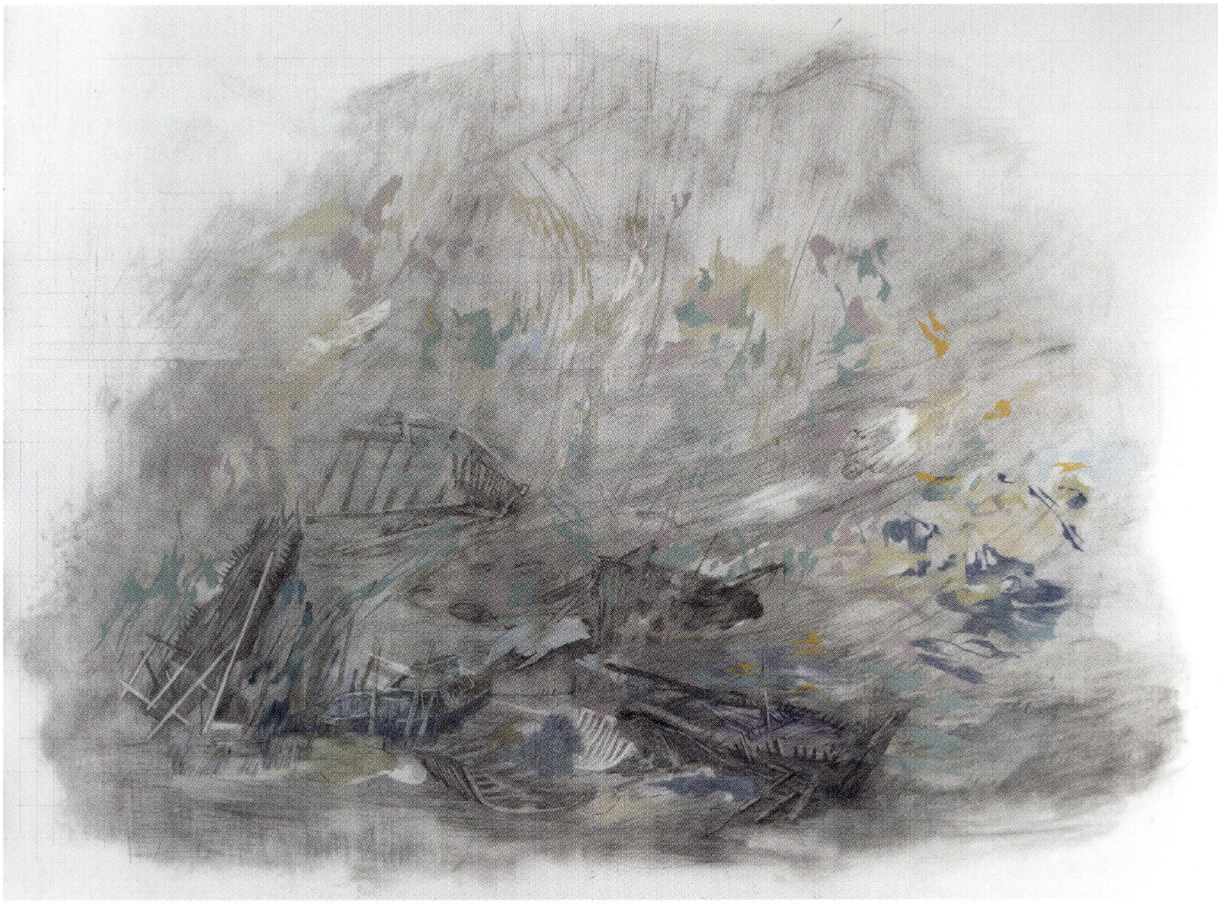

Ground Truth, 2024, Charcoal and Conté on paper, 140 × 180 cm

of destruction and reconstruction, damages and repairs.

In 2022, a major flooding event in Karachi and the surrounding region devastated local livelihoods in the delta area. This disaster revealed the precarity and vulnerability of these communities in the face of climate-related catastrophes. The government's lack of preparedness and limited capacity to manage such crises is evident. The artist suggests that in some ways, nature appears to be reclaiming what has been colonized by human activity. In the video work *Mapping Water* (2023), she is shown in her studio, drawing maps of various cities. She uses the highly fluid medium of watercolor to embody and give voice to the accumulation of water, allowing the material to evolve and express itself. The resulting series of drawings depicts water as a pervasive, transformative force. The maps become increasingly subsumed by water, culminating in a poetic yet dramatic gesture where the cartographic representations dissolve entirely into the aqueous medium. Through the interplay of drawing, material, and representation, the work grapples with notions of repair, vulnerability, and the dynamic interrelationship between human and non-human systems.

"It is fascinating to consider how artists have previously represented the ocean and what that has reflected about the collective understanding of the sea, from the emotive seascapes of Hokusai and Gaspard David Friedrich to imperial projects documented in maritime museums."

Ultimately, Khan's practice speaks to our collective perception and understanding of the ocean. For her, it constitutes an archive of diverse memories and trajectories, and she keeps investigating how we have collectively built its representation over time and across cultures. Yet, today, the question might rather be reversed: how does the ocean want to be represented?

Gideon Mendel

Born in South Africa in 1959

A photographer and environmental activist, Gideon Mendel began his career by documenting the apartheid regime in his native country, South Africa, attentive to political powers and social injustice. In the 1990s, he produced works related to the issue of HIV/AIDS, photographing patients in hospital ward at a time when the illness was still largely a taboo. For Mendel, it is crucial to engage with the pressing issues confronting the society in which he resides. Upon the birth of his children, he started contemplating the future world they would inhabit when they reached his age, focusing

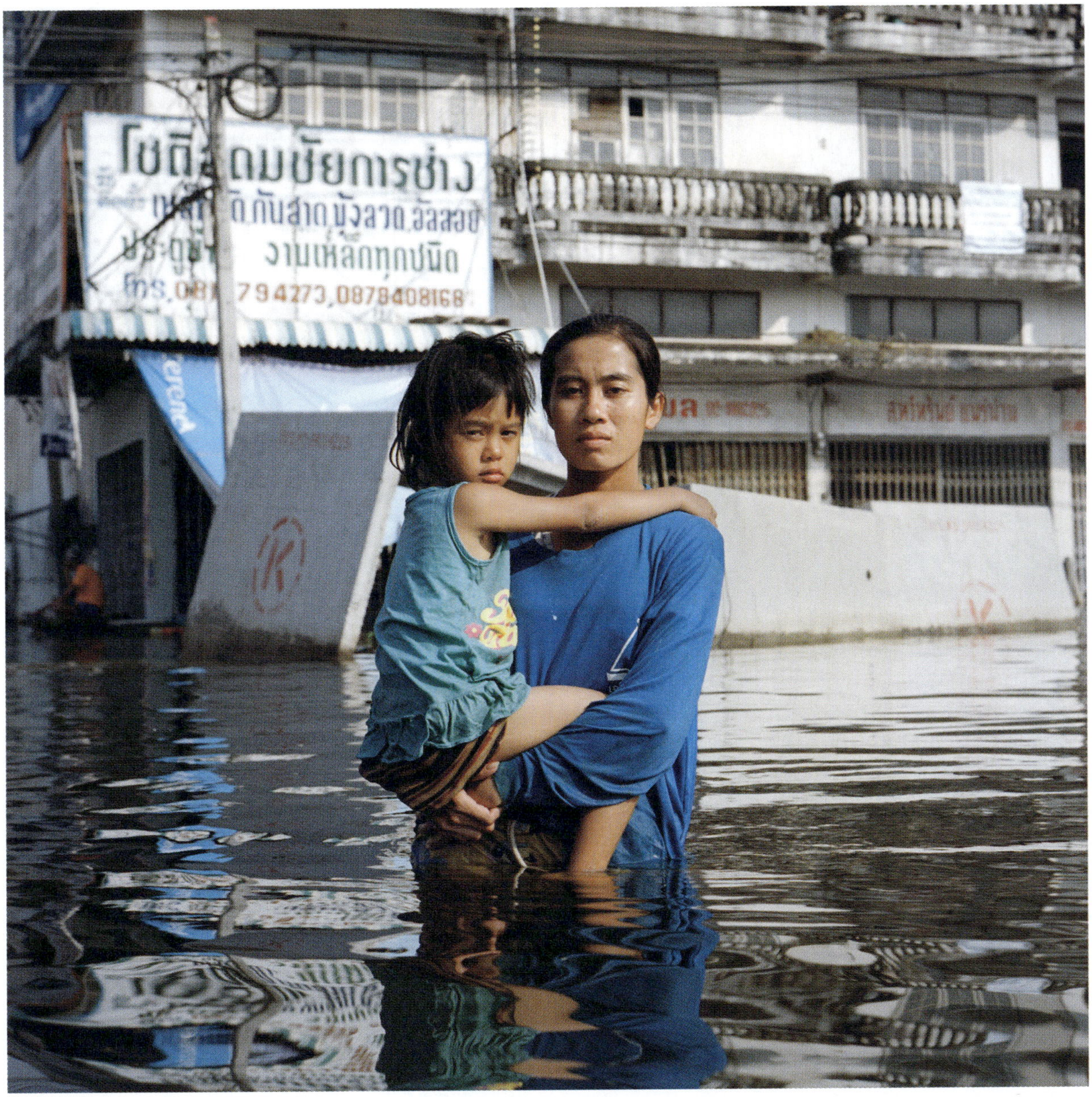

Submerged Portraits Series: Anchalee Koyama, Bangkok, Thailand, 2011, variable dimensions

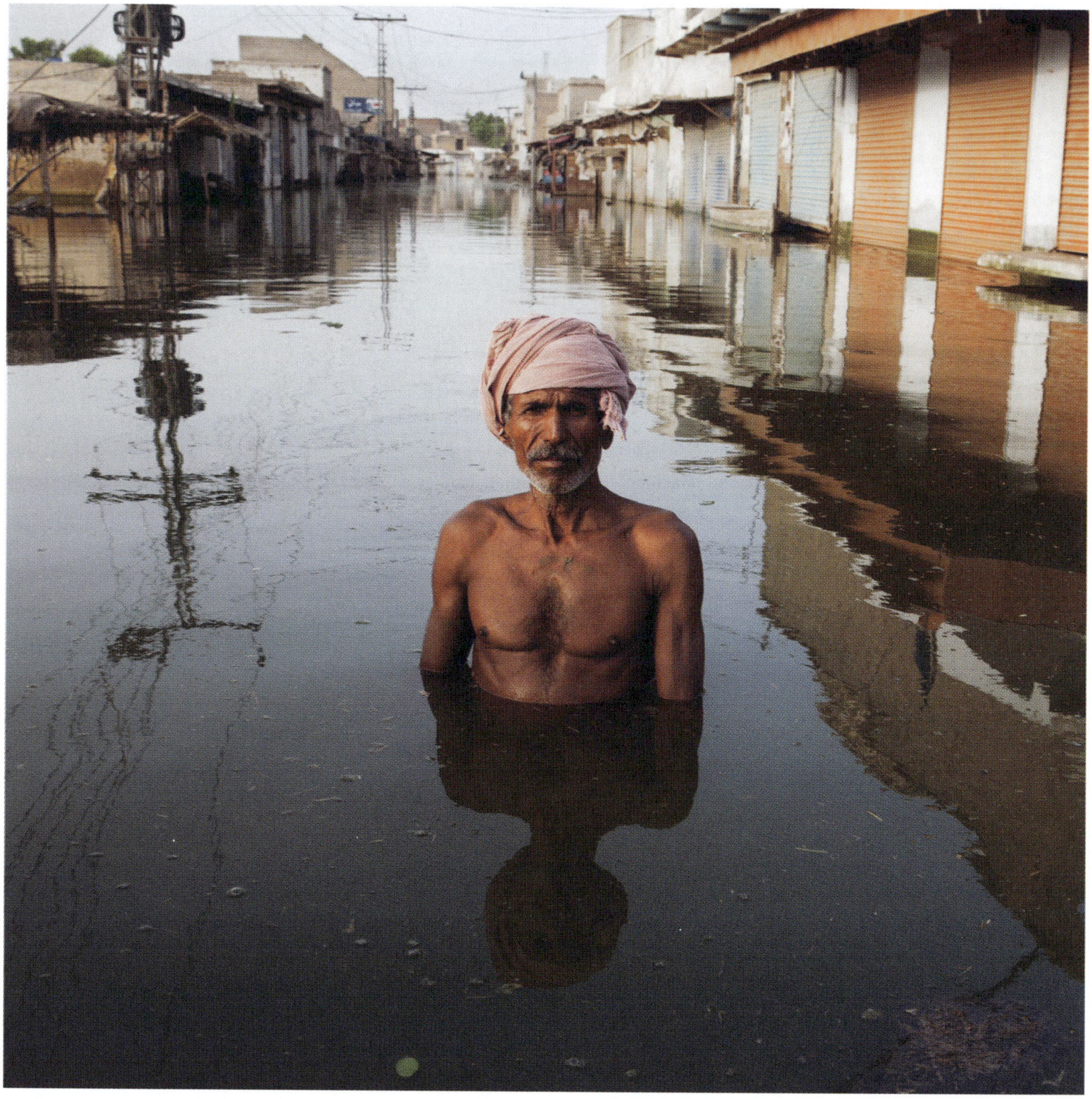

Submerged Portraits Series: Ahmed, Khaipur Nathan Shah Sindha, Pakistan, 2010, variable dimensions

on the potential impacts of climate change on their lives. This was in 2007, a time when the climate crisis was not widely discussed. When it was addressed in the media, the imagery was typically limited to polar bears and melting glaciers, phenomena that appeared distant and removed from daily life. "These images seemed so remote—how could people relate them to their own lived experiences?" recalls the artist. This disconnection motivated him to shift the visual representation of the climate crisis, aiming to make the issue more tangible and relatable to individuals. He sought to illustrate the direct and immediate effects of ecological issues on human lives.

This motivation led Mendel to embark on an extensive, long-term project documenting flooding events around the world. Over the years, his photographic approach has evolved from traditional documentary work to a more conceptual and activist-oriented practice. "Witnessing these disasters inevitably changes you," he says. Flooding became a central theme in his work, almost an obsession. When water infiltrates our private spaces, invades our homes and cities, it is no longer possible to deny its power and existence. The ocean ceases to be an invisible entity, and the water cycle, a mere elementary school lesson.

Mendel's photographs are aesthetically striking, meticulously composed, and visually compelling. However, the subject matter they depict is deeply unsettling. In his series *Submerged Portraits*, Mendel employs a systematic methodology to portray individuals whose homes have recently been devastated by flooding. These subjects stand, often partially submerged in water, gazing directly at the photographer—and by extension, the viewer. This direct eye contact is powerful and confrontational. For Mendel, it is essential not to depict these individuals as passive victims, but rather as dignified people. They do not appear powerless; instead, their expressions seem to ask the critical question: "Who is responsible?"

Since 2007, the artist has traveled across the globe, documenting these tragic events from Haiti to Pakistan, Brazil, and Thailand. While many of the floods Mendel captures do not occur along coastal regions, oceanic deltas are particularly vulnerable to flooding. The warming of the oceans has amplified atmospheric humidity through evaporation, resulting in intensified rainfall and increases in the severity and frequency of hurricanes globally.Ultimately, no one is immune to the effects of climate change. Mendel's series portrays individuals from a wide range of social classes, ethnic backgrounds, ages, and religious affiliations. As he notes, "Climate change makes no distinction," a statement that highlights the need for greater solidarity in addressing this global issue.

"I want people to look, look again, and then look once more." Mendel is deeply concerned about climate change skepticism and denial. His images force viewers to confront the stark reality of the situation. As he explains, "As we witness more and more alarming extreme weather events around the world, we also see increasingly aggressive denialism, often promoted by populist leaders; a global political system incapable of taking meaningful action; and petro-carbon corporations that resist even the most minor measures to reduce carbon emissions. In the face of this, I feel a personal responsibility to make this project resonate as powerfully as possible."

Mendel is also actively involved with Money Rebellion, a group of activists dedicated to exposing the underlying causes of the ecological crisis. The group engages in peaceful protests, sometimes utilizing Mendel's images printed on boards to raise awareness. In addition to his activist work, Mendel is deeply committed to educational initiatives. Recently, he collaborated with children living along the coast of Devon, England, an area particularly vulnerable to rising sea levels. Drawing inspiration from Mendel's photographic work, the children imagined how the landscape and their lives might be affected by climate change. Each child created a piece titled *When I'm 40*, expressing their vision of the future. Subsequently, Mendel photographed each child holding their artwork, standing in the outdoor places they like the most.

Paribartana Mohanty

Born in India in 1982

Paribartana Mohanty's artistic practice is strongly influence by his upbringing in Odisha, an eastern Indian state on the Bay of Bengal. He was born and brought up in the coastal district of Puri, and the sea was part of his everyday life. For him, it was linked to an idea of the sublime until the 1999 Super Cyclone that drastically changed his perception as well as the lives of the coastal communities, particularly the Nolia fishers. This cyclone was among the most destructive in the region, damaging more than 1.5 million homes, with casualties on humans and wildlife. It took place the same year the artist entered art school and left a deep impact on him.

Most of Mohanty's films and photographs deal with Odia-speaking people, living in the state of Odisha, and with the Nolia community, which is a Telugu-speaking marginalized fishing community from the same state. The artist is interested in their traditions, beliefs, and living conditions, which are on the verge of disappearing. The Bay of Bengal is especially vulnerable to the effects of climate change, with the coast being increasingly eroded. Consequently, villagers have to move inwards. They abandon their villages and their fields, unfit for crops due to the salinization of the water. The government is

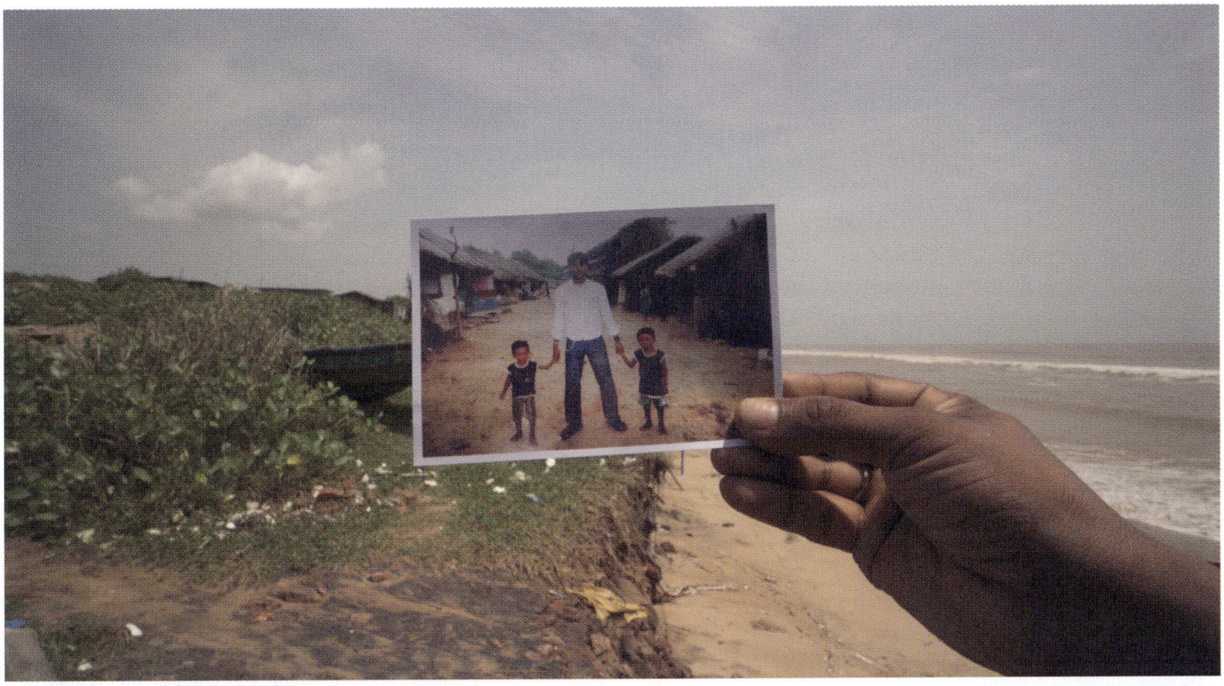

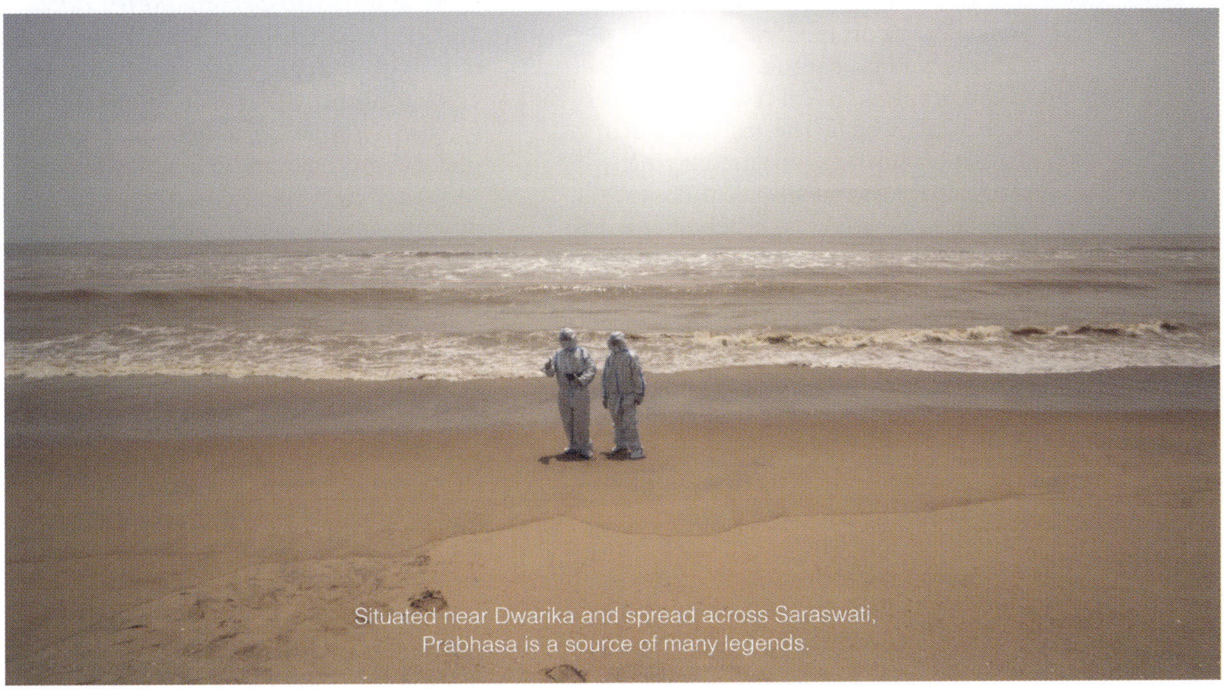

Ocean Mud Pickle, 2023-24, Video, 41 min

building rehabilitation colonies, but these new villages remain often empty: most families cannot pursue their artisanal fishing work and have no choice but to work in trawlers or construction industries in neighboring states or cities as daily wage laborers, leaving their newly donated-homes empty. Mohanty follows them as they face and resist these phenomena, recording their memory and questioning their perceptions of ecological disasters.

The video *Ocean Mud Pickle* (2023-24) focuses on the abandoned village of Satabhaya, one of the first Odia villages to be submerged in the sea. It begins with a strong storm that viewers can only hear: for a while, the screen remains black, forcing their attention to the sound and encouraging them to use their imagination. This darkness embodies the disappearance of the villages, and probably the sentiment of the people who feel that they have been swallowed by the sea. The video features musicians who are playing traditional local music called Daskathia, as they walk along the beach near a former temple, now in ruin. Images of the present day alternate with archival images showing the lives of the people in places now covered by sand and dunes. However, the video is not dominated by a sense of nostalgia: the empty spaces that appear now in maps become new spaces that the artist conceives of as dark wombs, places where everything can reunite and where, eventually, new beginnings might form. The title of the film refers to marine sediments, the mud from which everything originates.

According to Mohanty, any water body is sacred for the Odia people. In the vernacular language it is referred as "Ganga." There are plenty of stories and legends about the sea, often perceived as mother nature, which provides everything in plenitude. Along the coast, most of the people rely on the sea for their livelihood. One maritime ritual, the Boita Bandana, is still performed. It is a boat festival during which boats are decorated and paraded to commemorate the days when the Odia people would sail to Southeast Asian countries for trade.

The film ends with a wedding ceremony taking place in a rehabilitation colony, where villagers have been resettled. The couple comes from a low cast (they are Dalit migrant workers), and the ambiance is very joyful. For the artist, this is a way to suggest that life goes on, even among these marginalized communities.
The idea of defeating the consequences of today's ecological and socio-economic crisis is also at the core of *Four Songs of Apocalypse* (2021), a video that draws on local mythologies to imagine what Mohanty calls "post-environmental-disaster landscapes in the costal districts of Odisha." For him, this is a way to open the debate about climate change, a discourse that is otherwise rarely addressed locally. For instance, Mohanty rewrote a part of the *Mahabharata* to include climate references. In the Indian epic, Lord Krishna dies near the ocean. He combines this story with an Odia book of prophecies by the poet Achyutananda Das according to which a third world war will take place in 2032 followed by the fall of a meteorite in the Bay of Bengal, resulting in a tsunami and the devastation of the whole region. This is one of the oldest and most widely quoted literature on the apocalypse in

Odisha. Mohanty has heard its stories from childhood, and it has left a strong imprint in his mind. According to him, mostly uneducated religious people keep returning to it when natural disasters occur in the state.

"With these works, I want to tell the audience that 'ecology' is not only about humans; it is a non-human subject as well. It is about the 'queering' processes that all species go through in the universe."

Shneider Léon Hilaire

Born in Haiti in 1990

"My practice explores the complex and sometimes mysterious relationship Haitians have with the sea. Every civilization has developed collective beliefs, myths, and maritime legends. In Haiti, some of these beliefs remain alive and active. For example, misfortune can be warded off or eliminated through a sea bath known as 'the bath of luck.' When moving into a new home, it is not uncommon to cleanse the space with seawater to ensure it is filled with good energy."

Shneider Léon Hilaire grew up and studied in Port-au-Prince, in Haiti. His paintings are embedded in the cultural heritage of his country and strongly inspired by Haitian practices of Vodou linked to the sea. In recent years, he has undertaken a project to collect various local myths and beliefs, particularly those from the Côte de Fer, a coastal area in the southeast of Haiti where his family originates. For instance, one legend speaks of an underground cave along this coast that leads to the sea. It is said that growling sounds can be heard from the cave when it is about to rain. For the artist, this research is both a way to question the relevance of these beliefs and to preserve this cultural heritage, which is often transmitted orally.

In Haiti, the sea is a place of rituals–a space where magic happens, inhabited by spirits. Historically, part of Haiti's population originates from Benin, brought over during the transatlantic slave trade and French colonization. However, in Haiti, there is no Mami Wata, the powerful and ambiguous water spirit so prevalent in Africa. Instead, imported beliefs have been adapted and evolved. In Haitian Vodou mythology, Agwe is the master of the waters and marine creatures. His wife is a mermaid. "For us in Haiti," Shneider explains, "we do not fear the mermaid. It's not like Mami Wata, who is so ambiguous and seductive. Our relationship with these spirits is more peaceful. Damballah can bring wealth and luck." For example, in the painting *The Healing Mermaid* (2023), a couple prays to her to heal their sick child. With their bodies immersed in the sea and surrounded by candles, they entrust the child to the spirit of the waters, who leans over to heal him.

In general, fishermen pray to Agwe for protection and a fruitful catch. He is a central figure in local Vodou, and his worship is omnipresent on the island. In the painting *Marriage with Agwe* (2023), a woman is depicted uniting with him. These unions occur through revelation; according to the artist, it is the spirit who asks the woman for marriage. Typically, such unions are seen as very positive for the individual. There are specific laws governing these marriages, though: for instance, if the woman is already married, she must not see her husband on certain nights.

These artworks are painted using contrasting tones of black, blue, and white. For Shneider, this choice opposes the common stereotype of Caribbean art as being excessively colorful. It is also a way of emphasizing the unity between spirits, the environment, and human beings, all symbolically interconnected in the world of beliefs. Colors enable an invisible dialogue between every living being. Ultimately, these colors

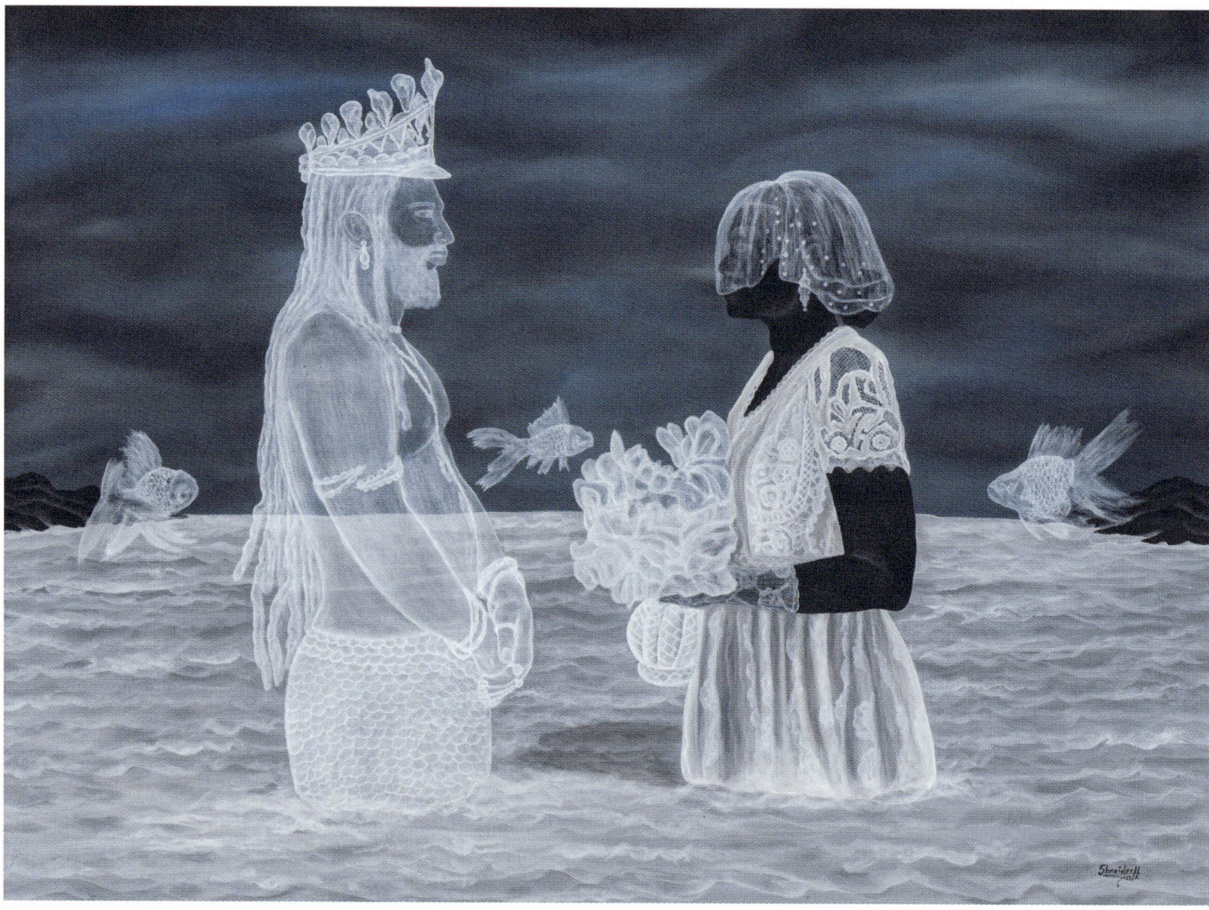

Mariage avec Agouet, 2023, Acrylic on canvas, 75 × 100 cm

are his personal choice, as the artist envisions the sky and the sea at night in these tones.

A large portion of Haiti's population does not know how to swim, and Shneider himself is no exception. He argues that this apparent contradiction–an intimate relationship with the sea but an inability to navigate it–does not stem from the country's colonial history, as is often claimed. Instead, it is more simply a matter of education, infrastructure, and the lack of political will.

Haiti is still a very poor country, with a highly complex political situation characterized by extreme levels of violence, corruption, and insecurity. A large part of the coastline is occupied by gangs, who build shacks by piling up garbage to reclaim land from the sea. "We no longer go to the beach near Port-

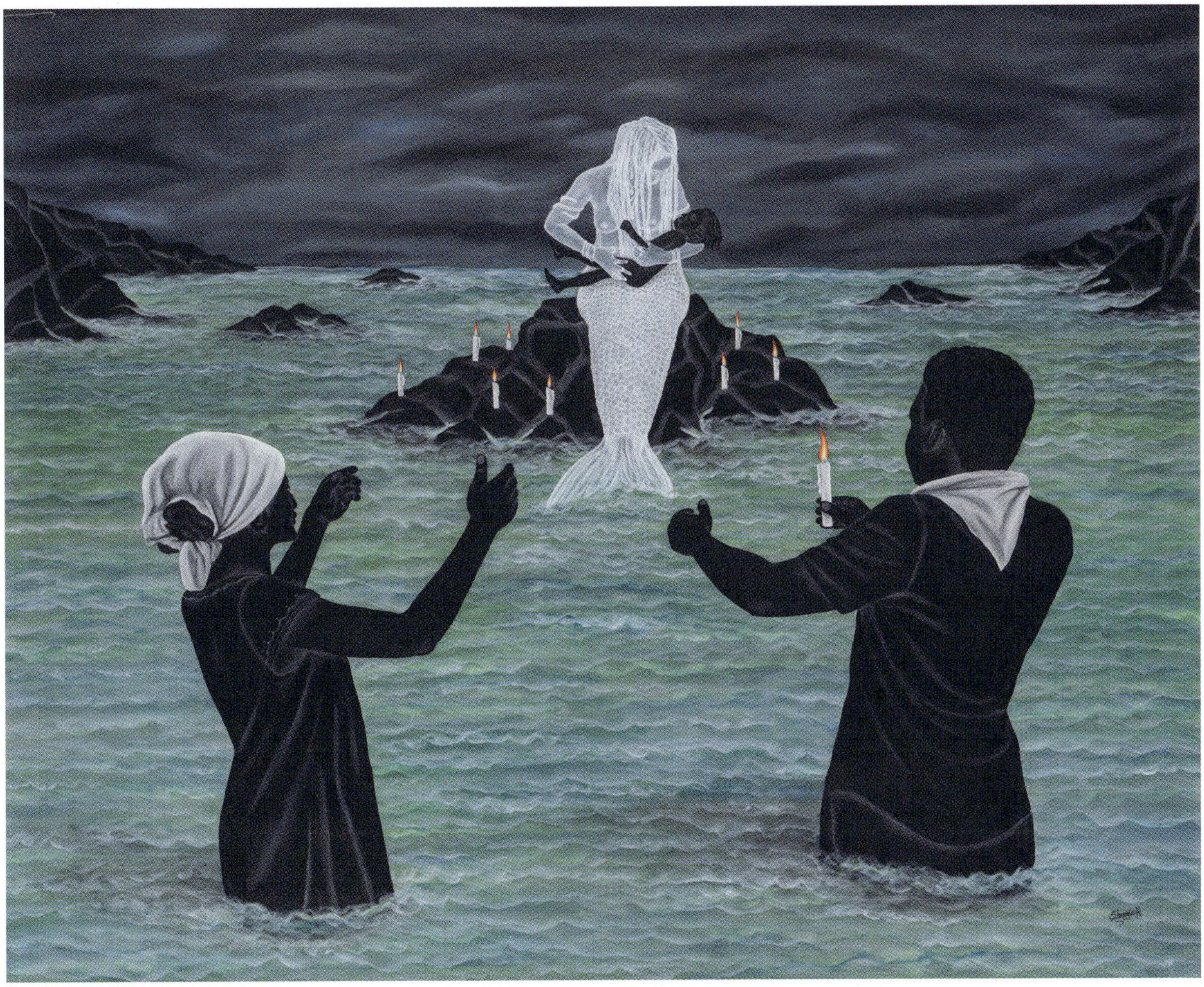

La sirène guérisseuse, 2023, Acrylic on canvas, 125 × 151 cm

au-Prince," Shneider says. As a result, the coast and its resources are neither valued nor protected. Pollution is also a major issue, with garbage piling up, clogging drainage systems, causing floods during heavy rains, and often ending up in the sea.

All these factors have distanced the population from the coastline, but they have not weakened their associated beliefs. Rituals do not require a physical presence at the shore–ceremonies can be performed at home. People imagine connecting with the spirit of the sea, often sprinkling seawater on the individual seeking protection or a favor. The spirits respond–they are alive and omnipresent. The fact that the mermaid and Agwe are depicted anthropomorphically illustrates how deeply their presence is interwoven into the everyday lives of Haitians, who actively coexist with them. In Haiti, the sea remains at the heart of their lives, dreams, and aspirations.

Judy Watson

Born in Australia (Waanyi) in 1959

Judy Watson's artistic praxis is grounded in her matrilineal lineage, the ancestral Waanyi aboriginal people, whose traditional territory is located in the Gulf of Carpentaria region of northeastern Queensland, Australia. Working primarily in printmaking and painting, and utilizing in-situ natural materials and pigments, she scrutinizes the local topography to reveal traces of concealed histories and markings of the current transformation of the coastline. She often displays her artworks unframed allowing them to seemingly float on the wall or undulate upon the floor.

The Waanyi are referred to as the "running water people," and the interconnection between water and land is paramount within Watson's culture and creative practice. In such arid landscapes, water is vital for all, and is perceived as a source of life. Two paintings, *Sacred Water* (2010) and *Sacred Earth* (2010), conceived as a complementary pair, offer a typical example of the artist's vision of the sea and the land as two foundational pillars of her cultural heritage.

moreton bay rivers, australian temperature chart, freshwater mussels, net, spectrogram (2022) offers another topographic view of the coastline, this time represented through diverse techniques and mediums in a reflection on the complexity and dynamism of the stakes involved in climate change. Working collaboratively with various women from her community, the artist has incorporated indigo dye, multiple types of paints, and wax on a large linen canvas, creating an original cartographic work. In addition to giving Australia's average air and water

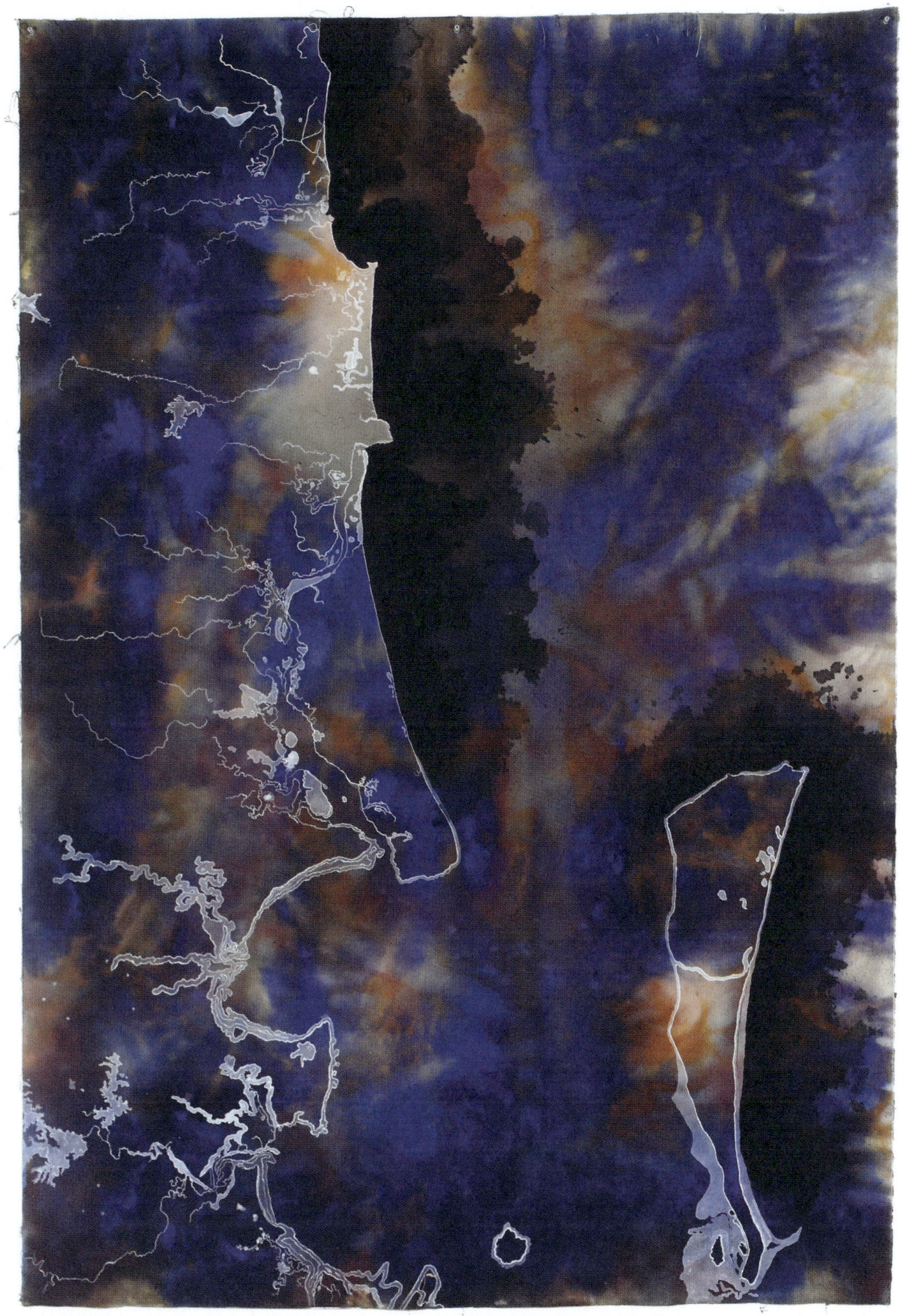

Bloom, 2009, Pigment, pastel, and synthetic polymer paint on canvas, 219 × 147 cm

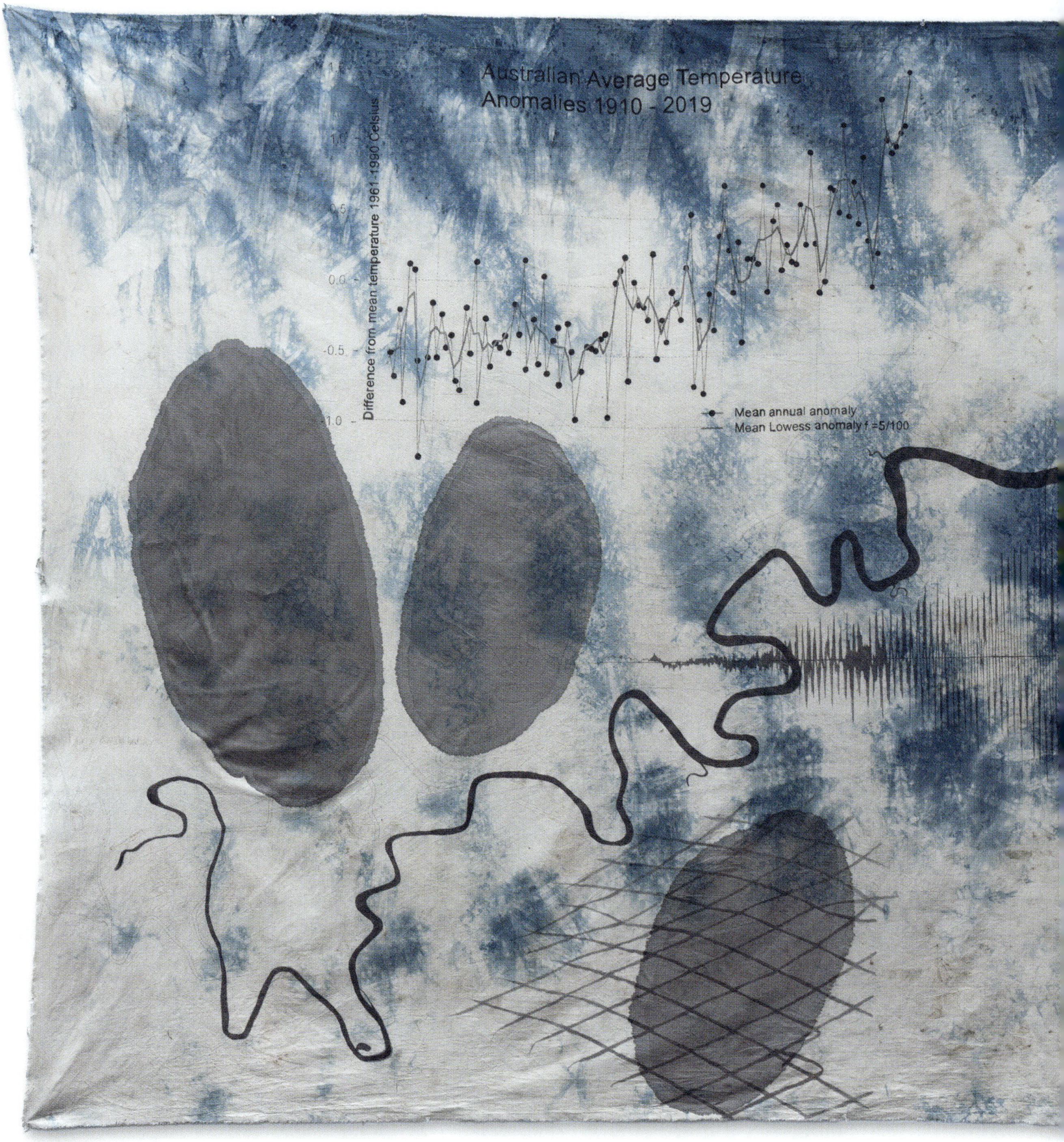
Australian Average Temperature
Anomalies 1910 - 2019
Difference from mean temperature 1961-1990 Celsius
Mean annual anomaly
Mean Lowess anomaly f =5/100

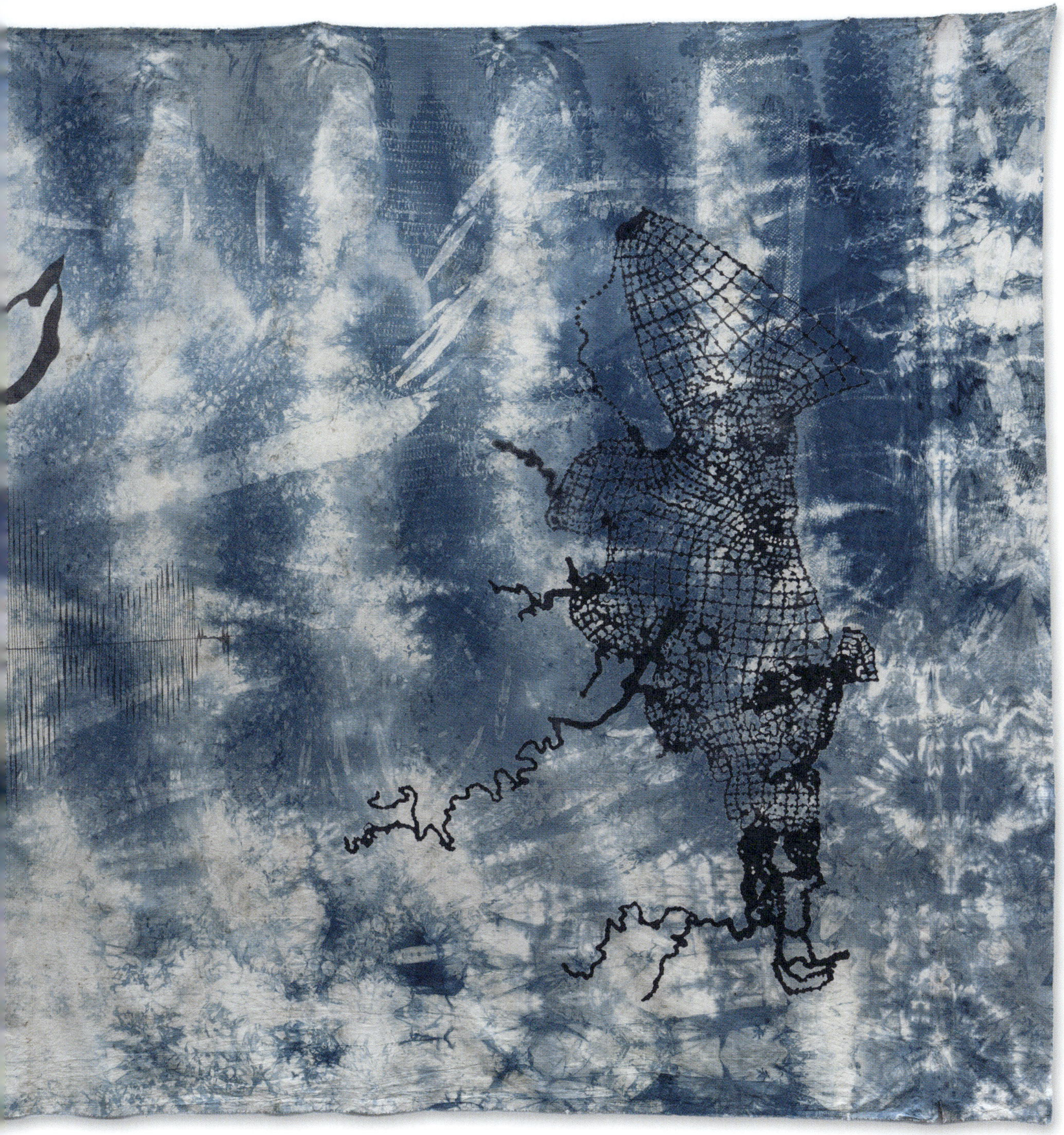

Moreton Bay Rivers, Australian Temperature Chart, Freshwater Mussels, Net, Spectrogram, 2022
Indigo dye, graphite, synthetic polymer paint, waxed linen thread and pastel on cotton, 247 × 488 cm

temperatures recorded between 1910 and 2019, the painting integrates indigenous knowledge and cultural symbols, such as the three large freshwater mussels painted on the left side of the composition. Known as "water beef," these marine species are an important dietary staple for the Waanyi people.

With her practice, Watson seeks to reconnect with her ancestral culture and to find a new way to restore, or re-invent, the now-broken link that, for a long time, used to bind her people with the sea and natural elements. Embracing and adapting to the effects of climate change is one thing, yet the coast is also directly affected by human industrial activities.

bloom (2009) is an example of the artist's response to a significant oil spill that occurred in 2009 in the Coral Sea, off the south-eastern coast of Queensland, when a container ship collided with the path of Cyclone Hamish, resulting in tons of fuel and ammonium nitrate being spilled into the sea. Her painting features an aerial perspective of the coastline sullied by shades of black. The contrast between the delicate lines of her drawing and these expanding stains invites viewers to contemplate the vulnerability of the local ecosystem, which is easily and increasingly exposed to degradation.

Absorbing Human Excesses

A volcano of cans
and tobacco papers
pukes in the corner,
melting faces
with tripletas
and Presidente freezers.

Cement gullies
coughing, clogged with bottles
and reused sunflower oil
like the feet that try
to reach the shore
but only step
on the people who were there
and won't leave,
not even in a million years.

Thaís Espaillat Ureña

1. Clay Apenouvon
Born in Togo in 1970

2. Edward Burtynsky
Born in Canada in 1955

3. Thaís Espaillat Ureña
Born in the Dominican Republic in 1994

4. Heba Y. Amin
Born in Egypt in 1980

5. Andy Hughes
Born in the United Kingdom in 1966

6. Denise Lira-Ratinoff
Born in Chile in 1977

7. Patrick Tagoe-Turkson
Born in Ghana in 1978

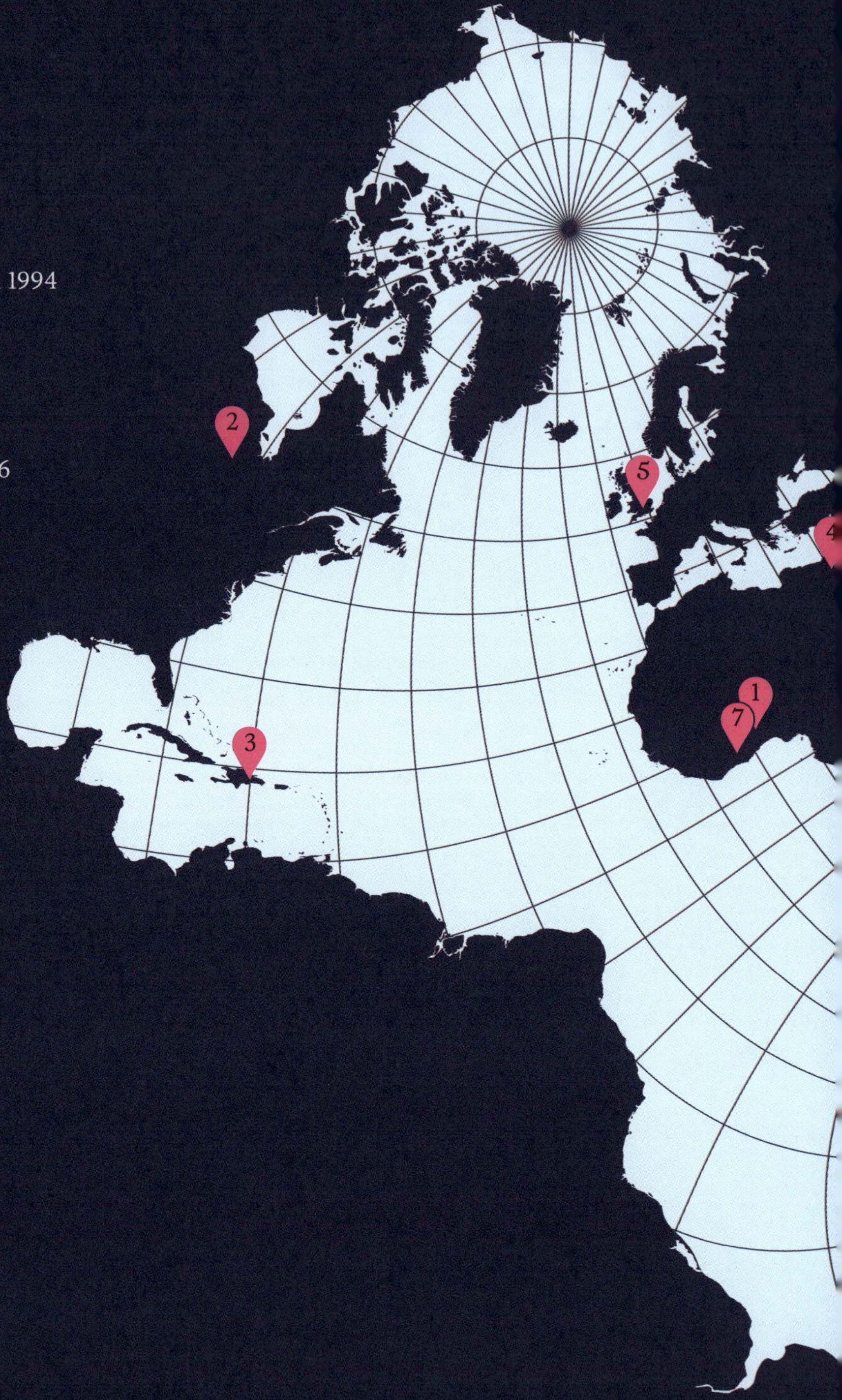

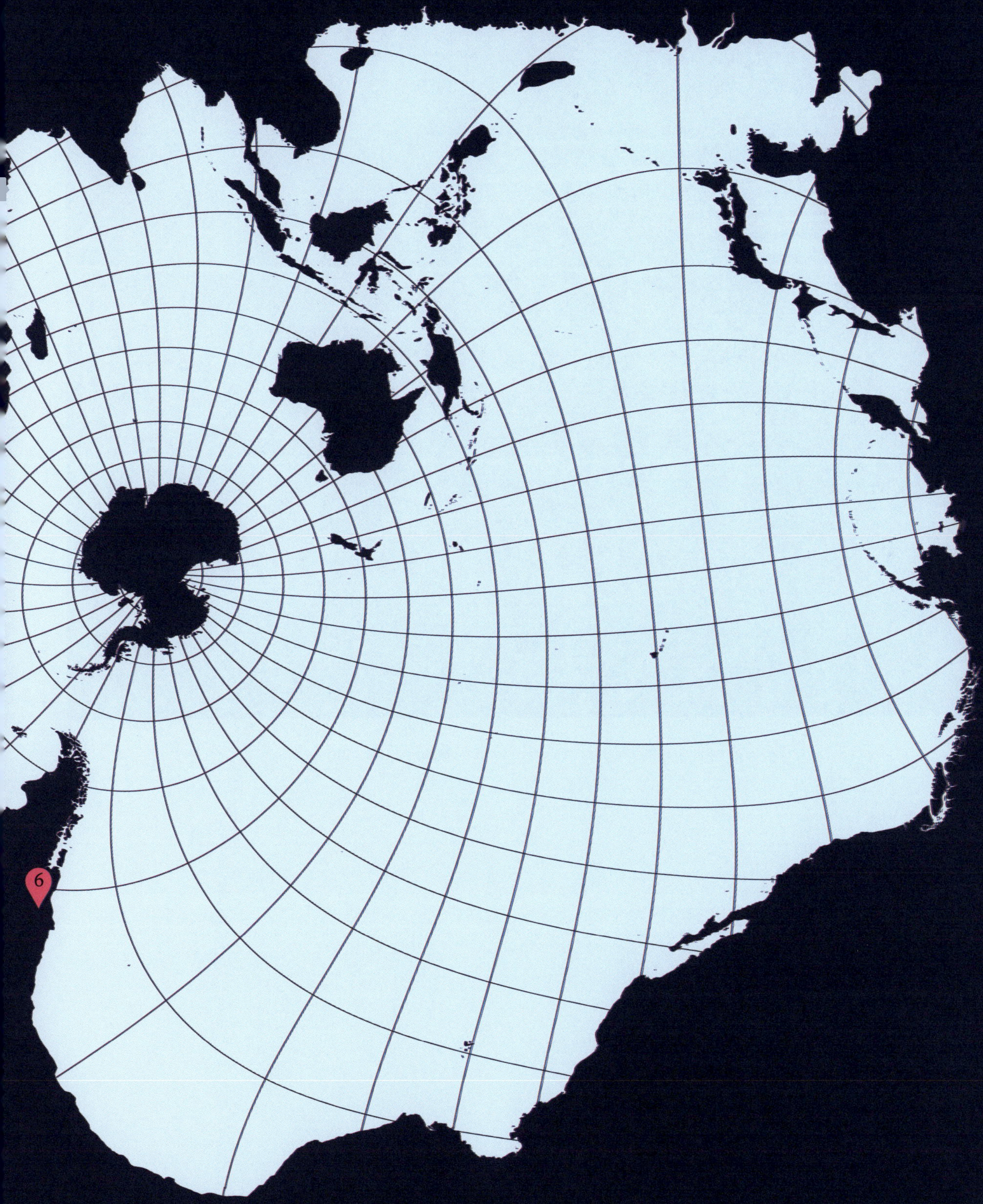
6

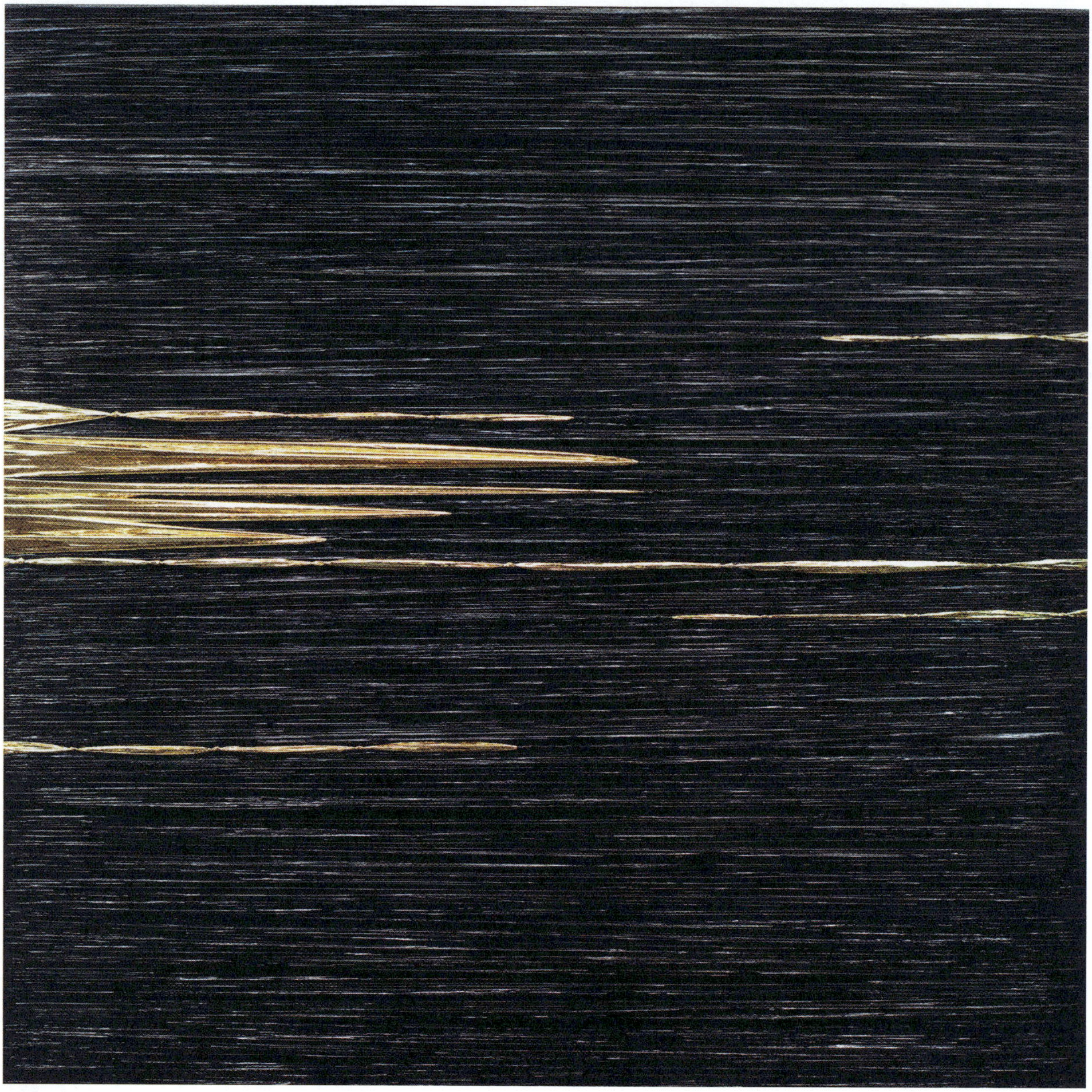

Lignes de survie, 2018, Black plastic film, survival blanket, 100 × 100 × 7 cm

Clay Apenouvon

Born in Togo in 1970

"The first time I saw the face of the sea, I was nearly overwhelmed by fear–the fear of being engulfed by that vastness. I clung to the palm trees, remaining far from the beach, observing the towering, formidable waves crashing onto the white sand, distant from me. I felt minuscule in the presence of something immensely large and limitless. In the days following this forbidden encounter, which my parents had warned me against, I experienced nightmares. I was overwhelmed by what my eyes had witnessed, yet I could only keep this secret to myself, despite being a child who often spoke a great deal."

Clay Apenouvon grew up in Togo, near the coast, and his early experiences and perceptions of the sea left a lasting impression on him. However, it was only decades later that the subject of the sea resurfaced in his consciousness, this time viewed through a tragic lens. In October 2013, the images of hundreds of lifeless African bodies washed ashore on the beaches of Lampedusa profoundly shocked him. This tragedy stemmed from a fishing trawler carrying approximately 500 migrants that caught fire near the Italian coast. Apenouvon reflects on how this event invaded his imagination. Seeking further information online, he found himself with profound questions about the circumstances, particularly the testimonies of survivors who recounted the harrowing experiences they faced on their journeys, including human trafficking in Libya and perilous desert crossings. Two years earlier, the war in Libya had erupted, a conflict ostensibly aimed at protecting Libyan civilians from the dictatorship of the Gaddafi regime, as the international community claimed. "Many of we Africans believe that this war was declared primarily to seize Libya's oil reserves, rather than for the official reasons presented," says the artist. "We can all observe the disaster wrought by this latest war declared in the name of democracy and human rights." Subsequently, he came across a satirical drawing titled *The Black Tide of Lampedusa*, which catalyzed the connections between Libyan oil, plastic waste, and the migrant crisis. Gradually, these elements coalesced in his mind, prompting him to create a noir film in which young Africans drown in a black tide, amidst widespread indifference in Europe.

In his installations, Apenouvon repurposes stretchable plastic film–commonly used in daily life for wrapping goods and food–into a compelling subject. This ubiquitous transparent film, derived from oil, is often discarded, contributing to the accumulation of plastic waste. Through this initial invisibility and toxicity, the artist effectually transitions the material from shadow to light. The black plastic, stretched across canvases or cascading like waterways, captures light beautifully and creates artificial landscapes that resonate with the migrants' aspirations for a better life. Thus, the face of the sea re-emerges as both a horizon of hope and a site of nightmares. →

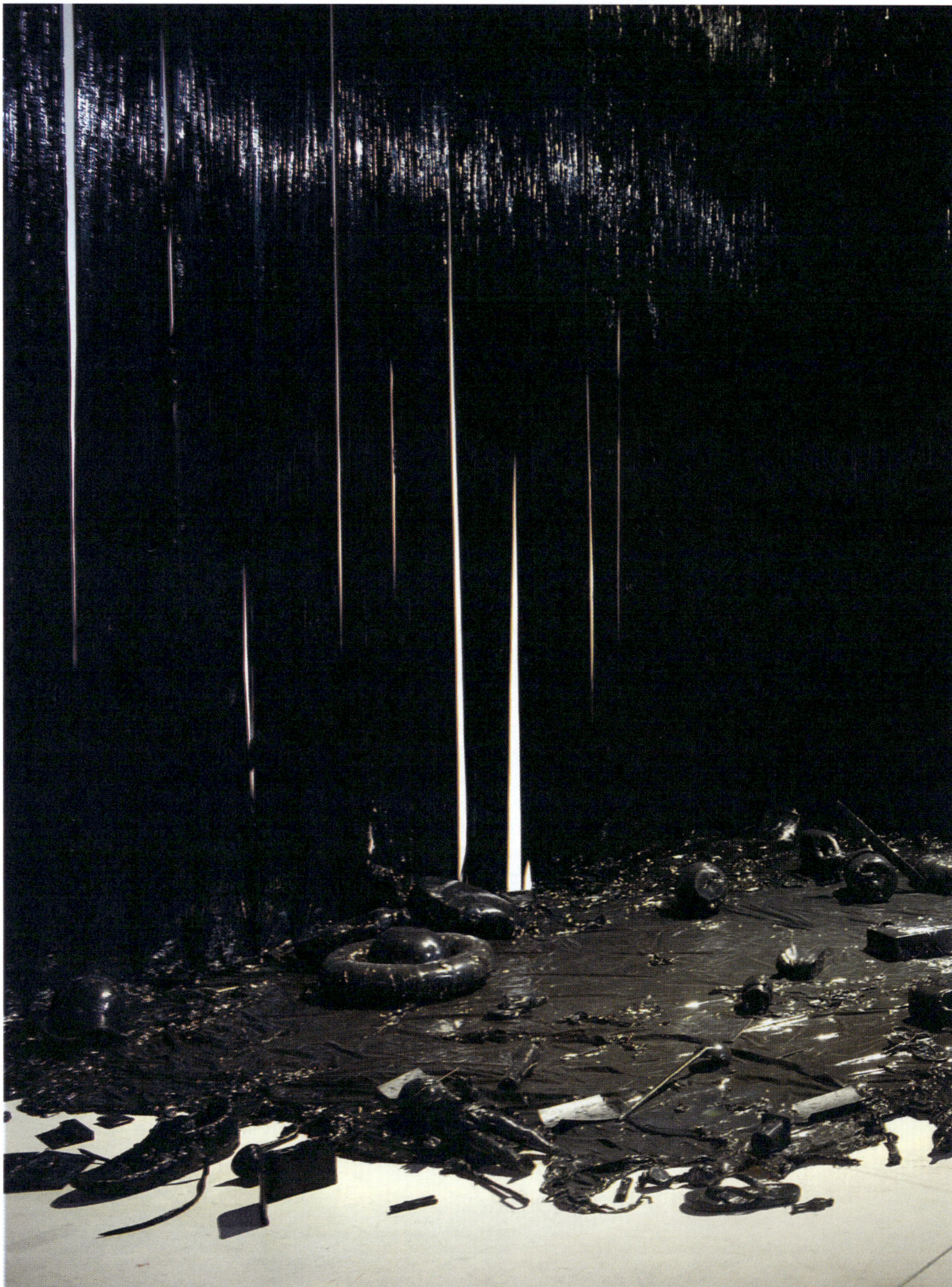

Film noir de Lampedusa, 2015, Site-specific installation, Black plastic film and objects

In parallel, Apenouvon utilizes discarded black plastic bags to address significant oil spills, such as the one perpetrated by the Shell company in Ogoniland, Nigeria, in 2012. His large-scale installations seem to engulf the entire space, stifling the air and suffocating the voices of those who attempt to protest against such extensive pollution. In all instances, the artist's work reflects a world dominated by plastic, often at the expense of the environment and human lives.

Edward Burtynsky

Born in Canada in 1955

"We come from nature. There is an importance to [having] a certain reverence for what nature is because we are connected to it . . . If we destroy nature, we destroy ourselves."

Edward Burtynsky's ecological critique and commitment are evident through his extensive artistic practice. For four decades, the Canadian photographer has created images that blend the sublime with devastated landscapes, resource extraction sites, and massive industrial infrastructures from around the world. Aware that the planet is approaching a critical tipping point, with human impacts on natural resources becoming nearly irreversible, Burtynsky initiated the Anthropocene Project in 2018. This multidisciplinary body of work combines artistic multimedia with scientific research to investigate humanity's imprint on the state, dynamics, and the future of the Earth.

Burtynsky's photographs can often be ambivalent, as he depicts industrial sites in a spectacular manner. For him, this approach highlights the contradictions and challenges we face as a species, especially in reconciling our needs and dreams of expansion with the planet's finite resources. As he explains: "These images are meant as metaphors for the dilemma of our modern existence; they search for a dialogue between attraction and repulsion, seduction and fear. We are drawn by desire–a chance for good living, yet we are consciously or unconsciously aware that the world is suffering because of our success. Our dependence on nature to provide the materials for our consumption and our concern for the health of our planet sets us in an uneasy contradiction. For me, these images function as reflecting pools of our times."

In the mid-2000s, Burtynsky produced a series of photographs focusing on oil, motivated by his desire to explore the industries linked to automobile production. "I began to think about oil itself, both as the source of energy that makes everything possible, and as a source of dread, given its ongoing threat to our habitat."

In 2010, following the largest marine oil spill in history in the Gulf of Mexico, Burtynsky traveled to the site to document the disaster using aerial photography. The spill, caused by the explosion and subsequent sinking of BP's *Deepwater Horizon*, is depicted in his photographs not only as a scene of devastation but also, more abstractly, as a transformation of the seascape. His images capture the massive lines of oil that stretch far beyond the horizon.

The sea takes center stage again in Burtynsky's *Pengah Wall* series (2017), photographed off the coast of Komodo Island, Indonesia. This time, he captured coral reefs and underwater landscapes like intricate tapestries—abstract compositions of textures and colors that reflect the richness of marine biodiversity. For Burtynsky, this series is also about hope, revealing the beauty that still exists, and underscoring the urgency of protecting it.

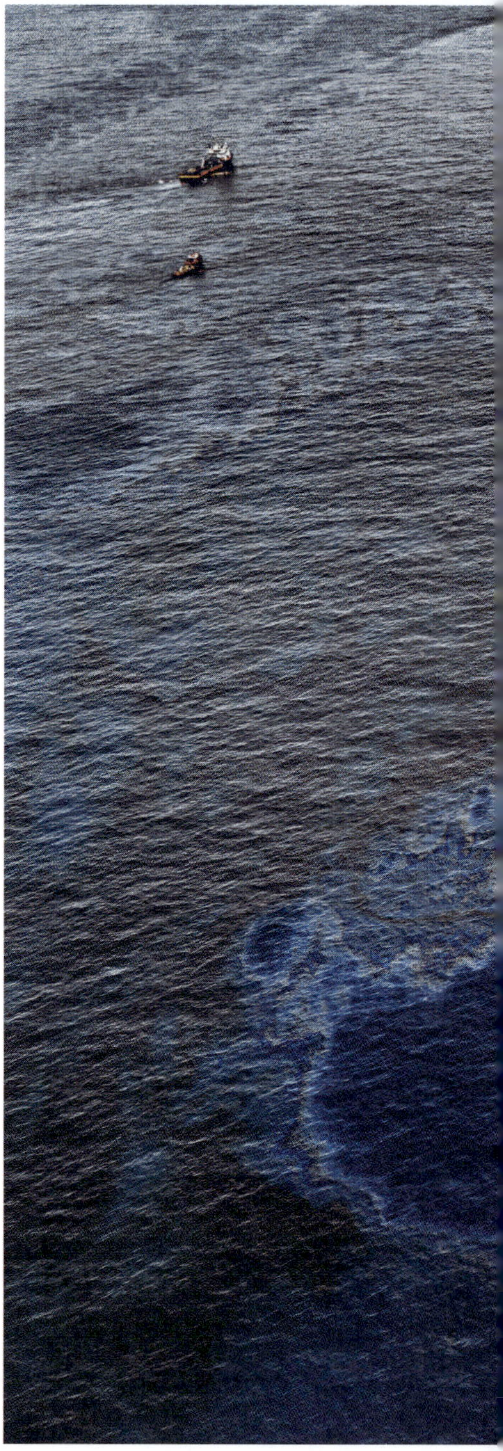

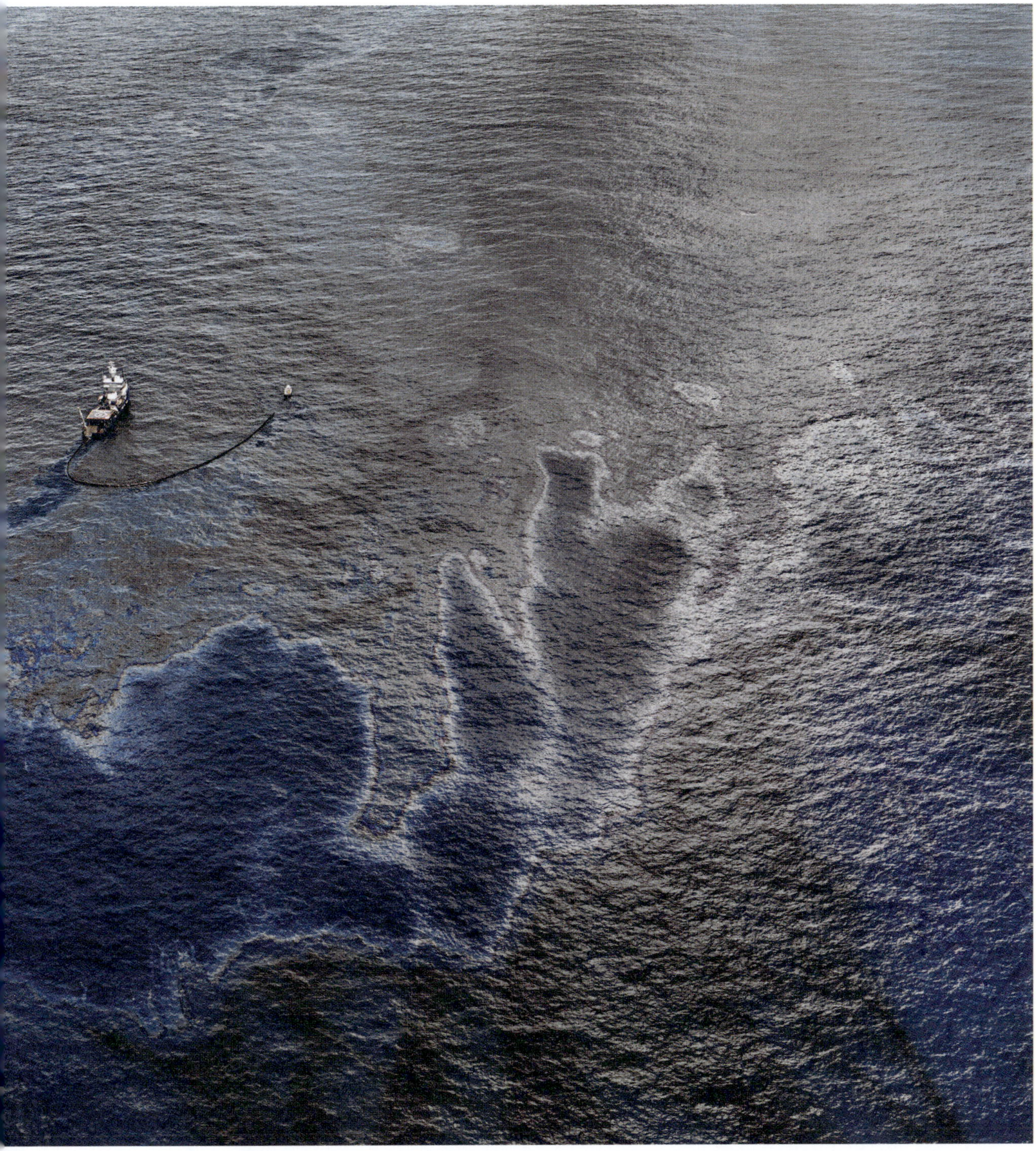

Oil Spill No. 4, Oil Skimming Boat, Near Ground Zero, Gulf of Mexico, 2010, C-Print, 121,9 × 162,6 cm

Thaís Espaillat Ureña

Born in the Dominican Republic in 1994

Thaís Espaillat Ureña's practice aims at pointing to the often overlooked yet undeniable realities of their home country, the Dominican Republic, situated within the Caribbean Sea, in the Antilles archipelago. Their work stands in stark contrast to the conventional, touristic portrayal of Caribbean destinations. Instead, it exposes the underlying consequences of economic growth driven by mass tourism, gentrification, and other ongoing colonial practices, including deepening social inequalities and the degradation of the local marine ecosystem.

Where Do These Waves Come From? (2019) is a transdisciplinary project that originated from the artist's investigation of the Malecón of Santo Domingo in March 2019. Despite being referred to as a "pier", the Malecón is, in fact, a prominent avenue that extends across nearly half of the city, running parallel to the coast. This makes it a significant site for exploring the dynamic interactions between the sea and urban life.

According to Espaillat, this project sought, through a loose interpretation of the scientific method, to identify intersections between the sea, poetry, and the manner in which we inhabit (and simultaneously destroy) the city and the Caribbean Sea, which serves both as a natural landscape and a receptacle for waste. During their fieldwork, Espaillat engaged in daily observations, filming recurring scenes at sunset over one month, photographing the same section of the horizon from Plaza Juan Barón, and taking detailed notes on a specific segment of the Malecón. From their photographic documentation, they produced a series of thirty cyanotypes on a single cotton sheet, which were developed on a bed of trash that washed onto the city's shores, symbolically absorbing the local pollutants and atmosphere.

"The stretch of the Caribbean Sea that I have lived near for most of my life is bordered by an avenue lined with semi-trailer trucks, abandoned buildings, and imported palm trees. It has never represented the idyllic, picturesque beach often depicted on postcards. Beaches, particularly along the southern and eastern coasts, are now lined with sargassum. This seaweed accumulates on the beaches that are truly public, whereas those appropriated by hotels are diligently cleaned by workers around the clock to ensure a pleasant experience for tourists. Mangroves, a critical component of the island's ecosystem, are being recklessly destroyed to make way for development projects, thereby removing a vital layer of protection against increasingly severe hurricanes and disrupting the habitats of countless species."

Findings (This Was Not The Future We Were Promised) is a speculative documentary video that illustrates the environmental degradation

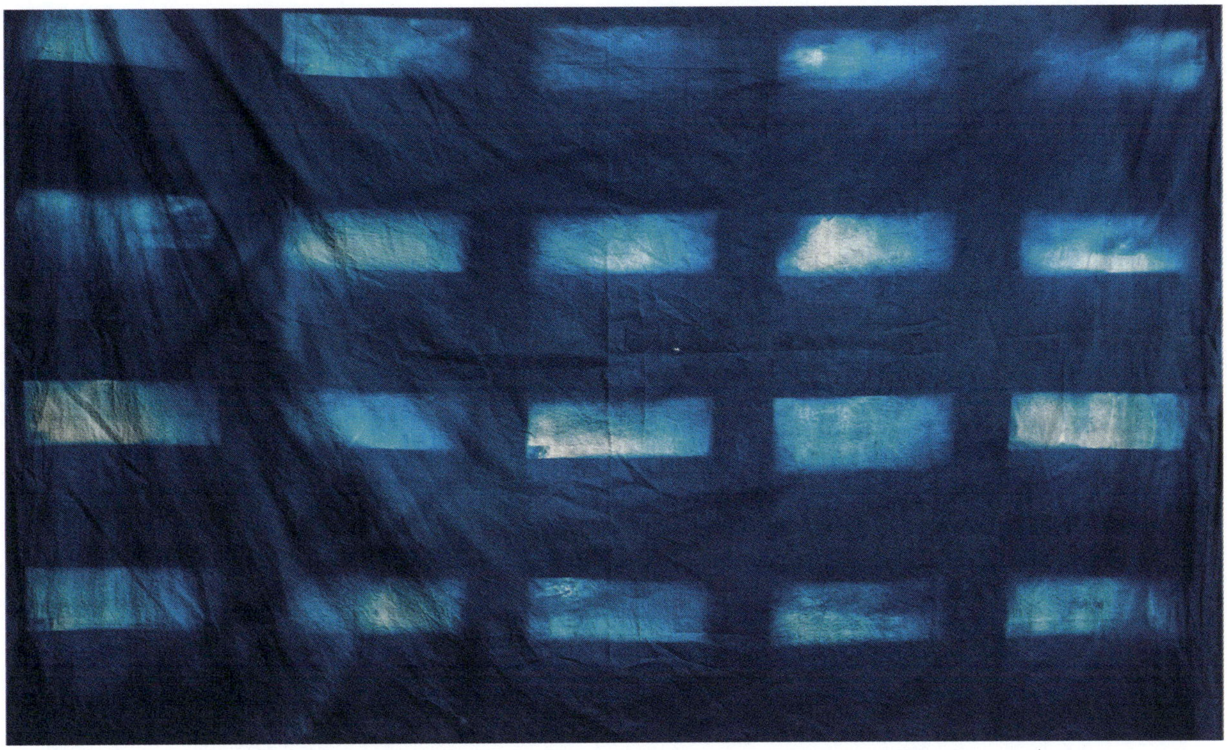

¿De dónde vienen estas olas?, 2019, Cyanotype on cotton, 140 × 110 cm

along the coasts of Santo Domingo and imagines what these would look like after "the end of the world." According to Espaillat, it is unfortunately common practice on the island to dispose of waste improperly.

The artist also attributes this collective denial of the coastline's deteriorating condition to the imposition of an imported colonial ideal of the landscape, which has led local inhabitants to turn their backs on the sea. "Santo Domingo, my hometown, has been constructed in a manner that disregards its coastal location. It is often said here that the city stands with its back to the sea. Yet, much of the city's waste ultimately ends up on the shores, so the sea cannot ignore us as the city appears to wish to ignore the sea. I can ignore neither the sea nor the city, and while I hold other, more pleasant images of the ocean, it is these foul-smelling, trash-filled waves that persistently haunt me."

Heba Y. Amin

Born in Egypt in 1980

With an artistic practice deeply grounded in extensive research, Heba Y. Amin's approach to the sea is mainly historical and political, informed by critical geography. In her ongoing research project, *Operation Sunken Sea* (2018-present), the Egyptian artist and scholar investigates mid-nineteenth century and early twentieth century techno-utopian proposals to relocate or drain the Mediterranean Sea. Among the projects of her investigation is a largely forgotten colonial proposal made in the 1920s, which aimed to lower the level of the Mediterranean Sea to facilitate the transformation and exploitation of the marine and coastal landscapes. This project, known as Atlantropa or Panropa, was proposed by German architect Herman Sörgel, who envisioned the construction of dams across the Strait of Gibraltar, the Dardanelles, and between Sicily and Tunisia. The intended outcomes of this massive undertaking included the generation of hydroelectric power, granting significant geopolitical influence to the controlling authority, the creation of new land for exploitation, and improved access to Africa to enhance trade.

For Amin, this "techno-utopian" project is not only emblematic of the geopolitical alliances and global politics of the post-World War I context but also serves as an example of humanity's often appropriative nature, driven by a quest for geopolitical dominance at the expense of the environment and local populations. By bringing attention to the complexity and richness

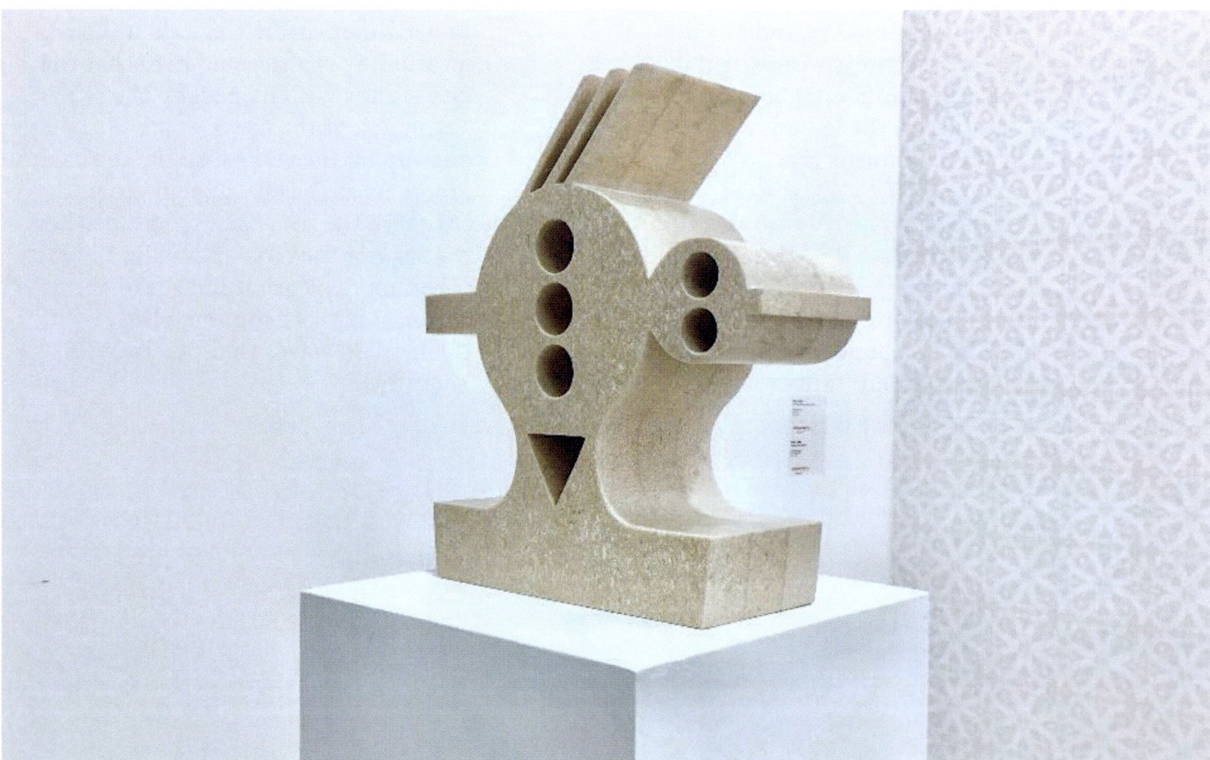

Visions of the Sea, 2018, Crema laguna stone, 43,5 × 38,6 × 15 cm

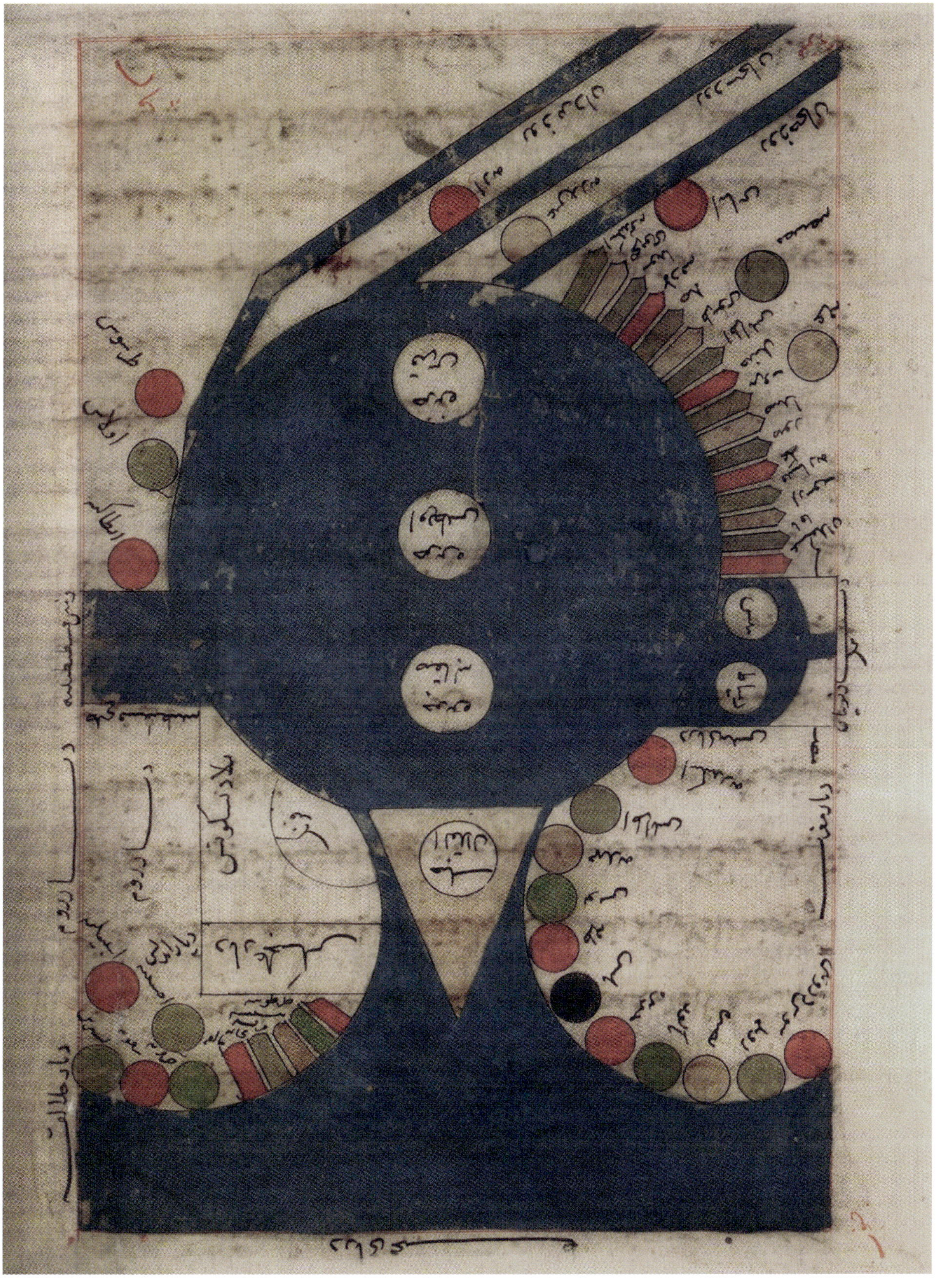

Persian geographer al-Istakhri's 10th century map of the Mediterranean Sea, c. 950

of the sea's economic and political potentials, Amin critiques the recurring utopian ambitions that frequently underpin such highly technical projects–projects that reflect a belief in humanity's ability to reshape the world according to its needs and desires. Notably, in the Atlantropa project, there was no consideration of the perspectives of local populations, nor of the profound ecological transformations and disruptions that would likely ensue. As the artist points out, the Mediterranean Sea might have been reduced to a giant hypersaline lake, while the newly created lands would likely have been too saline to support agriculture. This denial of human, social, and ecological realities remains a persistent mindset in many contemporary development initiatives.

Operation Sunken Sea manifests in various forms and is expressed through multiple works. In her video *Operation Sunken Sea: Relocating the Mediterranean, Inaugural Speech*, for instance, Amin imagines herself as the leader of a fictional nation, delivering an official address concerning a grandiose national project. Drawing on archival footage of actual political speeches, she conveys the aggressive tone, gestures, and body language typical of male authoritarian figures, while subverting these conventions with humor. The project's perspective is radically altered: this time, it is presented by a woman from an African country, and its purpose is to address contemporary political, social, and economic challenges.

Amin's sculpture *Visions of the Sea I* (2018) similarly embodies a radical shift in perspective. The form of the artwork draws inspiration from a tenth-century Persian map that represents the Mediterranean Sea as a positive space. This contrasts sharply with the colonial European vision of the sea as a negative or empty space. Notably, in this map, the Strait of Gibraltar is positioned as the central point of orientation. Through these various works, Amin invites viewers to reconsider perspectives inherited from colonial legacies, urging a more nuanced, respectful, and reparative approach to interpreting the world.

Andy Hughes

Born the United Kingdom in 1966

Born in 1966 in Castleford, Yorkshire, Andy Hughes was raised in close proximity to a coal mine, where, as a child, he would observe the smoke emanating from the facility from his window. Later, during his time as a student in Wales, where he took up surfing, the pervasive presence of pollution and the heavy industrial impact upon the environment became increasingly evident to him. He observed the waves and the sea covered in rising smoke and noticed the significant amount of debris floating on the water's surface. This experience sparked his enduring interest in the seascape, and from the late 1980s onward, Hughes began to systematically explore, through photography, the ways in which a society driven by excess was increasingly allowing plastic waste to accumulate along the shorelines.

Ahead of his time, Hughes embarked on journeys across multiple continents to document the plastic debris that had washed ashore on various beaches and littoral zones. Each of his photographs is accompanied by detailed captions, specifying the exact location of the shot,

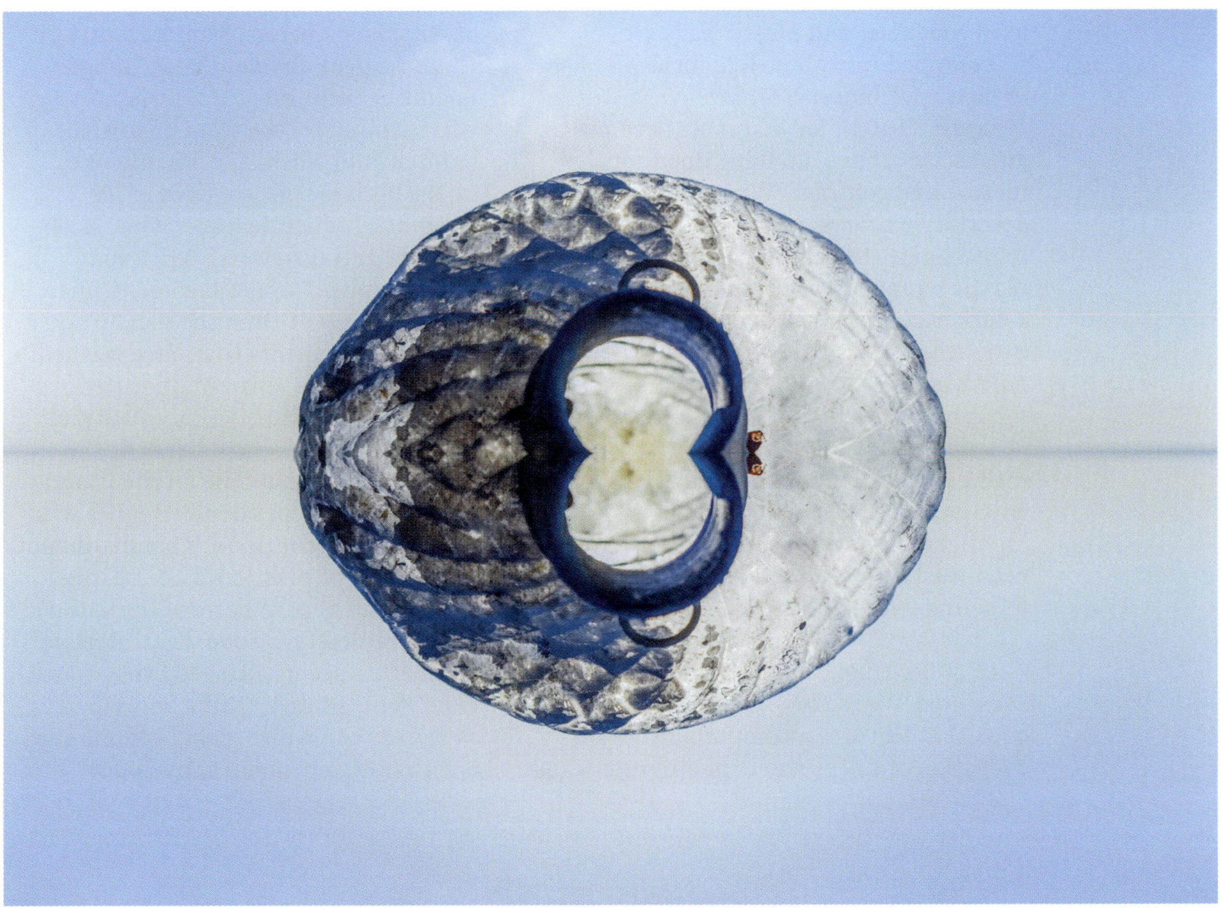

Sky Egg, Gapado Island, South Korea, 2022, Digital C-Type Photograph, 76 × 100 cm

as well as the time and contextual background. Additionally, Hughes often provides personal commentary, reflecting on his reactions to some of the more disturbing or unusual encounters. In a consumerist society where material consumption reigns supreme, his artistic perspective critically examines our relationship with material artifacts, which are often venerated. The marine litter is photographed from an intimate perspective, filling the entire frame of the image. These objects–once mere waste–are transformed into subjects, almost magnified. Through this process, Hughes introduces a form of ambiguity in how viewers are expected to engage with these detritus.

Captured in large format, objects such as abandoned oil drums, used condoms, and plastic bags are transformed into alien-like forms or objects of inquiry. *Sandwich Wrapper* (2004), for instance, becomes an abstract shape, striking due to its texture and contrast. Without its descriptive title, it would be difficult to recognize the familiar triangular plastic wrapper, which emerges from a dark background, still wet from its journey across the sea. In *Plastic Particles* (1990), Hughes shifts his focus to the almost invisible forms of pollution that are gradually infiltrating beaches and natural environments. In particular, he highlights tiny coal particles that have been washed by the sea for decades, mingling with the sand.

In 2006, Hughes published *Dominant Wave Theory*, a seminal book that features a comprehensive collection of his previous photographic works. The book operates almost as a portrait gallery, displaying a wide array of trash, brand logos, forms, and textures of plastic debris. The sheer scale and accumulation of images reflect the increasing accumulation of pollution in the marine environment and natural world in general. Hughes also collaborates with scientists, studying the impact of human waste on the marine and coastal environment. In 2013, for instance, he joined the project "GYRE: The Plastic Ocean," traveling to Alaska with researchers and artists.

Over time, Hughes' artistic practice has evolved towards a more conceptual format. Through his lens, plastic debris is transformed into enigmatic forms, with artificial colors that verge on kitsch aesthetics. More recently, Hughes spent six months on Gapado Island, South Korea. His series *Sky Egg* was inspired by the work of British painter Paul Nash, specifically one of Nash's later pieces *Flight of the Magnolia*, which features what Nash termed "aerial flowers." With his photographs, Hughes created similar "flowers," albeit from the plastic bottle caps and fragments that have washed ashore from the surrounding seas. Gapado is also home to the Haenyeo, the traditional women divers of the region. During his time on the island, which is near Jeju Island, Hughes was struck by the continuous bombardment and arrival of plastic waste. Despite the island's goal to become the world's first completely carbon-neutral island by 2030, it seems a tragedy that waste from South Korea, China and Japan, as well as the wider global community, is unlikely to be neutral by 2030.

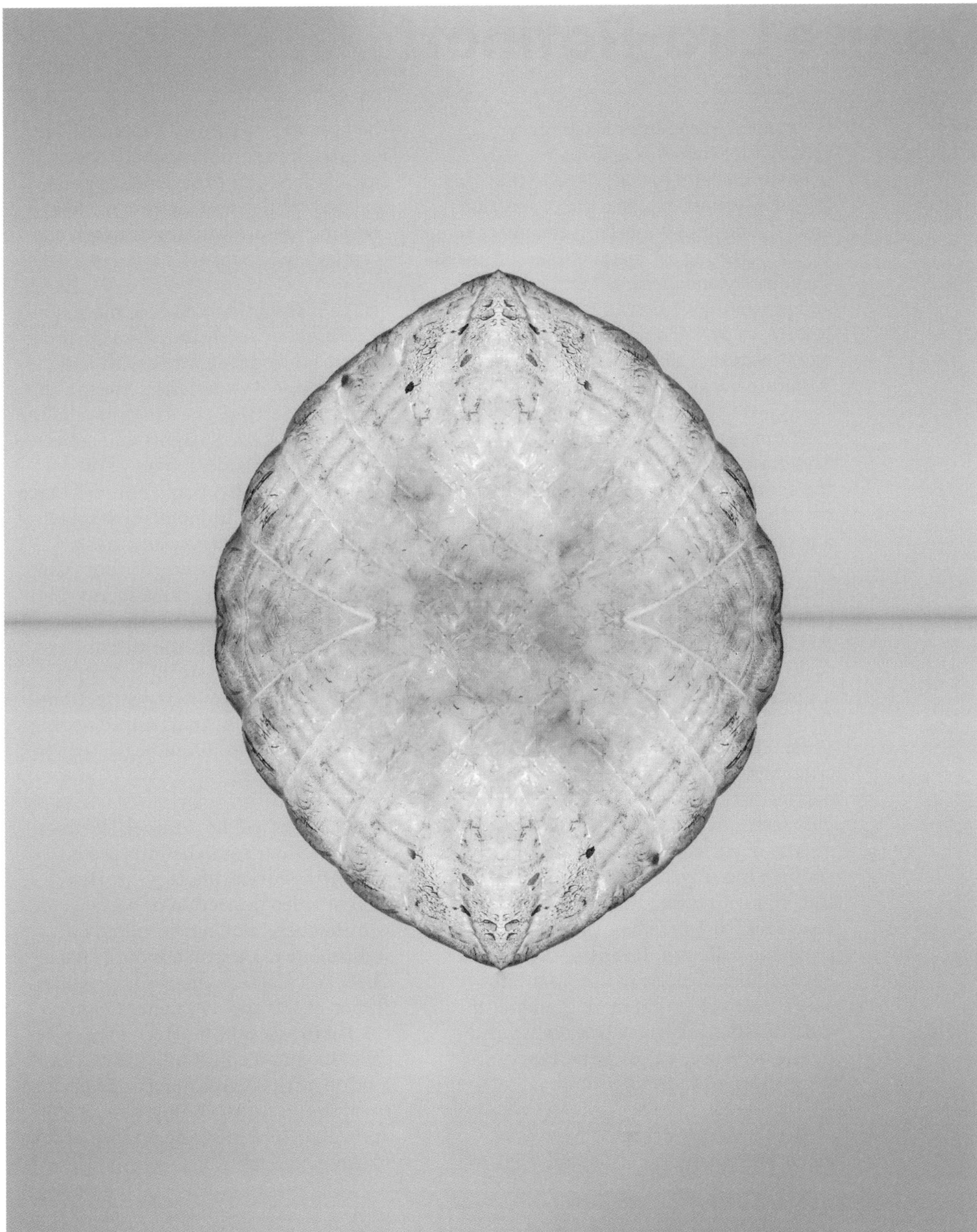

Sky Egg Flower, Gapado Island, South Korea, 2022, Digital C-Type Photograph, 100 × 80 cm

Denise Lira-Ratinoff

Born in Chile in 1977

"I am a Chilean interdisciplinary artist with extensive experience in environmental issues linked to the Ocean. My work explores the connections and relationships between Humanity and Nature, creating art installations & films that display and present their closeness and proximities. I first think of myself as Nature, as part of an order that exceeds me. If Nature is sick, I get sick with it. If Nature burns, I burn with it."

For more than a decade, Denise Lira-Ratinoff has been focusing on the ocean from different perspectives, creating sensory artistic experiences and a personal language that she describes as "thinking through feeling, a language that goes right to the body." *Chronometer* (2018-2019), co-created with her husband Patricio Aguilar Diaz, is a large-scale installation comprising a giant architectural maze constructed entirely from single-use plastics collected around Chile. The accumulation of plastic debris stretches from floor to ceiling, with only narrow passages accessible to visitors. This restrictive environment induces a sense of suffocation as one navigates through the labyrinth of waste, evoking the plight of marine life, such as fish and cetaceans, that perish in the ocean due to plastic pollution. It is also a synopsis of the dramatic noise pollution in the ocean. "Knowing that more than one of these plastics may have been consumed by the audience, we ask what legacy we want to leave" the artist wonders.

"I believe that art must have an active role in education, and that the best way to convey a message is through experiences. With this installation, you experience anguish and fear while entering this narrow corridor. We are walking through our own legacy."

The installation allows a maximum of five visitors at a time, transforming the artistic journey into a physical and emotional experience. Within this dystopian setting, the songs of whales, recorded by Lira-Ratinoff and collected from oceanographers, emerge as a poignant call to action and resistance. Additionally, a chronometer displaying the current time serves as a stark reminder of the urgency of addressing climate change and radically altering our consumer habits. For the artist, this element grounds the work in the present, emphasizing individual responsibility for our current actions. Above all, Lira-Ratinoff aims to bring visibility to issues often neglected in our daily lives.

At the end of the "tunnel," visitors arrive at what could be interpreted as a metaphorical shore, featuring a large video projection of waves continuously moving, seemingly indifferent to the surrounding chaos. The water in these images is pristine, filmed at 150 and 450 meters high on the Pacific Ocean. It offers a space for calm and meditation. The constant motion of the waves appears to cleanse everything, transcending time, events, and space, and bringing a sense of hope to the installation.

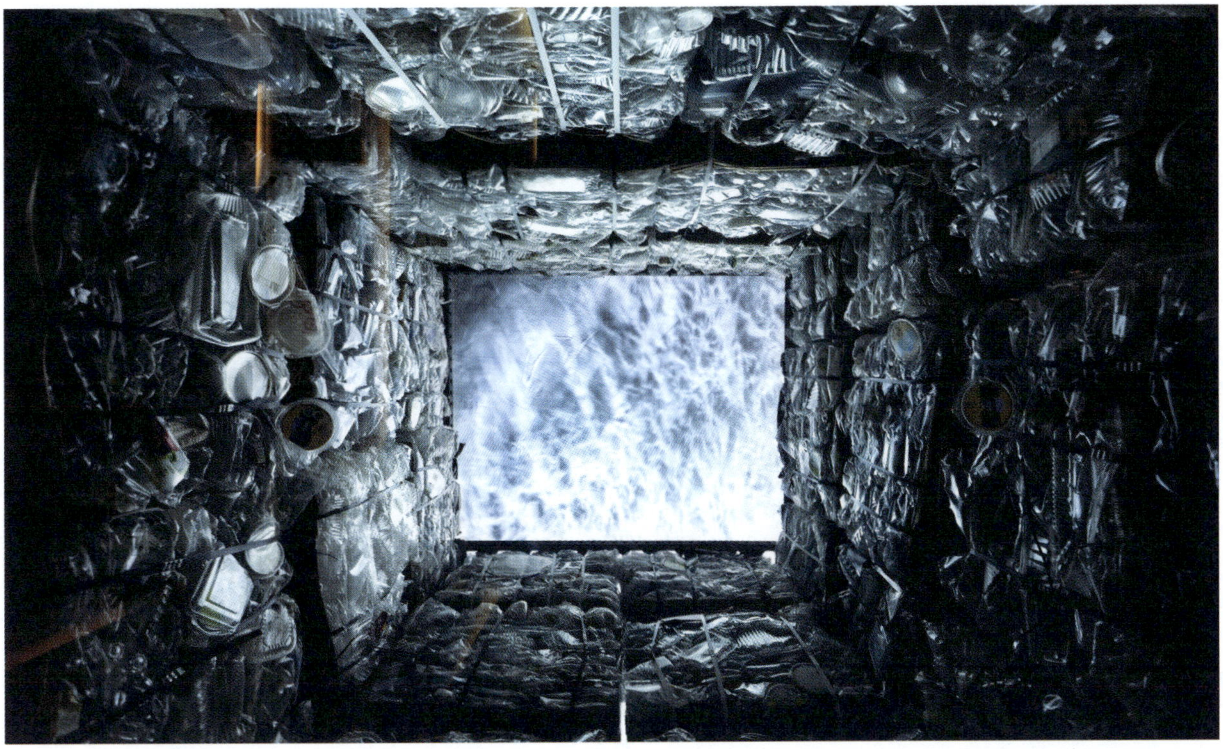

Chronometer, 2019, Site-specific installation made of 2,300 m of intertwined steel rafters, 1,400 bales od solid residues, variable dimensions

"Standing next to the ocean, I think of how we, as humankind, are dumping into it a totally uncertain destiny. We are about to make barren and condemning it forever to the silence of its magnificent inhabitants. We are ready to give voice to the ocean's grief."

The 11'30" video *Threshold* (2019) is projected onto a floor covered with 3,000 kilograms of salt. It features the enigmatic movement of waves with their ebb and flow, accompanied by a powerful soundscape that includes the chants of whales. Gradually, human sounds begin to interact with the marine songs, altering their frequencies. On the floor, a woman tries to respond to the various rhythms as if lost.

This large-scale installation and performance highlight the acoustic contamination caused by human technologies that infiltrate oceanic spaces, profoundly disorienting its inhabitants. It serves as an invitation to become more attuned to aquatic life, which is often imperceptible to human ears. The video employs a monochromatic palette, with large dark spaces that open like voids, symbolizing the death of marine species and the resulting emptiness in marine environments. For the artist, the whale's chant resembles a moan. However, as viewers become mesmerized by these songs, they may develop a more empathetic understanding of marine life.

Patrick Tagoe-Turkson

Born in Ghana in 1978

"My connection to the sea goes back to my childhood, growing up near its shores in Winneba, Ghana. My family's livelihood depended on fishing, instilling in me a deep respect and understanding of the ocean's importance. From playing with plastic debris to participating in traditional sea-related rituals, my bond with the sea has been lifelong. About twenty-five years ago, I began noticing alarming changes–rising water levels, increased pollution, and communities being affected. This sparked my determination to address these issues through art, using plastic waste collected from beaches as my medium.

"In Ghana, where I live, we have a deep and fundamental connection to the ocean because fishing is our primary traditional vocation. When I was growing up, the elders in our community emphasized the significance of preserving and conserving the ocean so that it might continue to support our livelihoods. We believe in the presence of sea gods and goddesses who govern the ocean, and as a result, we undertake rituals and sacrifices in their honor, hoping for a plentiful harvest. These customs are strongly ingrained in our society, with fishermen pouring libations and performing rituals before embarking on their voyage and then returning home. Fishermen do not go fishing on Tuesdays. Our respect for the ocean is also mirrored in our annual festivals, which are lively celebrations of diverse art forms and performances honoring these deities.

"Historically, we have been encouraged to avoid actions that could harm the ocean because we consider it sacred. Strong beliefs and taboos around the ocean instill fear and respect in the community, thereby preventing pollution. However, the landscape is shifting. The growth of technology, combined with the effect of modern religion on our traditional values, has resulted in a reduction in the great affinity and reverence we previously felt for the ocean. The discovery of oil and the widespread issue of plastic pollution of the ocean have had a huge influence on the fishing industry. Many families that previously relied only on the sea for a living have now switched to other trades.

"This transition elicits a range of feelings. On the one hand, there is a sense of loss of the traditions and our strong spiritual connection to the water. On the other hand, it demonstrates our community's tenacity in adapting to new economic realities. The ocean still remains an important part of our identity and legacy, and we must strike a balance between accepting modern innovations and protecting the sea's cultural and natural integrity.

"In Ghana, marine ecological issues mirror those of many African nations. Destructive fishing practices, degradation of the habitat, pollution from coastal development, and improper waste disposal threaten our marine ecosystems. Fishermen often find more plastic than fish in their nets.

Despite challenges, efforts such as protected marine areas, fishing close seasons, and environmental education campaigns are under way. Artists like myself are also playing a role by transforming beach-collected plastic into art. Yet, more action is needed to ensure the long-term health of Ghana's marine environment.

"I define myself as an activist artist, leveraging the power of art to challenge perceptions and inspire action on environmental issues. My strategy as an artist is multifaceted. Through workshops and public art interventions, I aim to raise awareness and foster dialogue about marine conservation. By transforming ocean plastic into beautiful artworks, I seek to highlight both the hidden beauty of our maritime world and the urgency of addressing environmental threats. My tapestries, woven from upcycled flip-flops, serve as symbols of resilience and renewal, illustrating the potential for beauty amidst environmental degradation.

"I also frequently encounter unexpected challenges when I work on the beaches of Takoradi, Ghana. When I first visit a new fishing community, for instance, the inhabitants are usually both welcoming and skeptical. As an outsider, they are wary of me until I introduce myself to the chief and the elders. This introduction is important because each group has its own set of norms and conventions that govern their interactions with the ocean. For example, during a beach performance I gave, I covered myself in a white cloth and faced the water. This pose, unbeknownst to me, offended he community's sea deity, so I had to sacrifice two white goats to appease her. Regardless of such cultural obstacles, community support can be overwhelming. There have been times when the chiefs and elders have rallied the entire community to help with my beach cleanup activities, even providing refreshments for participants. However, firms whose activities have a direct influence on the ocean frequently pledge support but fail to deliver, leaving us to rely on self-funding from the revenues of my paintings.

"One of the most difficult issues is managing and transporting the collected plastics. The financial burden of these operations is significant, as all beach cleanup actions are self-funded. Furthermore, the Ghanaian government's efforts to safeguard coastal towns from erosion by covering beaches with massive pebbles unintentionally trapped plastic debris. This has made our cleanup efforts more challenging and raised our operational expenditures as we are compelled to relocate our operations further south. Besides, there is an urgent need for studios and workshop facilities in these communities to instruct young people on how to turn collected rubbish into successful businesses. Establishing plastic recycling studios and workshops would provide jobs and improve the lives of the youth in these communities.

"Overall, the community enjoys seeing its waste transformed into beautiful and expensive art. This metamorphosis not only helps to clean up the environment but also instils a sense of pride and hope."

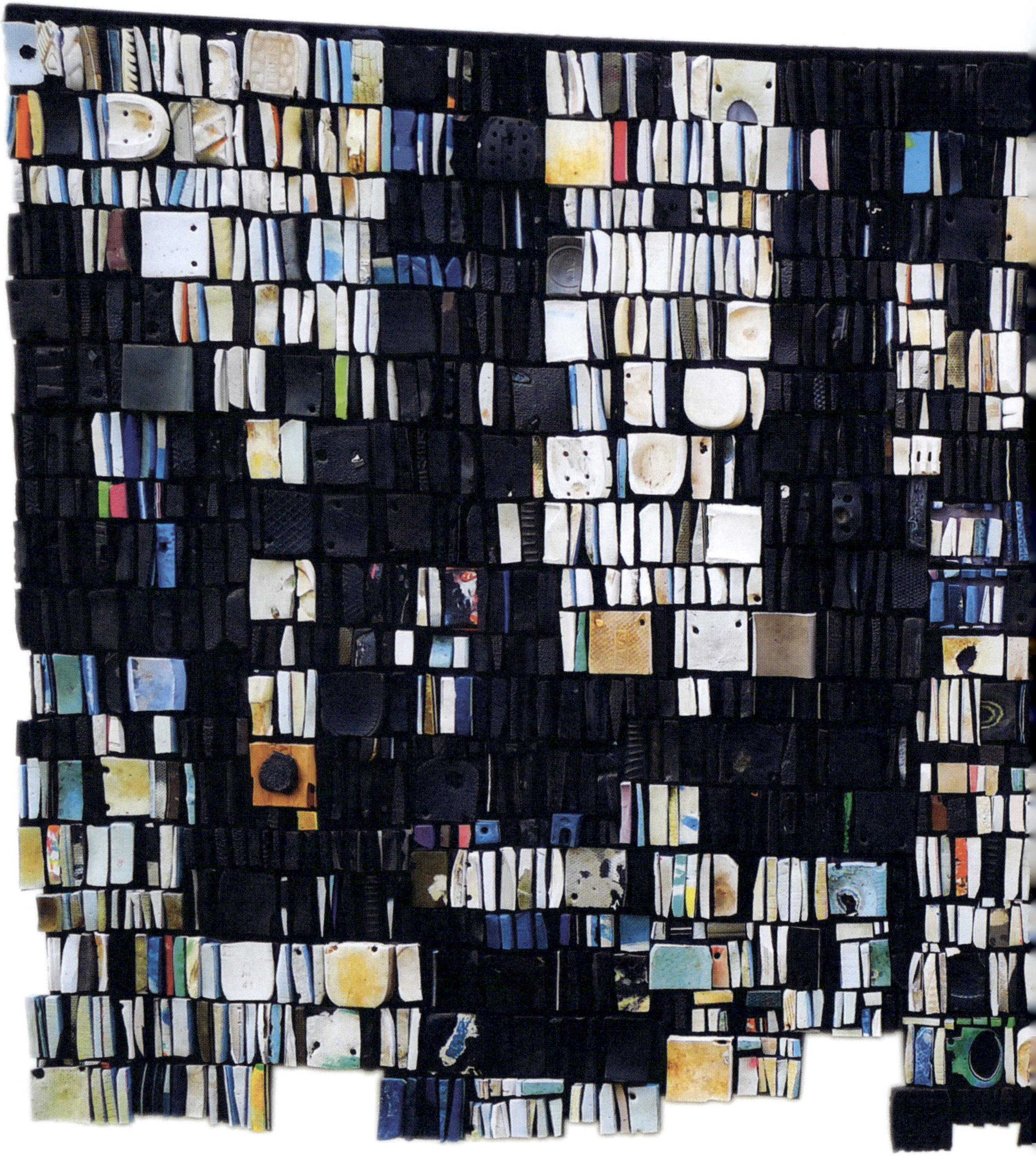

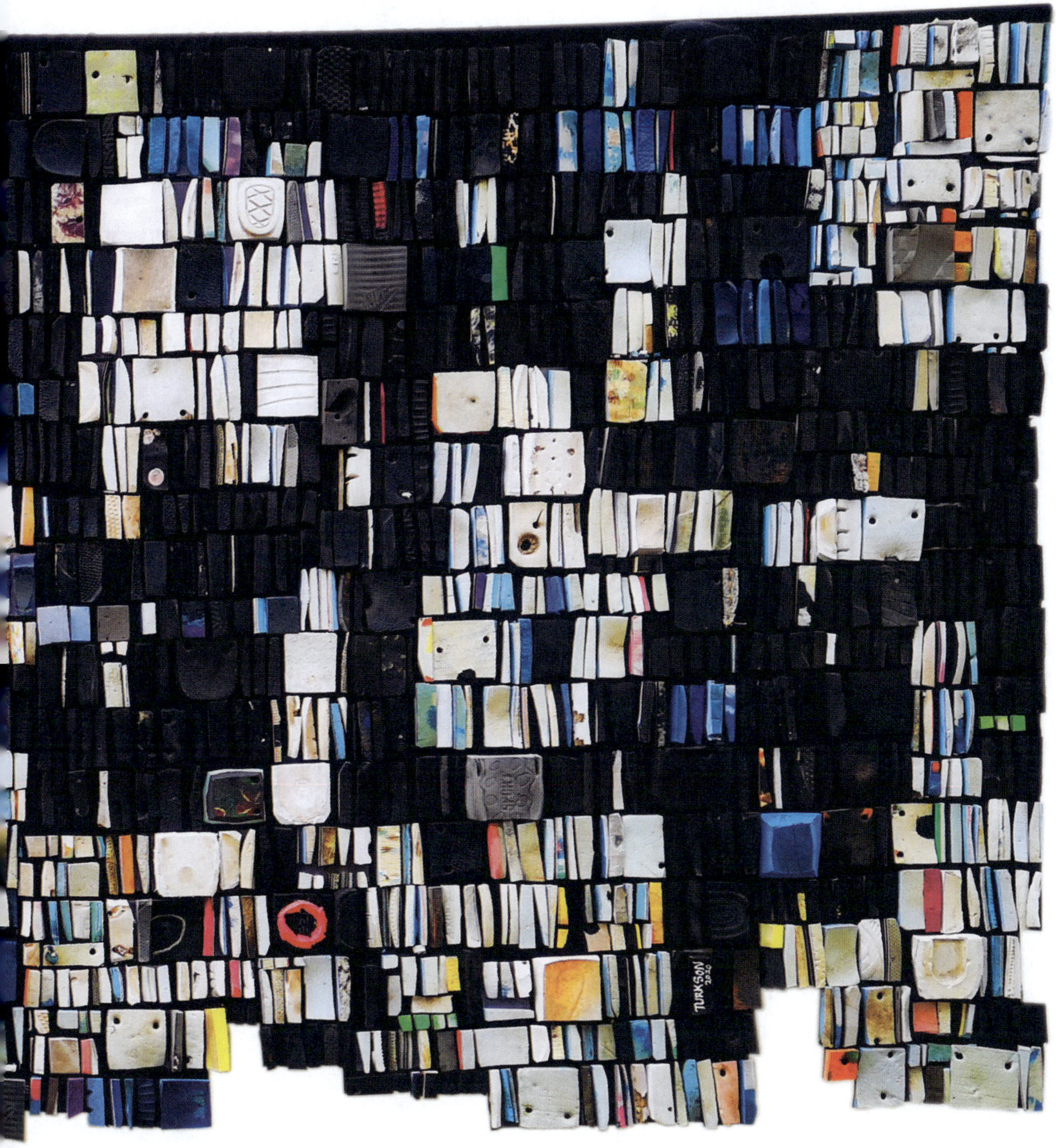

Enyifra na Tsiew (blind and deaf), 2020, Found Flip-flops on Suede, 242 × 127 cm

Carlos Manuel Rodriguez
CEO of the Global Environment Facility (Washington)

The ocean of my youth no longer exists; all the predators have disappeared. The beaches where I used to go surfing are now polluted beyond recognition, mainly by plastic pollution. Many of our local breaks have disappeared due to new hotels and highways built nearby, and above all due to rising sea levels. And the fresh fish we used to easily catch have migrated to cooler waters far away. This is my ocean, but it does not have to be yours too.

I consider myself incredibly fortunate to have spent my life fighting to protect this beautiful and amazing planet. Although I am a lawyer by training and a politician in practice, I am a conservationist in my heart and a surfer in my soul. I have always felt most at home among the waves, feeling my body slowly connect with the rhythm of the swell, the currents, and the tides.

From Playa Hermosa (Costa Rica) to Uluwatu (Indonesia) and many other beautiful breaks all around the world, that feeling of connection with the ocean can be found everywhere. It is not just found in the waves. Swimming, snorkeling, and even just a fun day splashing around on a clean beach should be available to everyone. For the many of us who are lucky to have found our connection to the ocean, you know—once you find that stoke that exhilaration, it never leaves you, and it always calls you back for more.

Yet we should all remember this is not our ocean or our planet. We share this beautiful world with an amazing wealth of flora and fauna. And for the ocean, we are only guests. When you dive underwater, you are entering the home of an amazing diversity of flora and fauna. There are marine mammals and cephalopods with intelligence that rivals that of humans. There are fish stocks that feed entire communities and drive national economies. There are coral reefs and mangrove forests that protect whole countries. And there is much we do not know about our undersea world. It is our duty to be precautionary and proactive in protecting all living and non-living marine resources in a sustainable way.

The fight we face to protect the planet and restore our oceans is both personal and global. Every one of us is fighting for our own reasons, but we are all fighting. My sincere hope is that our fight is just getting started. And we have every reason to be optimistic because I see the fight strongest in our youth–in our children and grandchildren. These next generations of leaders are growing up with aspirations to become ocean scientists, climate negotiators, and marine park rangers. Their passion, their connection, and commitment to saving the planet is core to their identity in ways we will never fully understand.

Many say that our actions over this decade may decide our fate on this blue planet. We cannot wait for youth to forge their own career paths over future decades before they have an impact. No, they are already here, committed, enraged, and ready to contribute in impactful ways. We need to listen to what they have to say now. We need to stand aside and follow their example. This is not about our future. It is about theirs.

The ocean of my youth may be no more, but the same does not need to be true for future generations. We must ensure they have the skills to succeed. And succeed they will. The ocean of the future will be full of our children's own memories and their own ocean stoke. In fact, the ocean of the future may well save the entire planet.

A Heritage to Regulate

The water understands
Civilization well;
It wets my foot, but prettily,
It chills my life, but wittily,
It is not disconcerted,
It is not broken-hearted:
Well used, it decked joy,
Adorned, doubled joy:
Ill used, it will destroy,
In perfect time and measure
With a face of golden pleasure
Elegantly destroy.

Ralph Waldo Emerson

1. Emma Critchley
   Born in the United Kingdom in 1980
2. Ana Teresa Fernández
   Born in Mexico in 1981
3. Inuk Silis Høegh
   Born in Greenland, Denmark in 1972
4. Lee Tzu-Tung
   Born in Taiwan in 1990
5. Winnie Soon
   Born in Hong Kong in 1979
6. Charles Lim
   Born in Singapore in 1973
7. Uili Lousi
   Born in the Kingdom of Tonga
8. Tita Salina and Irwan Ahmett
   Born in Indonesia in 1973 and 1975

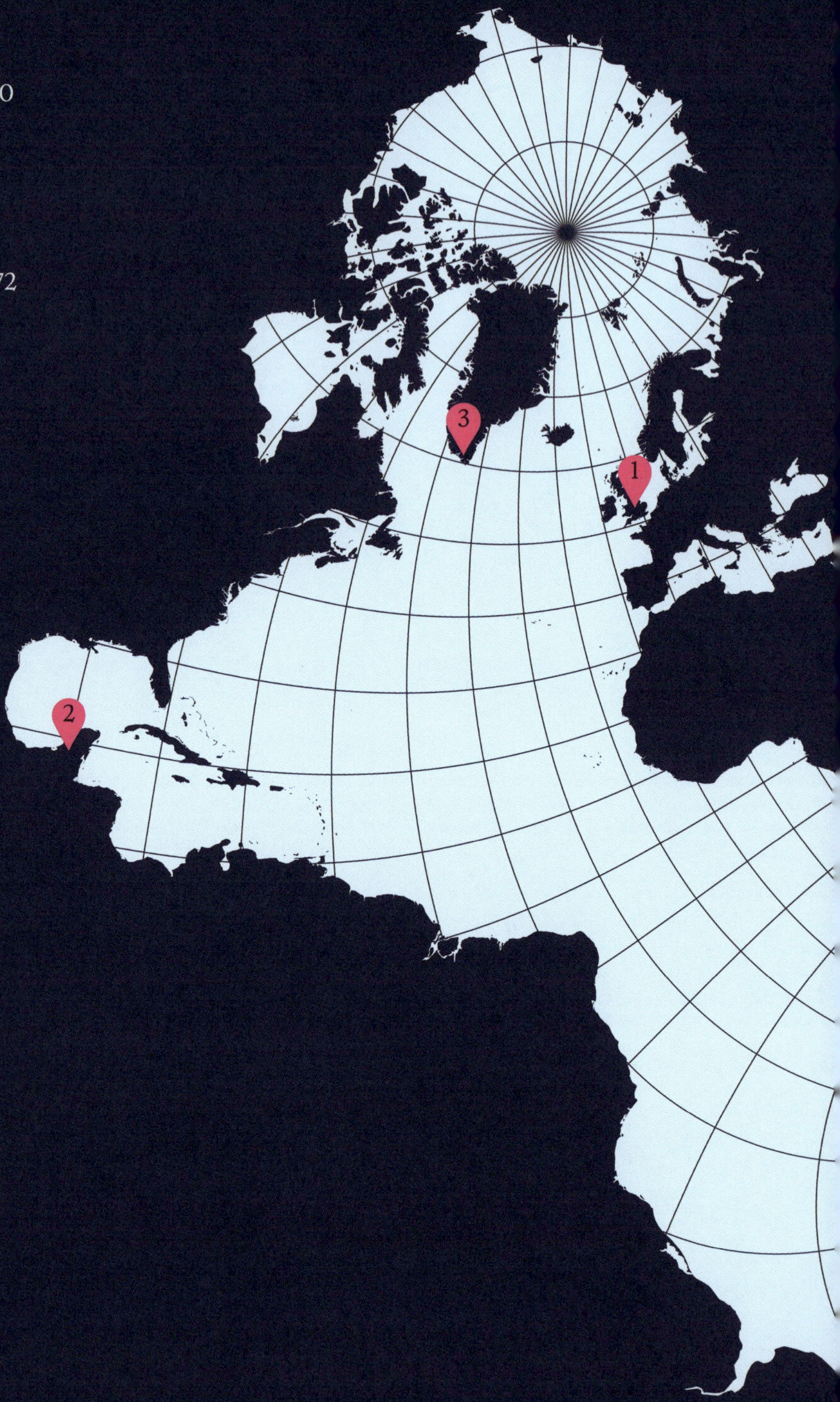

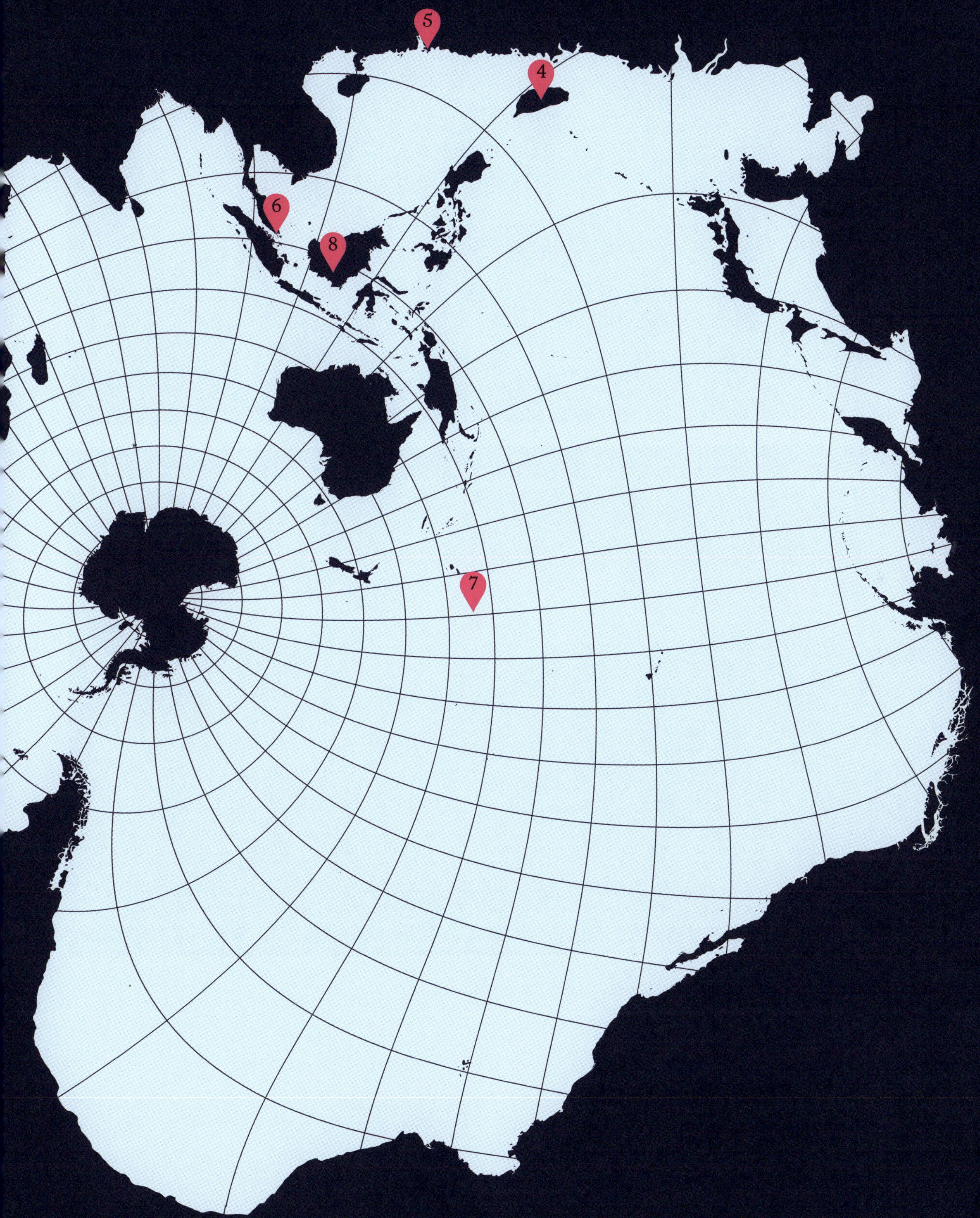
5
4
6
8
7

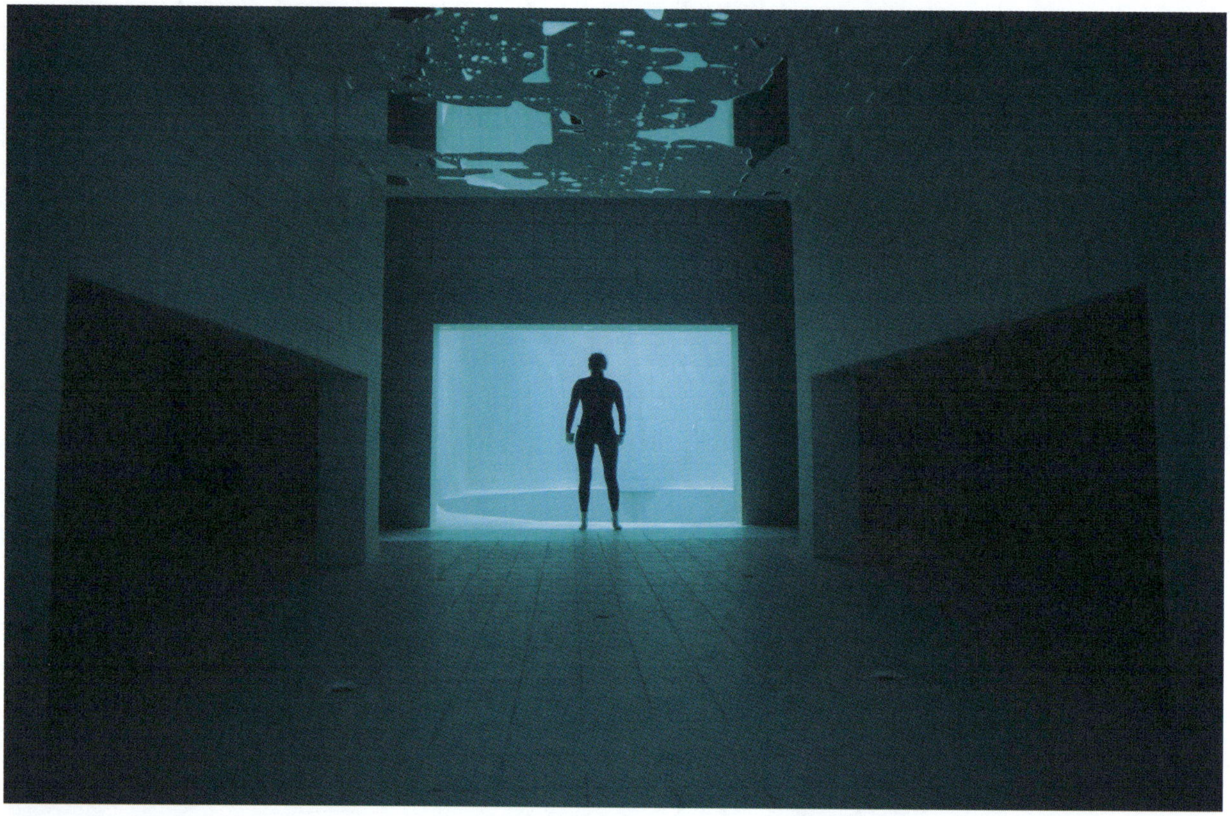

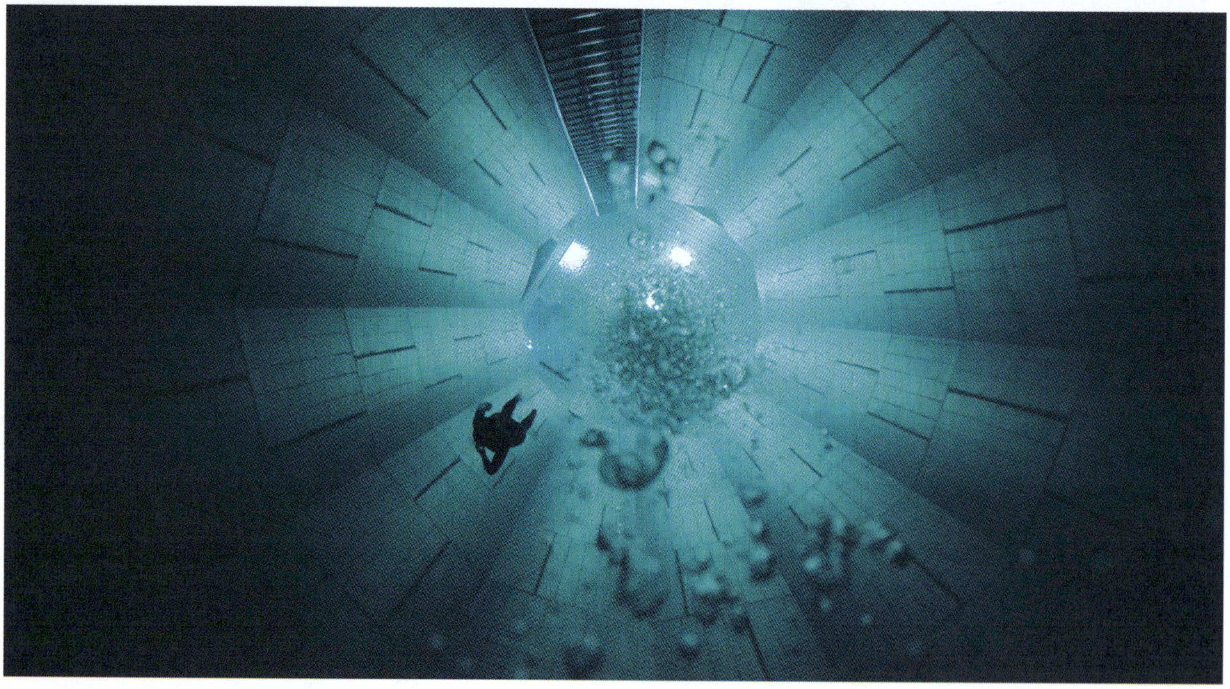

Common Heritage, 2019, Video, 25 min

Emma Critchley

Born in the United Kingdom in 1980

"Being immersed in water is a powerful scenario that resonates not only with me as an artist but unites us all; it is something we have all experienced. The thick liquidity of the space brings to the fore our interconnected, symbiotic relationship with the environment we inhabit, we are part of, we are. Yet, our relationship with the ocean is in crisis." (*The Routledge Handbook of the Ocean Space*, ed. K. Peters, London & New York: Routledge, 2023.)

With her films, art installations, and photographs, Emma Critchley creates constructed environments that intertwine scientific realities with imaginative narratives. Her artistic practice aims to question and embody our relationship with the ocean. Often perceived as a vast, illimited and often abstract entity, or merely considered as a location for resources, the ocean is difficult to apprehend. The artist believes that art and fiction are paramount to nurture people's imagination. It enables them to better understand the ocean and grasp today's challenges of climate change. "I feel we need to find a connection with the ocean to care about it, to protect it," she said. The artist defines her practice as politically engaged and through herwork aims to create a space where people can reflect on their own positions.

Single Shared Breath (2011) is Critchley's first artwork that directly addresses environmental issues. The short film features two freedivers exchanging a single breath—each must use some of the breath before passing it back. Over the two-and-a-half-minute duration, what at first resembles a kiss becomes more a reflection of dependency and mouth-to-mouth resuscitation. Critchley considers the breath as a thread that connects our internal bodies with the environment around us, part of the terrestrial world that we take with us when entering water.

This first film is silent but, increasingly, Critchley has been working with music and sound, collaborating with sound artists and musicians. She likes to remind us that sound is the most important sense underwater. Some of her works explore underwater acoustic pollution, such as *The Space Below* (2020), an immersive large-scale sound installation, made in collaboration with artist Lee Berwick, that amplifies the current effects of human noises. Her practice also addresses the melting of glaciers, for instance *Witness* (2021), and the rising sea level with *Do You Know Nothing? Do You See Nothing? Do You Remember, Nothing?* (2018).

One of the most pressing marine issues that Critchley wishes to emphasize is deep-sea mining. In her film *Common Heritage* (2019), she sheds light on the fantasies we construct around exploration, and draws into focus how "these romanticized places are in fact borders of conquest, annexed for geopolitical territory appropriation and mineral resources." →

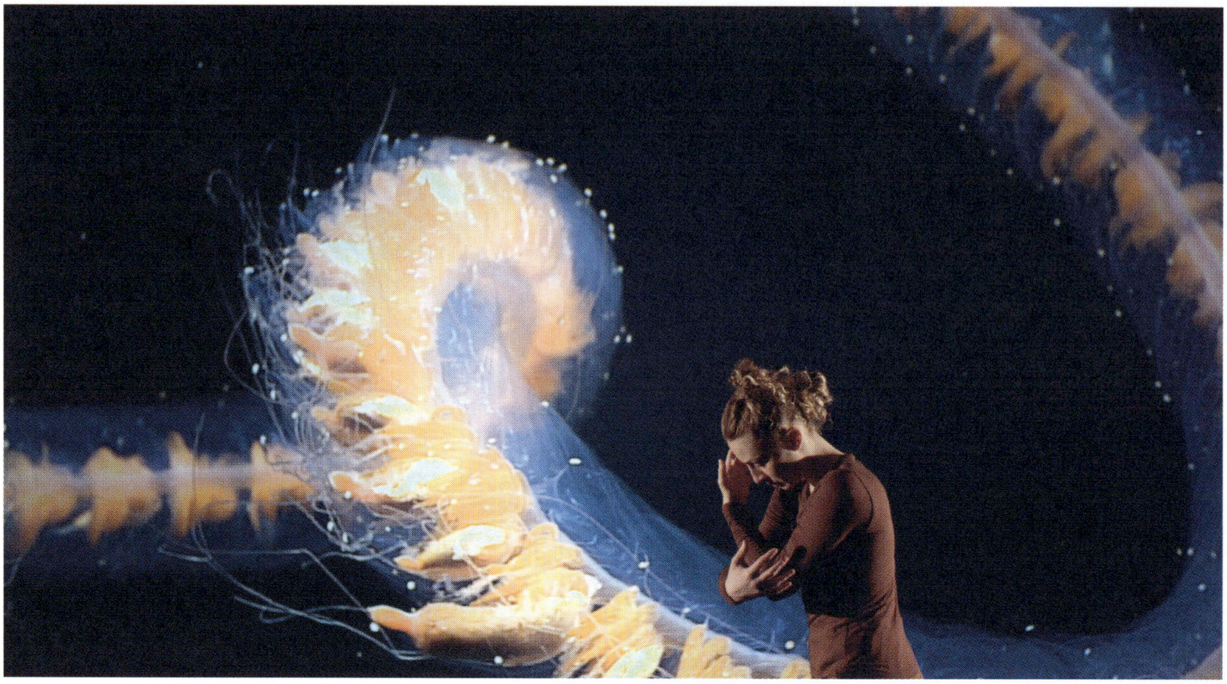

Sirens II & III, 2023, Video, 3 min 20

The film references the 1967 General Assembly of the United Nations, where Arvid Pardo, the representative of Malta, raised the critical question of seabed jurisdiction, proposing that it should belong to the "common heritage of (hu)mankind." The film's title directly alludes to this speech, which was pivotal at its time and changed the course of ocean governance. Critchley points out, however, that it is not without fault, as it assumes the minerals are resources that can be exploited and doesn't consider the cultural value of the ocean.

Throughout the film, archival footage shows us speeches, press conferences, and declarations by international states, revealing the tensions and contradictions over nine years of annual meetings, the time it took to draw up the Common Heritage of (hu)mankind principle and the Law of the Sea Treaty. Although the United Nations Law of the Sea Convention was adopted in 1982, it took over a decade for it to come into force, and it should be noted that, the United States has not signed it. The idea of regulating the high seas is relatively recent, as prior to the 1950s, and even dating back to the seventeenth century, the area beyond a cannon shot (approximately three nautical miles) was considered a realm belonging to all without any regulations or legal frameworks–an idea referred to as *mare liberum*.

When Critchley embarked on this research journey, deep-sea mining was scarcely discussed. However, today, the urgent need for a legal framework to protect the deep sea and its largely unknown seabed arises once again, with Norway approving commercial-scale deep-sea mining within their national jurisdiction in early 2024. In *Common Heritage*, the film contrasts indoor footage of bureaucratic debates with scenes from the depths of the ocean. Critchley draws a parallel between this new conquest and a form of modern colonization. She raises the critical question of whether we should truly wish to venture into these realms and repeat history. The film suggests strong parallels with colonial conferences, when Western states debated the future of their colonies without inviting the colonies themselves to participate in the discussions.

Ana Teresa Fernández

Born in Mexico in 1981

Ana Teresa Fernández grew up in Tampico, in Mexico, not far from the sea. As a child, she could smell the refineries from the coastline, but she did not realize what that meant until much later. Rather, she was enjoying the beach, surfing, always in the water. For a long time, her art did not address ecological issues, focusing instead on the questions of borders and migration. From the American-Mexican border to the Mediterranean Sea, her paintings, videos, and installations mainly aim to embody the plight of refugees and their painful and often deadly struggles at sea. From this perspective, the marine environment is depicted as a place of both hope and suffering, a transitional and liminal space between countries, between life and death, between the lightness of floating bodies and the heaviness of their burdens.

A few years ago, Fernández had a wake-up call during a conference about climate change when she heard that sea levels were predicted to rise up to 1.8 meters by 2100, a height that exceeds her own. She realized, however, that this data remained abstract to her, just like all numbers linked to ecological issues. She searched for the best way to transform this abstraction into a sensory and empirical experience that everyone could feel and immediately understand through their bodies.

On the Horizon (2021) is a public art installation consisting of sixteen acrylic resin cylinders filled with seawater. The artist conceived them as "sea bodies," and they measure exactly six feet, the projected sea level rise. The work is installed on beaches, near the shore, where the sea and the land meet. People can touch and play around them; they offer a way to physically grasp what is happening without any form of dramatization. Rather, viewers are invited to take the time to ponder the situation and consider how each of us can respond to this phenomenon and to climate change at large. →

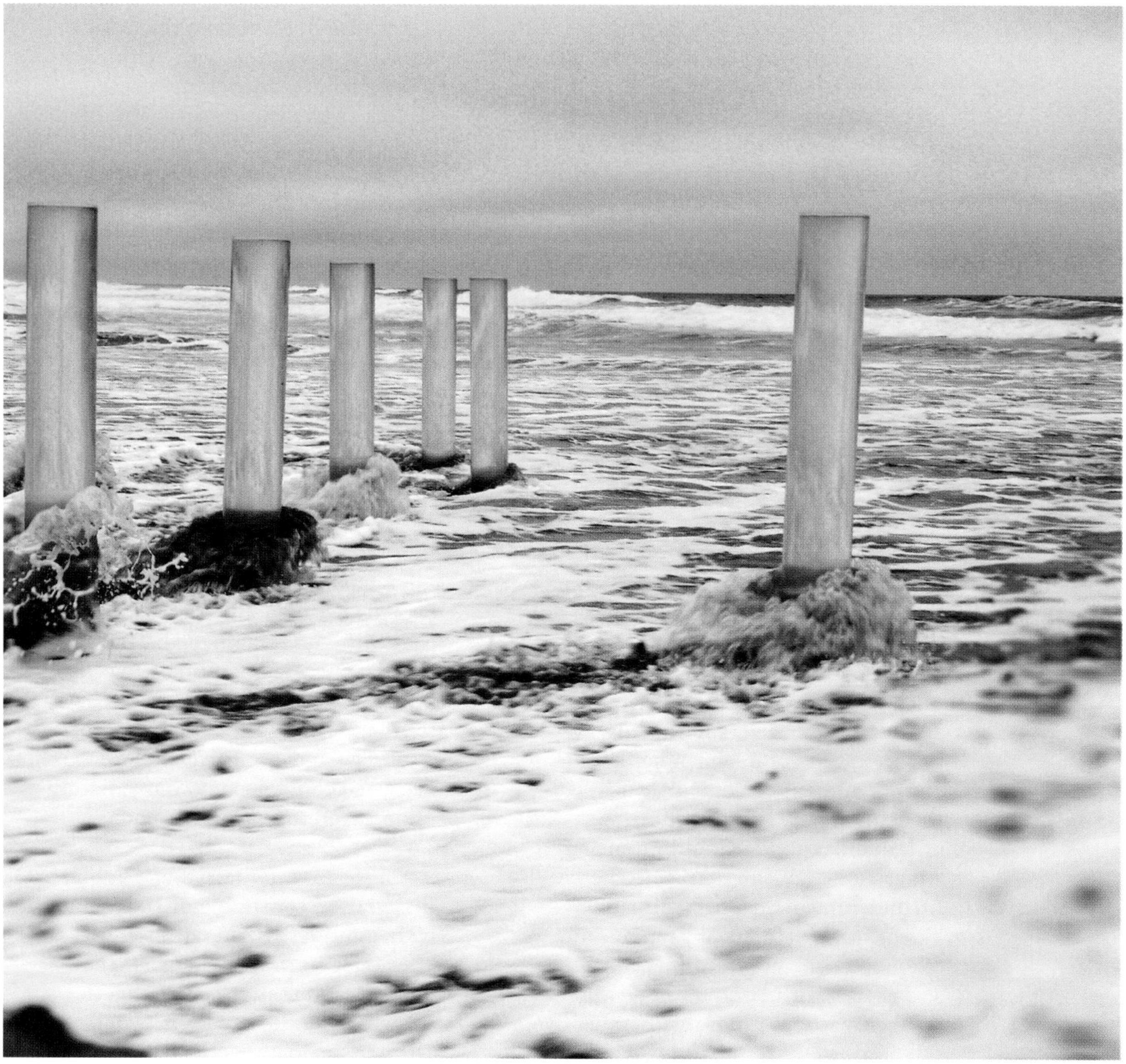

On the horizon, 2021, Site-specific installation

The columns are a window into our future, into ourselves.
It is the story of us.
Standing still is not an option.
We must all carry the water.
(Artist's statement for her exhibition *Listen Louder*,
di Rosa Center for Contemporary Art, 2023-2024)

For *SHHH* (2023), another public art installation, Fernández directly addresses our collective behavior regarding the rise of the sea level, particularly our silence and lack of attention to its effects on coastal communities. The title itself points to the denial we often adopt when considering climate change and its consequences. "Many people may feel less concerned about the rise of the sea level because they do not live near the sea. However, it is a pressing issue for many islanders who are already losing their land, homes, and ways of living," she explains. The linguistic diversity of small islands is especially important, with some endangered languages already spoken by very few individuals. With the displacement of coastal communities, these indigenous languages risk extinction. This represents a significant loss for humanity, yet the topic is seldom addressed.

The installation is conceived as a living monument that advocates on behalf of these threatened communities and languages. It is constructed with 1,800 golden acrylic mirrors that continuously move with the wind and reflect sunlight, responding to and trembling with the environment.

Referring to Joseph Beuys' theory, Fernández refers to her installations as "social sculptures" to remind us that there should be no border between art and life. She calls for urgent action.

Inuk Silis Høegh

Born in Greenland, Denmark in 1972

A multidisciplinary artist based in Nuuk, Greenland's capital, Inuk Silis Høegh aims to challenge the stereotypes that surround his native Greenlandic and Inuit culture. His sculptures, films, and installations are deeply rooted in contemporary Greenlandic politics and highlight the radical transformations the territory faces due to rapid modernization and the effects of climate change.

In 2013, Høegh installed a gigantic iceberg sculpture outside the National Gallery of Canada in Ottawa to address the pressing issue of global warming. Titled *Iluliaq*, which means "iceberg" in Greenlandic Inuktitut, the sculpture was not an actual iceberg but a giant print covering the building. Inside, though, one could hear recordings of melting glaciers emitting pops and sounds as if the artefact was also melting. The artist had previously created a similar work in 2009 for the United Nations Climate Change Conference (COP15). In his ongoing computer-generated series, *Audio Abstractions*, Høegh converts the sounds of the Arctic into colorful abstract spectrograms printed on aluminum, accompanied by corresponding audio

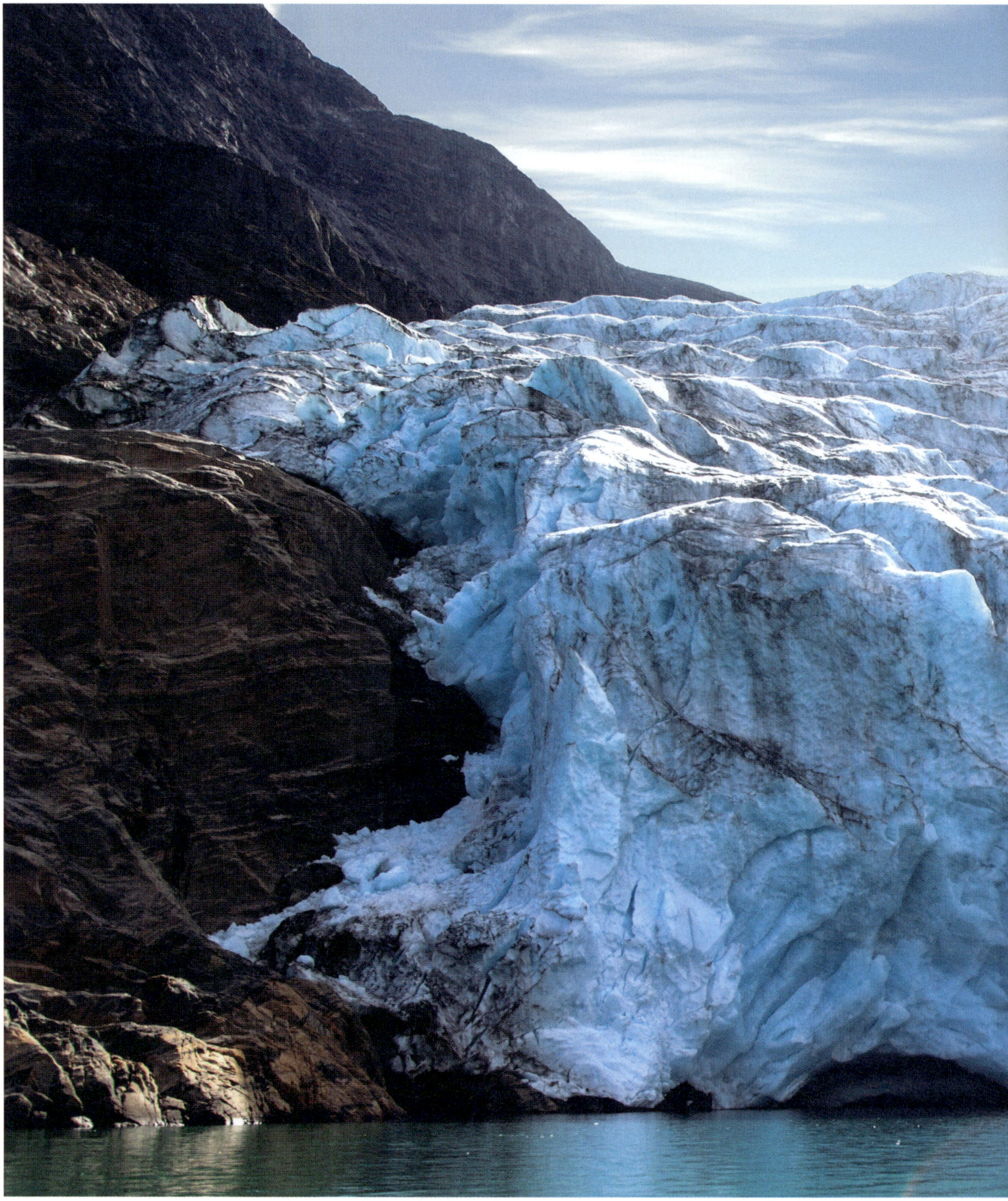

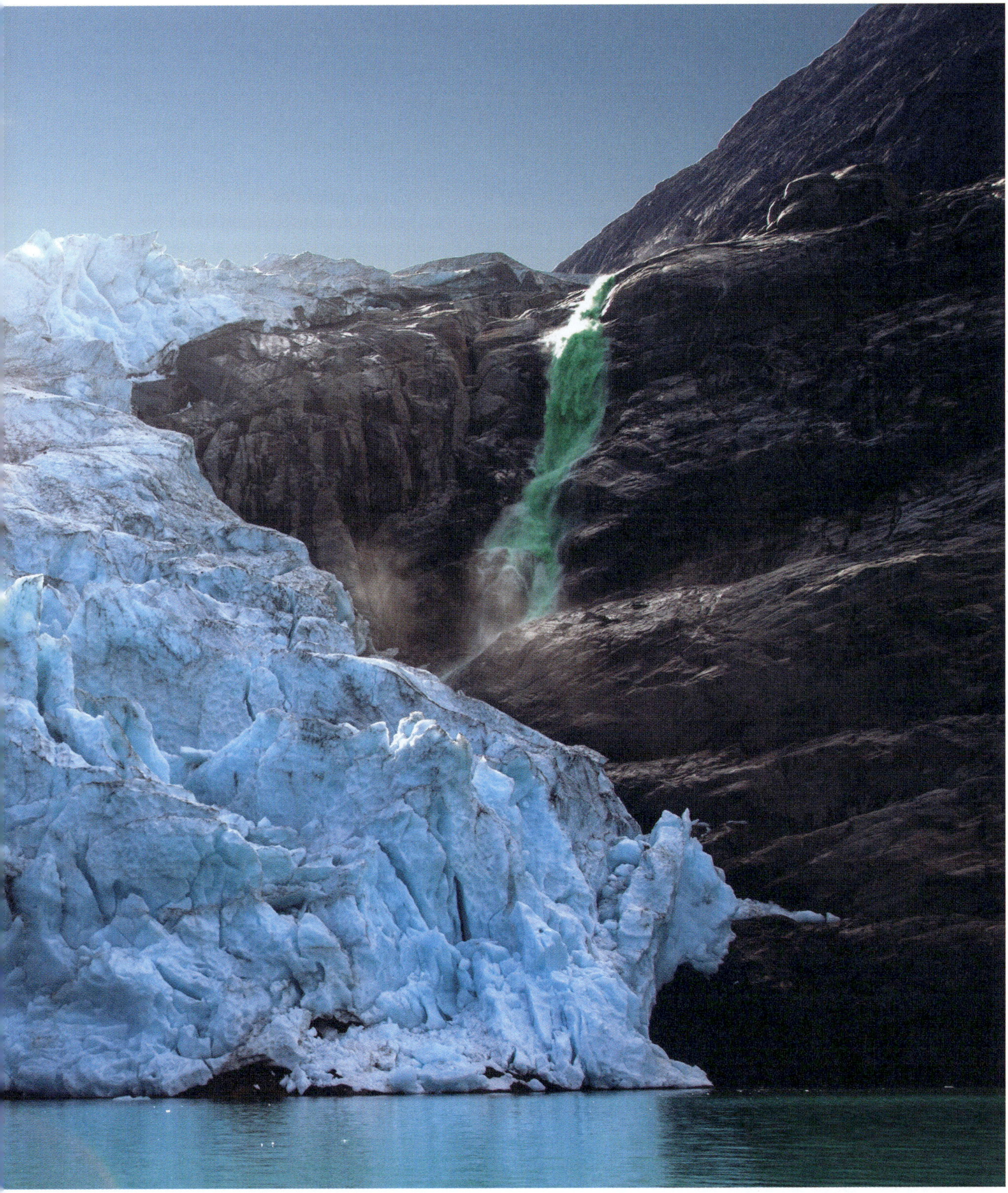

Water, image from The Green Land, 2020, Film installation, 34 min

experiences such as male humpback whale songs and the cracking of ice. Each time, the artist likes to play with different artistic languages and mediums to connect different feelings and create original synergies.

Arctic sea ice is diminishing at a rate of almost thirteen percent per decade, and even the oldest icebergs have shrunk drastically. Some predictions suggest that the Arctic could be ice-free by 2040. The melting of ice is a complex issue with multiple global effects. In the Arctic region, one consequence is that the reduction of ice facilitates navigation and drilling activities.

The origin of the thirty-four-minute film *The Green Land* (2021) stems from Greenland's 2013 vote on uranium mining. Although Høegh opposed mining, he acknowledged that a mine could provide access to rare earth elements, which could benefit green technologies. The film captures this tension, reflecting the current ecological controversies surrounding the development of certain technologies that justify non-green policies.

The film's starting point was a land art project in Nuuk and further north around Manitsoq on Greenland's west coast. The artist and his team created installations representing the four elements—water, fire, earth, and air—to reference alchemy and to picture the landscape of Greenland from a spiritual perspective. The glowing green color used to transform these elements was chosen because it is a perfect embodiment of the contrast between "green ecology" and toxicity. Working on-site was part of the performance, as the team had to continuously adapt to the weather conditions. For Høegh, this approach gave agency to natural elements over the human desire to control everything. The stunning beauty of the film's images prompts viewers to question the consequences of human impact on nature. There is no clear boundary between the artistic act and the natural elements, which merge to create a dreamlike landscape. The film suggests finding a way to continue cohabiting and transforming the world together. Alternatively, the beautiful spectacle may be the last performance of humanity before the human race withdraws from the scene. Høegh leaves viewers with their doubts and questions.

For the film's soundtrack, Høegh collaborated with sound artist Jacob Kirkegaard, who recorded sounds at each location. In her conversation with the artist, Mai Misfeldt describes a moving moment during these recordings: "For Inuk Silis Høegh, one experience in particular proved decisive for his understanding of what sound means to our experience of the world. He and Jacob Kirkegaard were in a boat on the ice fjord; it was eerily quiet, only a faint crackle from the ice could be heard. Kirkegaard dropped his microphone deep into the sea, and Høegh was allowed to listen. "I burst into tears when I put on the headphones. It is a fairy tale cavernous world down there, there was so much going on, rippling water, little distant explosions. It really took me by surprise." (in Mai Misfeldt, "A conversation with Inuk Silis Høegh," *The Green Land*, Nordatlantens Brygge, 2023)

Through his work, Høegh underscores the importance of preserving our natural world while navigating the complexities of modernization and climate change. His poetic and critical perspectives reactivate today's debates about ecological policies and green investments in Greenland and the broader Arctic region.

Lee Tzu-Tung & Winnie Soon

L. T-T.
Born in Taiwan in 1990

W. S.
Born in Hong Kong in 1979

Lee Tzu-Tung is a Taiwanese conceptual artist whose work integrates anthropological field research, political engagement, and economic activism. Winnie Soon, a Hong Kong-born artist and researcher, focuses on data circulation, invisible infrastructures, and the culture of coding practices. Together, they have developed a participatory project that poses questions about the rights of the ocean and the possibility for individuals to buy, own, or extract even a milliliter of the South China Sea.

Hailing from Taiwan and Hong Kong, where numerous communities have historically depended on the sea for their livelihoods, the duo underscores the essential connection that has historically linked these populations to maritime environments. They draw upon the work of anthropologist Epeli Hau'ofa, who introduced the concept of a "Pacific Way" and the term "Ocean of Archipelago" to describe communities residing across the Asian

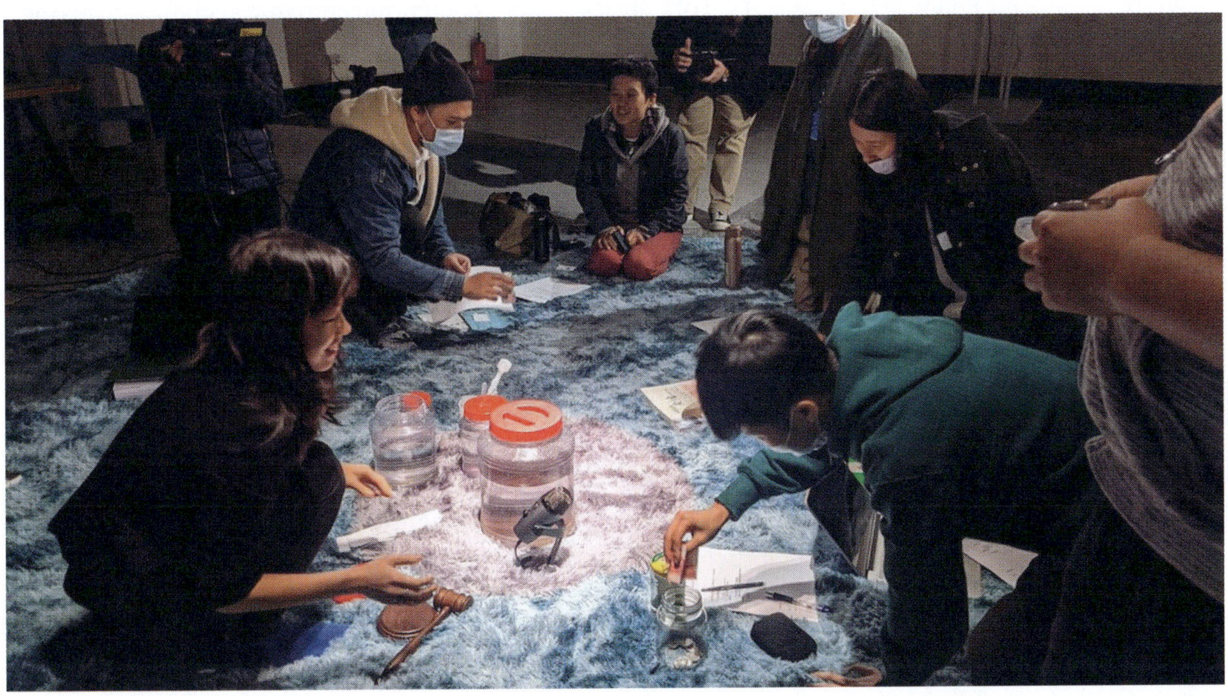

Forkonomy(), 2019-2021, Contract workshop, performance, bash code, contract, NFTs

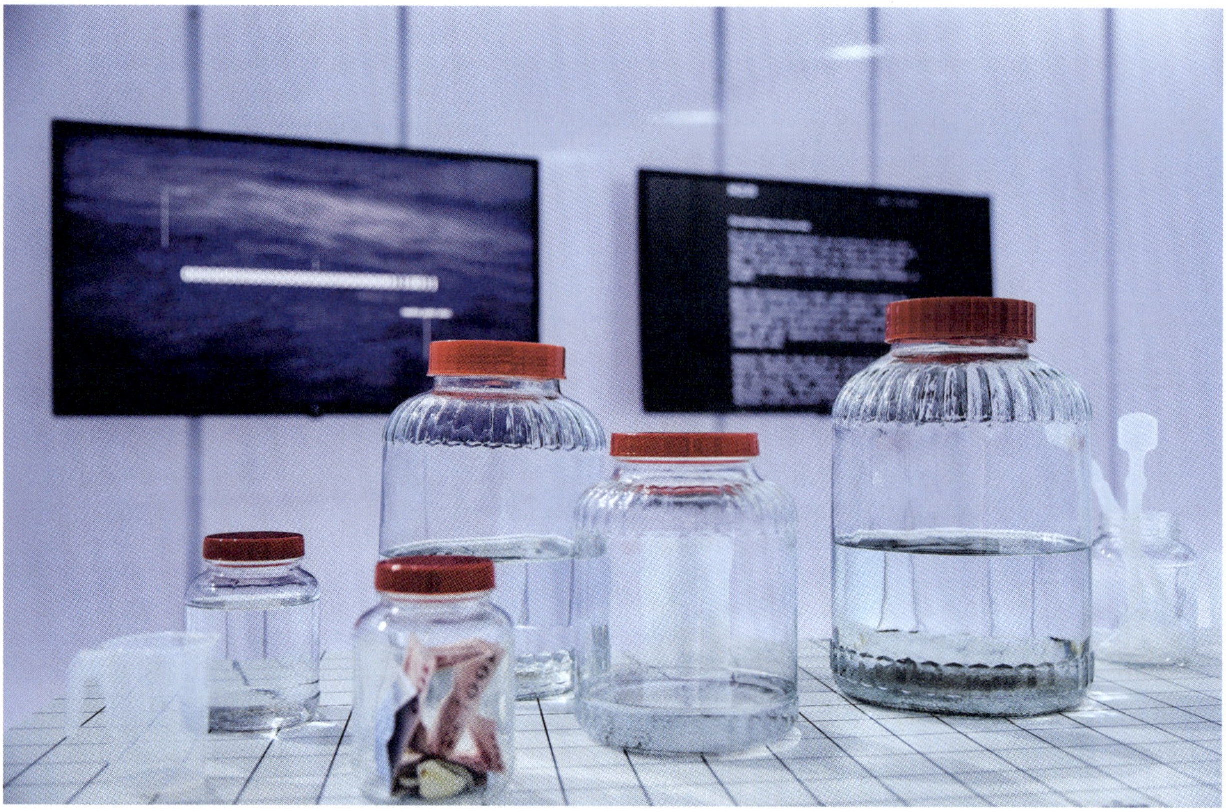

Forkonomy(), 2019–2021, Contract workshop, performance, bash code, contract, NFTs

and Polynesian realms. Hau'ofa posits that the sea serves as a connective medium rather than a barrier, fostering a sense of shared space among diverse islands and archipelagos. However, in the context of the South China Sea, despite efforts by various ethnic groups within this "Ocean of Archipelago" to maintain respectful relationships rooted in their respective traditions and histories, escalating tensions and territorial disputes are evident.

The South China Sea is notable not only as one of the world's busiest waterways for international trade but also as a region rife with contentious territorial claims involving Brunei, China, Malaysia, Taiwan, the Philippines, and Vietnam. Multiple Asian governments assert sovereignty over rocks, reefs, and other geographical features, as well as the underwater natural resources therein. Rather than conceptualizing boundaries and demarcated lines on maps that partition these aquatic spaces, the artists advocate for a "queering" of the South China Sea, encouraging a reimagining of its shared nature that emphasizes connectivity and kinship. For them, the notion of queerization entails embracing the fluidity of identities, enabling individuals to reinterpret existing boundaries and foster a sense of communal belonging.

Conceived as a collective laboratory, *Forkonomy()* (2020–present) is an ongoing project that invites participants to engage in reflective practices through various modes of participation, discussions, and contract formulation. In the initial iteration, workshop participants established a price and drafted a contract, determining that a milliliter of the South China Sea should be co-owned as a "cooperative," each participant committed to assuming ecological and economic responsibilities for the sea in a collective and cooperative manner. The second iteration shifted focus to non-fungible tokens (NFTs) on a blockchain network, minting the *Forkonomy()* contract on a decentralized protocol that enables more people to buy the contract. The transactions relating to the South China Sea are recorded, thereby of South China Sea, thereby further exploring the concept of ownership.

"What does it mean to mine and to co-own one milliliter of the ocean from the South China Sea? By employing free and open-source software and decentralized protocols, we set up the participatory project as a communing ship for people of the Pacific who care about and want to queer hierarchies, ownership, and gendered labor division, as well as fight against the constant threats to maintaining a high degree of autonomy over the land and the sea. *Forknomy()* shall sail the economy and autonomy into a queer ocean of freedom and the sea of common-wealth."

Charles Lim

Born in Singapore in 1973

Charles Lim is a former professional and Olympic sailor. His artistic practice is deeply influenced by his intimate knowledge of the sea and informed by field research and experimentation, drawing, photography, and digital video. His long-term *Sea State* project (2005-ongoing) attempts to break the usual tropes associated with the sea, namely its sublime or romantic side, and to reveal it as a highly economic, political and strategic space.

The name of the series, *Sea State*, is borrowed from the field of oceanography. It refers to the general condition of the sea and varies from 1 to 9 according to the wave height, wind, and swell conditions, with 1 corresponding to a calm sea and 9 to an exceptionally rough one.

Lim grew up in a village by the sea and the ocean has always been part of his life. He used to consider the sea like his office, but one which mirrored his inner thoughts. →

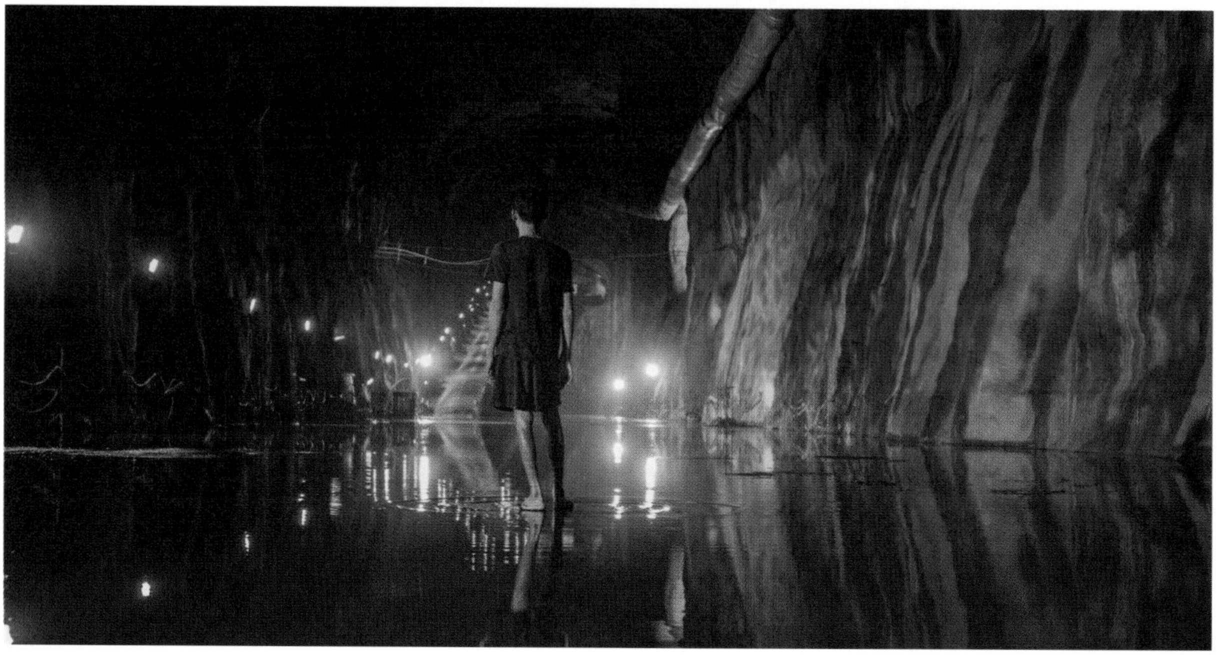

Seastate 6: Phase 1, 2015, Video, 31 min

However, today, he sees it from a more critical position. The turning point, for him, was his investigation of the Internet cables running under the sea in Singapore.

At that time, he was an avid online gamer. Towards the end of the 1990s, there was a prevailing myth surrounding the Internet–a belief that it possessed an almost mystical power to level the playing field, eradicating gender disparities and liberating individuals from their physical bodies. It was a realm of virtuality, where concepts like cloud computing and invisible networks held sway. However, one day, the Internet ceased to function in Singapore, leaving him perplexed and curious about its inner workings. While there was no definitive explanation, he discovered that the issue originated from a section of the world's longest undersea cables, near Taiwan. Several potential causes emerged: an earthquake in the seabed vicinity, a storm that forced a ship onto the undersea cables, or even the possibility of sharks damaging the cables due to their attraction to electromagnetic sensors.

These incidents underscored the vulnerability of these cable systems and, by extension, the Internet itself. Most significantly, they shattered the illusion of the Internet as an ethereal and boundless network, revealing its reliance on physical infrastructures and its tangible existence.

Simultaneously, Lim realized that since these cables lay on the ocean floor, his perception of the sea–a

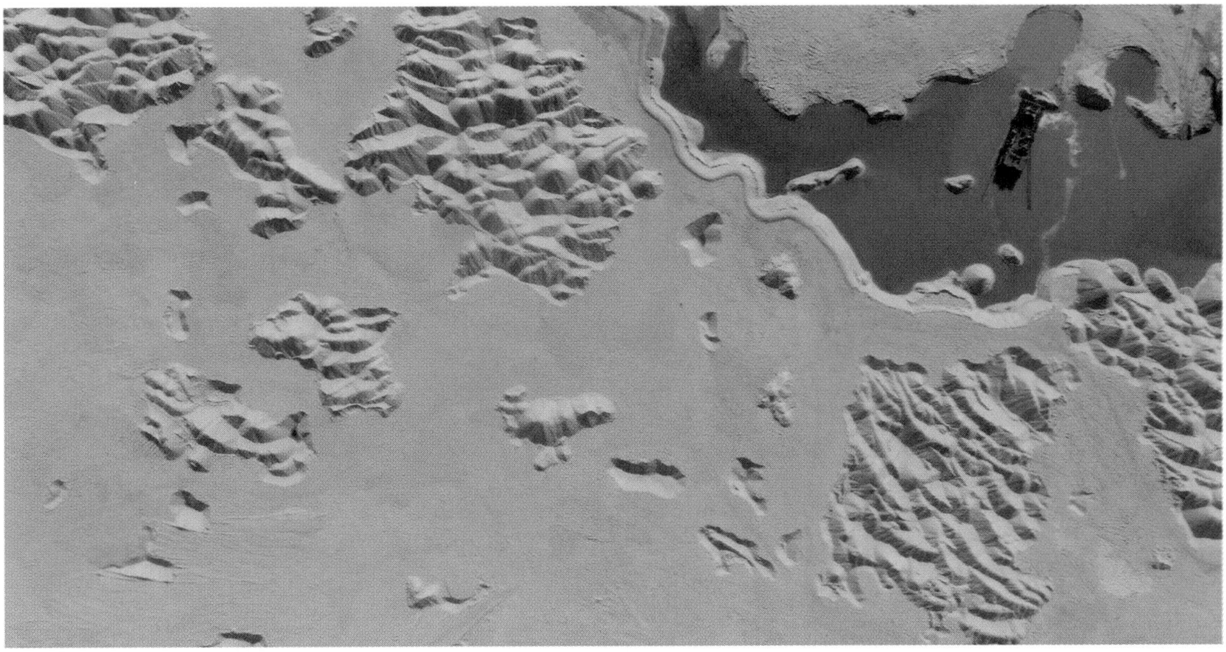

Seastate 9: Proclamation (drag), (drop), (pour), 2018, Video, 22 min 34

realm he believed he knew intimately as a sailor–might be incomplete. This revelation marked the inception of his research as he began to critically examine the sea with fresh eyes.

Lim's video *Alpha 3.9: silent clap of the status quo* (2016) results from this investigation. It shows more than two hours of archival footage from the bottom of the sea where Internet cables are laid. The work is striking in its simplicity but functions very well in giving tangibility to what is usually conceived to be floating in the air. It took years for the artist to access this secret material. The exact locations of the cables have been blanked out for they reveal highly political issues.

In parallel, Lim embarked on a long-term exploration of the sea as a geopolitical and economic realm. In his work *Sea State 6* (2015), he delved into the subterranean caverns situated 130 meters below sea level in Singapore. These caverns served as infrastructure for the petrochemical industry, housing vast quantities of liquid hydrocarbons. Established in 2014, they were the first underground oil storage facility in South-East Asia.

The artwork comprises two juxtaposed videos. The first video, titled *Capsize and Phase 1*, possesses a metaphorical essence, capturing the artist sailing and capsizing. As viewers, we feel immersed in the water, experiencing the disorienting sensation of everything being turned upside down. The second video takes on a more documentary-style approach with fictional elements. →

It is filmed within the caverns and follows a group of individuals transporting a sailing boat. The footage offers a glimpse into the cavernous space with its astonishingly high ceilings. The constant drip of seawater, accompanied by a dramatic musical score, creates a sense of oppressiveness. Here, Lim aims to convey the sensation of being submerged deep beneath the sea, surrounded by its vastness, and the lurking fear of entrapment.

These caverns are not the only ones constructed in Singapore. The country also houses subaquatic storage facilities for the military's munitions and explosives, spanning an area equivalent to approximately 400 football fields. This additional detail underscores the intricate, hidden connections between geography and politics, which Lim's artistic practice seeks to unveil. Despite the humble undertones of the primary gesture, the artist's desire was to reveal the complex multi-layered reality of the seascape.

As a sailor, Lim is used to translating the invisible into imaginative forms. A striking example lies in his ability to envision the wind, despite its intangibility. Though it is unseen, he can conjure its presence through the ripples of the sea, sensing its touch upon his neck or envisioning it as a force propelling the clouds. He can even discern its scent. Likewise, Lim presents his works in series, inviting individuals to forge their own connections and feel what they address. Just as the wind engages in a dialogue with the ocean's surface, his main sheet, or the clouds, the elements within his work converse with one another.

For Lim, today's most pressing issue in Singapore is the revival of land reclamation, with the "excuse" of fighting the rising sea level. In 2017, Mother Nature in Cambodia sued Singapore for sand-dredging activities, and Singapore lost the case. As a result, reclamation slowed down and few countries banned sand exports to Singapore. However, according to the artist these bans are no longer effective, and the government is resuming reclamation projects in the framework of a "Climate Change Defensive" policy. Since its independence in 1965, Singapore has increased its territory by more than twenty percent through land reclamation. Lim is also interested in questioning the ownership of this reclaimed land: to whom does it belong, since it was taken from the sea?

Lim hesitates to define himself as a cultural activist but acknowledges that there is no other way to criticize, especially in the context of Singapore. His artwork never tells people what to think, however. Rather, he hopes to shed light on certain situations. These days, he takes people with him on boat tours. Most of them do not know anything about the sea. They see it from a distance, and tend to look at it through the trope of the sublime. For him, it is time to dig deeper.

Uili Lousi

Born in the Kingdom of Tonga

Uili Lousi is a visual artist and climate change activist from the Kingdom of Tonga, an archipelago located northeast of Australia. The islands of this Pacific nation are relatively flat and particularly vulnerable to the effects of climate change. In fact, Tonga is considered the second most climate-vulnerable country in the world. Rising sea

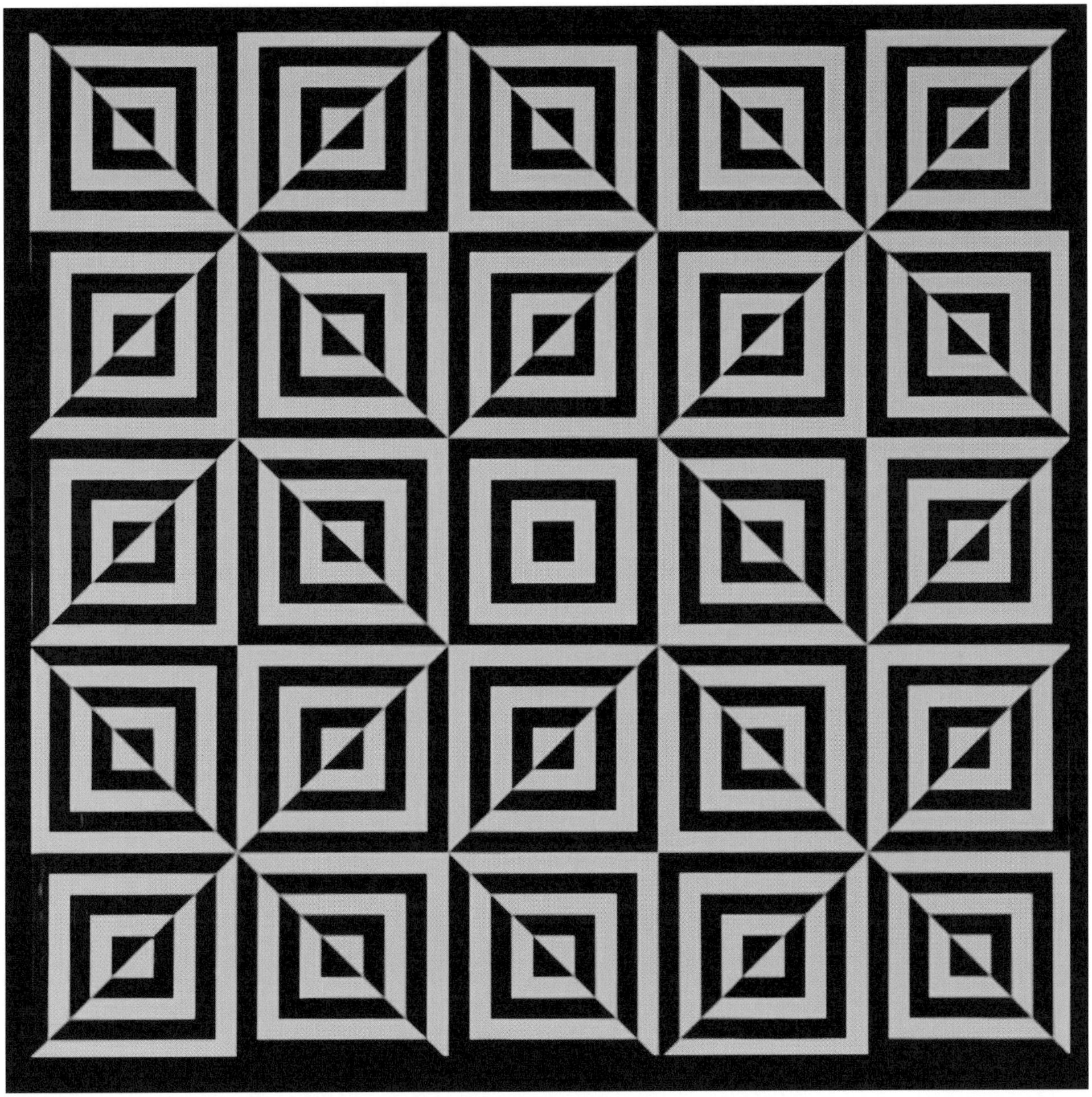

The Dancing Fish in Motion, 2016, Acrylic on canvas, 101 × 101 cm

levels, saltwater intrusion into freshwater supplies, and an increase in hurricanes and inundations are among the country's most pressing environmental challenges.

Lousi's artistic practice is deeply rooted in the ecological crisis he witnesses daily. As he explains, "My artwork is about the universal flow of energy, the interconnection of people, and the harmony of the world. Currently, nothing is in balance because we have been disrupting the natural order." His paintings serve as creative responses and potential solutions to mitigate the impacts of climate change. The geometric, abstract and often repetitive forms in his work depict colorful networks that crystallize a vision of flux, common vibrations, and interdependency. Many of his paintings are done on Tapa cloth, a non-woven fabric made from the bark of the mulberry tree in Tonga, which lends his work a fluid texture and a subtle luminosity.

Since 2015, Lousi has participated in the Conferences of the Parties (COP), the international conventions on climate change. He also serves as the President of OHAI Tonga, a nongovernmental organization dedicated to raising awareness about the impact of climate change on island nations.

"Too many artists adopt the perspective of victims, seeking someone to blame. That is not my approach," Lousi says. "I aim to advance ideas of kindness and collaboration. We need to search for common solutions together and think globally. Tonga is just one example on the map."

Tonga is situated along one of the migratory routes of humpback whales, which travel from Antarctica to warmer waters. Recently, Lousi introduced the concept of the "Whale Corridor," advocating for a systemic approach to protect whales across the oceans. He believes it is essential to address marine issues and environmental challenges from a holistic perspective in order to recognize the complexity of planetary interconnections. His initiative includes creating a whale corridor that would rejuvenate the whales' marine ecosystem from the Arctic to Antarctica and, of course, including Tonga.

In 2024, whales in Tonga, as was the case in New Zealand, Tahiti, and the Cook Islands, were granted legal personhood and inherent rights. Lousi's Whale Corridor proposal seeks to build on these efforts, preserving and regenerating the marine ecosystems on which the global health of our environment depends.

Tita Salina & Irwan Ahmett

T.S.
Born in Indonesia in 1973

I.A.
Born in Indonesia in 1975

Tita Salina and Irwan Ahmett are an artist duo working in Jakarta. Their performative and often community-based artworks are deeply rooted in the context of the Indonesian capital, a city struggling with rapid urban growth and pollution. The megalopolis is located in Jakarta Bay, a large water-based region where commercial and industrial activities take place alongside very vulnerable communities living below the poverty line. The place is also strongly impacted by extreme pollution and rising sea levels as it is located at the mouth of a watershed that drains thirteen river systems into the ocean.

"Our government keeps claiming that Indonesia is a modern country with many islands, all united, but in reality, it does not care about the coastal communities nor about the sea."

In Jakarta, most people do not visit the nearby seashore due to severe pollution and increasing privatization of the coastline. The artists themselves rarely went there until they began investigating the coastal communities in 2015. Their interest was piqued by the large amounts of waste carried by the city's rivers into the ocean. This led them to initiate what they

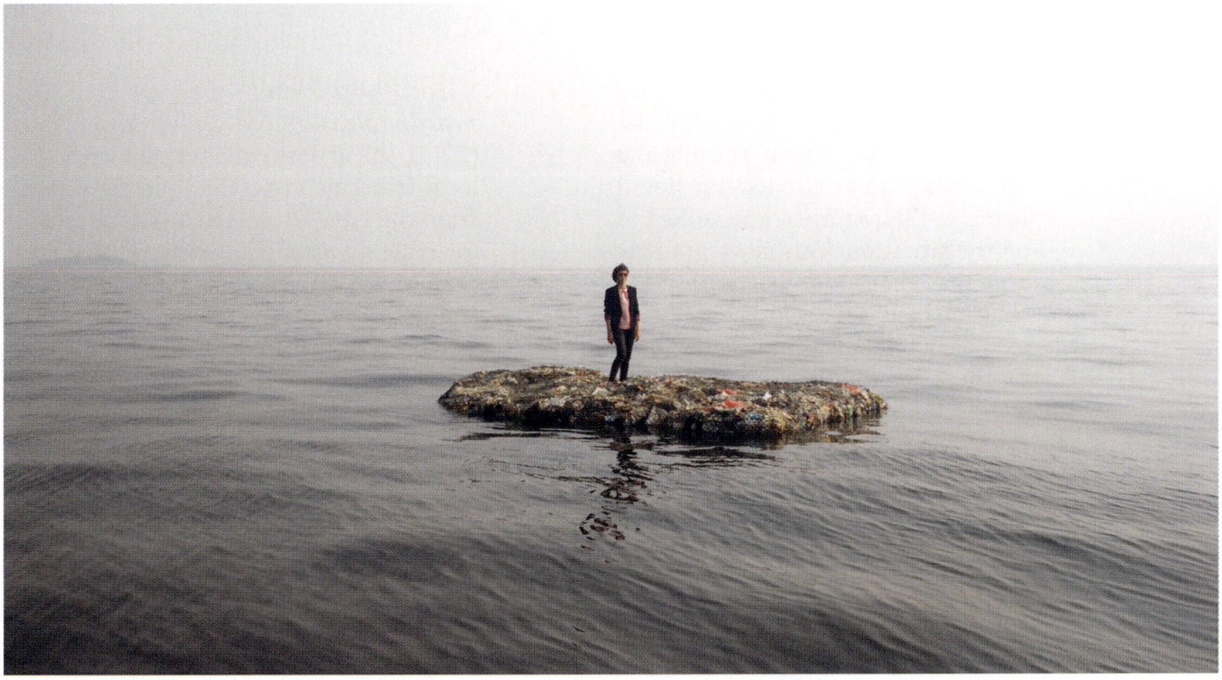

1001st island – The most sustainable island in archipelago, 2015,
Plastic waste, fishing net, rope, floats, bamboo, LED lights and oil barrels; Video, 14 min 11

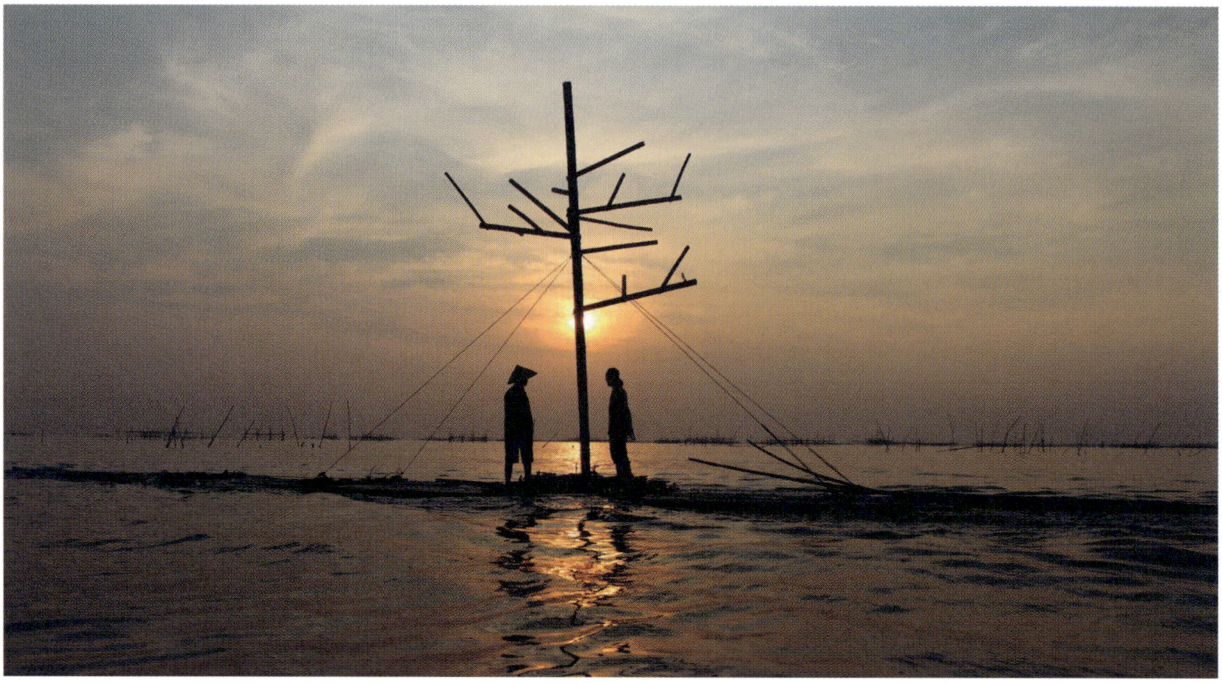

Harvest from Atlantis, 2019, Video, 11 min 38

call *Pilgrimage to the North* (started in 2018), an annual two-week walk along the coast. During these walks, they meet and talk to local residents, listen to their stories, and reflect on the situation. This slow, grassroots approach focuses on bottom-up solutions and serves as their response to the complex, layered issues faced by these communities. It is also their way of building strong, trusting relationships.

For Ahmett, this connection to the sea has become deeply personal and physical. "I feel it in my blood, in my tears, in my whole body. After starting to walk along the coast and meeting the local communities, my life has changed dramatically. I will never be the same," he says.

For *1001st island - the most sustainable island in archipelago* (2015), the artists collaborated with local fishermen to collect plastic waste from the bay. They used a fishing net to create an artificial island, which was then towed into the bay and released, symbolically adding a new, almost indestructible island—due to the durability of plastic—to the Thousand Islands north of Jakarta. The process was filmed by a drone and shows Salina standing alone on the floating island.

Working with these local communities can be challenging: the artists do not wish to speak on their behalf and are still searching for the best ways to collaborate. According to Ahmett, the fishermen and coastal villagers are very open-minded.

Their imagination about the sea is magnificent, but they do not take care of it and are now traumatized by its destructive power. This paradox is something the artists repeatedly encounter: the people love the sea and depend on it for their livelihood, yet they harm it by neglecting its sustainability. However, moralizing is pointless; their immediate priority is survival due to their extreme poverty. "We are not activist, but artists," the duo asserts. They strive to build networks, share knowledge, and foster discussions, especially when these communities are overlooked. They remain cautious about any form of artwashing or greenwashing.

For *Harvest from Atlantis* (2019), Salina and Ahmett followed coastal villagers who farm green mussels, documenting their activities over a seven month period. The title references a myth suggesting that Jakarta was built on the site of the lost Atlantis—an irony given the city's ongoing situation. The area has been sinking for a long time because it is built on a wetland. Each year, the villagers add layers of mussel shells to reclaim the land. The soil is highly unstable, and this practice is illegal. Despite severe pollution, green mussel cultivation persists in Jakarta Bay. Entire families rely on this resource, collecting, cleaning, and shelling the mussels before cooking them.

Today, resolving the issue of the sinking of Jakarta has become a government priority, leading to the construction of a sea wall along the northern coast of Java. For the artists, this wall poses significant problems. It is not designed to meet the needs of coastal communities and is highly destructive to the environment. Named Garuda to glorify the project, it tends to benefit developers rather than the local population. On-site, the artists observe fishermen who now must climb the wall to fish, while many villages continue to be submerged by the sea. When villagers lose their homes, they receive no compensation from the government. Some situations are dire; even cemeteries are engulfed by the sea.

The artist duo is now working on a new project centered on the villagers' dreams and hopes, collecting their stories. Perhaps due to the challenging circumstances, an old legend has resurfaced: the myth of a hidden underwater treasure, a remnant from maritime trade and colonial times. Local people strongly believe in this legend, and it may offer them hope during these difficult times.

What applies to a basin of course holds true also to oceans. Our seas are polluted, threatening the lives of everyone and everything that depends on them. And ignoring the situation is not the solution. As UNOC (Third United Nations Ocean Conference) works on how to conserve and sustainably use the oceans, seas, and marine resources for human development, I make an appeal to work together to take care of our common pool, and see the seas as "our seas" again.

We all have a relationship with our oceans, seas, and life below water.
For me, the relationship runs deep. I grew up in the Camargue (France), in the Rhone Delta where the Mediterranean Sea meets lagoons and marshes. My childhood there, and my family's continued commitment to wetlands and the unique species they harbor, made me intimately aware of why and how many generations before us came to see the Mediterranean Sea as "our sea" (or "Mare Nostrum", as the Romans said).

Oceans, seas, and rivers are the lifeblood of humanity. Without the seas and the rivers that lead to them, there would have been no human civilization. There would have been no Mesopotamia without the Tigris and Euphrates. There would have been no Egypt without the Nile, no Greece or Rome without the Mediterranean Sea.

The same is true around the world. The cradles of civilization, and the social capital they developed, all crucially depended on water. Some civilizations, like the ancient Peruvian one, even developed specifically around marine resources.

Today, billions of people live by the oceans and seas, and continue to both enjoy and depend on them. That is true in a narrow sense for tourism, and in a broader sense for resourcing.

But will the polluted sea still be able to fulfill its recreational role for tourism, and its vital role for fishermen and seafarers? Will a warmer, acidified ocean be able to continue to cool the planet, or play its critical role in the natural water cycle as effectively? How brutally will it restitute the huge amount of energy provided by global warming?

We are acutely aware of the adverse effects of the deteriorating quality of water of the oceans and seas on biodiversity, and the ecosystem services it supplies. Many of us know, from memory and history, how important, even vital, the contribution of the sea has been and continues to be to human development.

We find it deeply troubling to see the oceans and seas under such severe threat. When we lose nature in the broadest sense, when climate change gets out of control, and when we overexploit natural resources, we bite the hand that feeds us. To harm the world's natural capital is to harm our human, social, and financial capital.

Businesses, in particular, have a crucial role to play in reversing this trend. As key actors in the global economy, they have the capacity to drive innovation, implement sustainable solutions, and mobilize resources at scale. The private sector must not only mitigate its environmental footprint but actively contribute to restoring marine ecosystems. By investing in circular economy models, reducing plastic waste, and adopting sustainable supply chain practices, companies can be powerful agents of change. Moreover, through partnerships with governments, NGOs, and research institutions, businesses can help bridge the gap between conservation efforts and economic development. A thriving ocean economy does not have to come at the expense of marine health. On the contrary, it can be a catalyst for long-term prosperity and resilience.

It is only when we allow nature to thrive that humanity can thrive. And it is only when all society's actors–including business, government, and civil society–work together to achieve these common goals that we will achieve a brighter, more sustainable, and more inclusive future for all.

A Site of Wonder

He looked into the water and saw that it was made up of a thousand thousand thousand and one different currents, each one a different colour, weaving in and out of one another like a liquid tapestry of breathtaking complexity.

Salman Rushdie

In the Caroline Islands, it is said that the sun follows quite an irregular path in the sky until it settles, in the early evening, in an orange tree from which it throws its fruits into the ocean to scare away the sharks before making its final descent into the sea. The people of Peleliu are known for having often tried to catch the sun while it was in the orange tree: they searched to prolong the length of the day.

Palauan folk tales of the Palau islands

1 Alimi Adewale
Born in Nigeria in 1974

2 Soly Cissé
Born in Senegal in 1969

3 Ale de la Puente
Born in Mexico in 1968

4 Faris Ridzwan
Born in Malaysia in 1993

5 Rebecca Rutstein
Born in the United States of America in 1971

6 Kim Yip Tong
Born in Germany in 1991

7 Juria Toramae
Born in Morocco in 1983

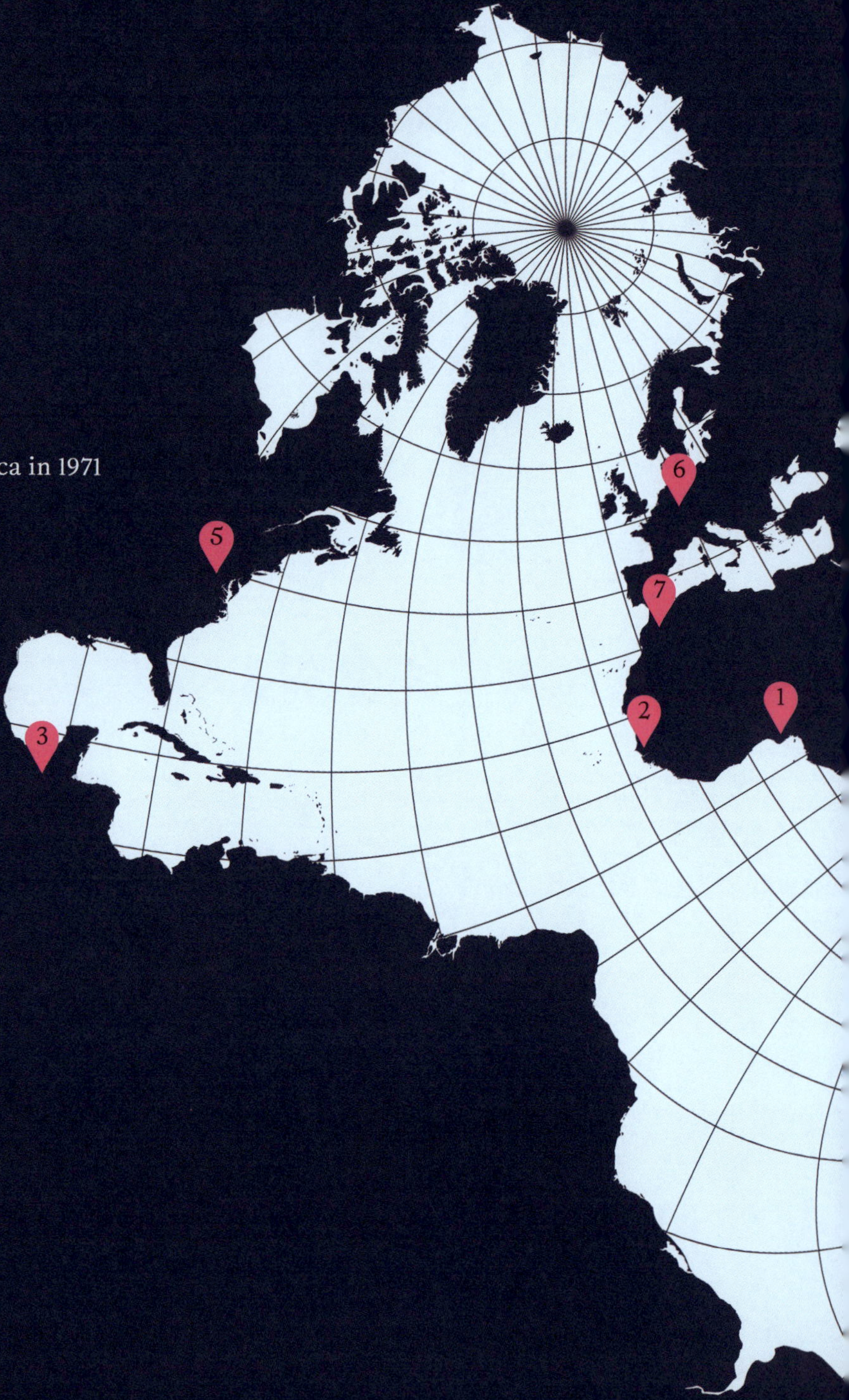

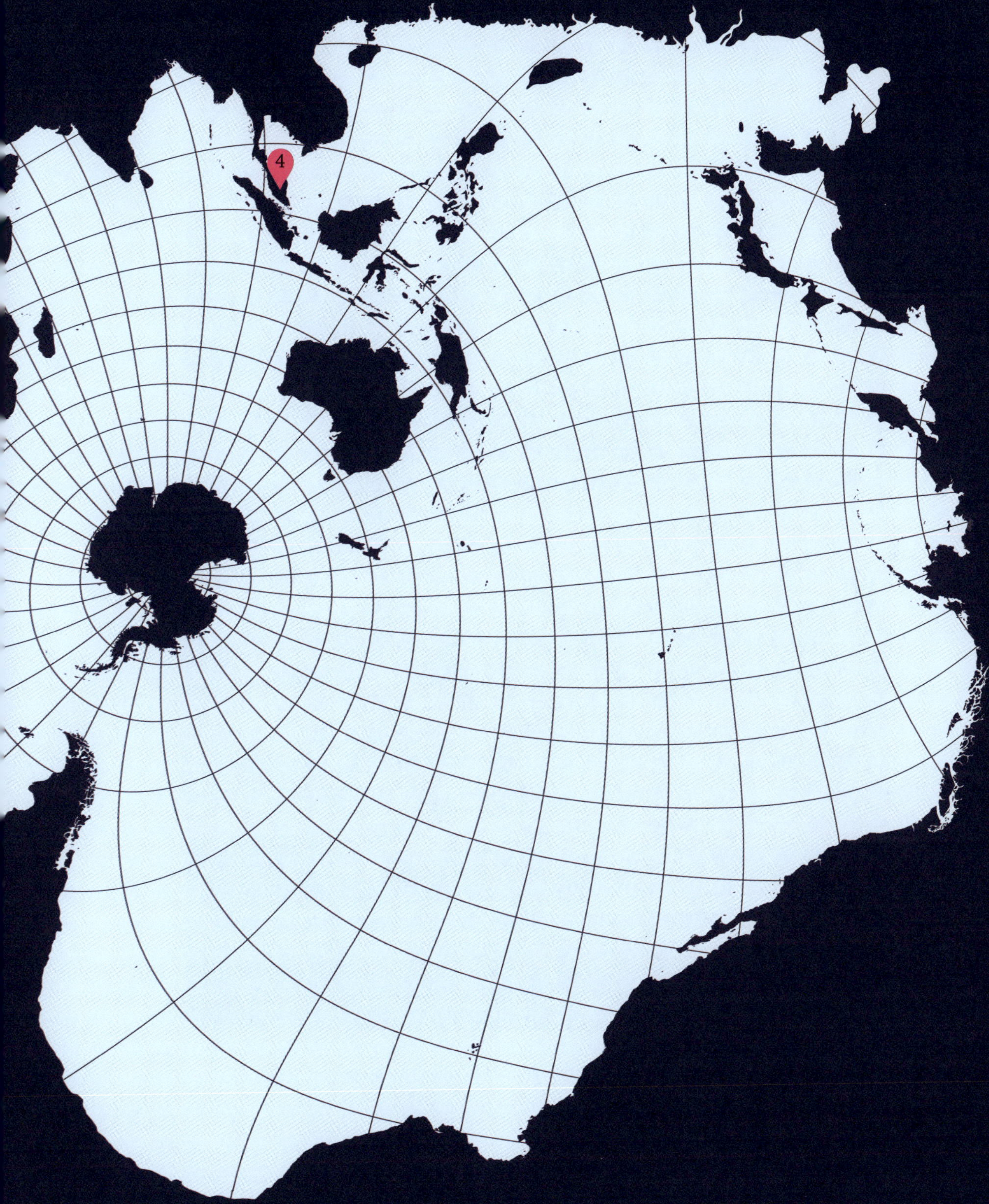
4

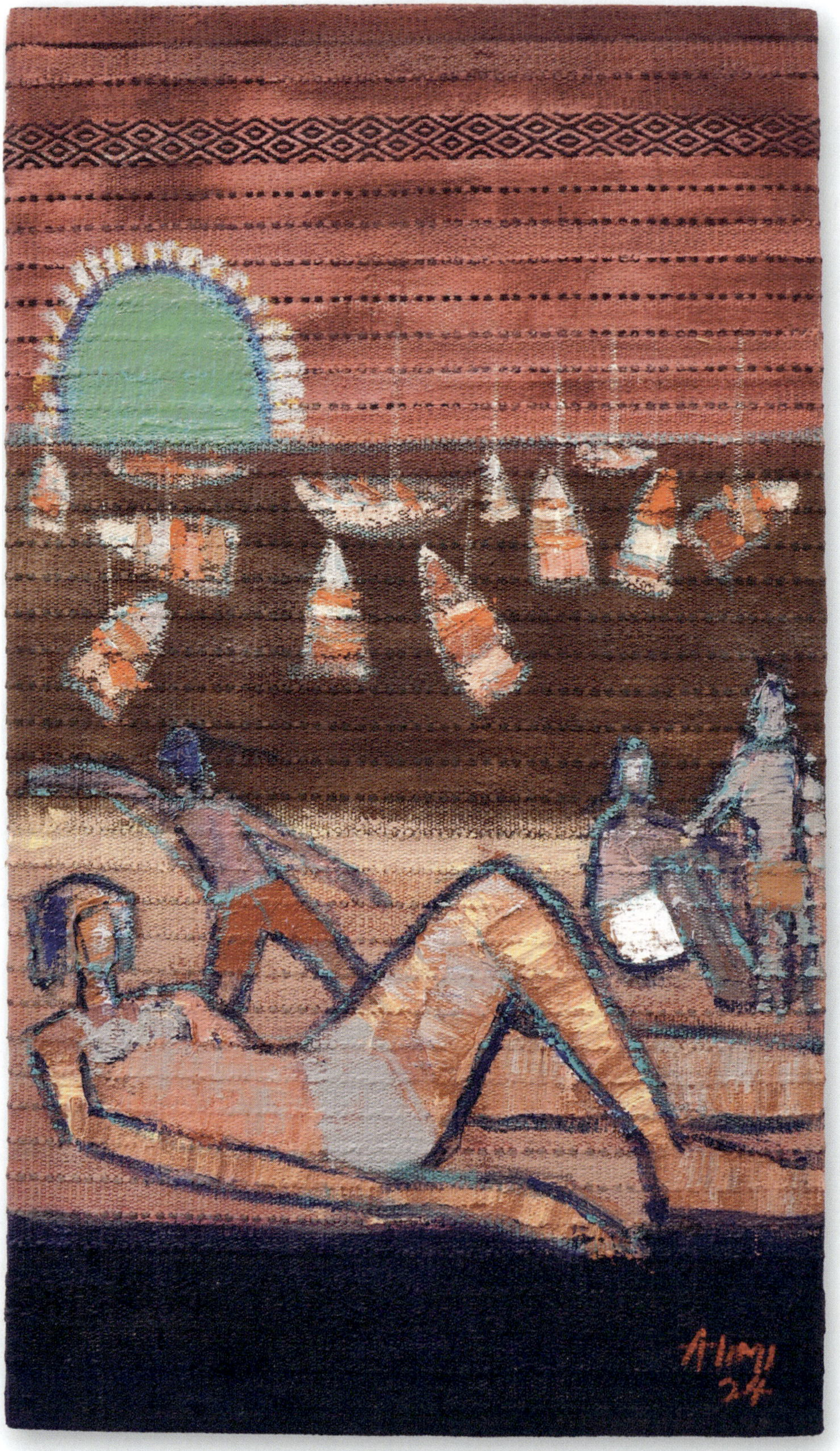

Coastal Reverie, 2024, Acrylic and dye on kilim, 142 × 80 cm

Alimi Adewale

Born in Nigeria in 1974

"Being close to the shore deeply influences my work–its vastness and unpredictability mirror the creative process. The marine environment provides endless inspiration, not just aesthetically but also conceptually, as a metaphor for resilience and transformation."

Seaside Reflection (2024) could be described as a classic marine landscape with an endless sea bordered by the horizon of the sky, a multitude of boats, and a gigantic pale sun. Yet the impression is entirely different: the hulls of the boats are depicted either from the side or from above. They seem lost on a strangely calm sea. They resemble fish searching for something on the surface. Perhaps they are drawn to the hooks piercing the sun, attracted towards somewhere else? The painting thus exudes a deep ambiguity, a tension between calm and unease.

This same ambivalence can be felt in *Coastal Reverie* (2024). As its title suggests, everything here is an invitation to dream. A languid woman lies on the sand. Behind her, a child plays. It is almost a family portrait. In the distance, the sun sets over a tranquil sea. Sailboats are anchored, and one can almost hear the clinking of the masts. In *Seaside Reflection*, Alimi Adewale has an almost paradoxical way of representing the sea, using earthy colors. The artist's palette ranges from brown and dark blood-red to deep blue, painting expanses of water that seem abstract. The fact that he paints on kilim reinforces the materiality of the work and gives it a surprisingly sensual texture.

Thus, men, the sea, and the beach are closely intertwined, layers woven into and against one another, in symbiosis. It is this solidarity that Adewale seeks to highlight. "Through making art and other forms, I aim to explore and highlight the tensions between humanity and the marine environment, emphasizing both our dependence on and responsibility to it."

Adewale's studio, in Lagos, faces the sea, and he observes both its beauty and the pressing issues faced daily by the local communities, such as pollution, overfishing, and the impacts of climate change. "While there is growing awareness, many ecological concerns still don't receive the urgent attention they deserve. In Nigeria, as elsewhere, education and advocacy are critical to building this awareness. We have a few environmentalists championing these concerns. Art can play a transformative role in rethinking human relationships with nature. By creating work that provokes thought and emotion, we can inspire dialogue and action."

Soly Cissé

Born in Senegal in 1969

Soly Cissé grew up along the coast of Senegal, and the sea has always been an integral aspect of his life. For many years, he participated in surf competitions, experiencing various tastes of waves and surf spots around the globe. His artistic practice questions Senegal's socio-cultural context and explores the tensions between globalization, modernization, and the evolution of traditions. While studying at the École Nationale des Beaux-Arts in Dakar, Cissé focused on ecological issues, believing that art should foster reconciliation between people and their environment. His paintings celebrate the beauty of nature, which he regards as a precious resource requiring preservation and transmission to future generations. "We are responsible for these gifts," he says, "and we must preserve them for our children."

Cissé's series of fish paintings illustrates both the aesthetic appeal and the pressing challenges confronting marine ecosystems. "A few years ago, I concentrated on advocating for my country's natural resources, particularly in relation to their exportation. Consequently, I sought to highlight the endangered species of certain fish and to explore the marine realm." Through thick, vigorous, and often vibrant brushstrokes, he captures the dynamism of aquatic life that is endangered by overfishing, climate change, and pollution. For his series *Poissons* (2019), he employed a combination of acrylic and tar. These artworks present an explosion of vivid colors, conveying the beauty of fish frolicking against a bright marine backdrop. However, upon closer inspection, one observes that these backgrounds are devoid of life, and the fish, disproportionately large for the confines of the canvases, appear trapped and suffocated within this barren space. The tar encircles them, potentially encroaching upon their forms.

This sense of entrapment is further emphasized in another series, painted solely with charcoal and sanguine and thus characterized by dense shades of black, red, and white. In *Poissons 1* and *Poissons Rouges 1*, the fish are depicted as completely compressed, leaving no room to breathe or escape. For Cissé, the accumulation of fish within the composition symbolizes the overexploitation of coastal resources by Western trawlers. The darker hues evoke marine pollution, representing despair and serving as urgent reminders for public awareness campaigns at both national and international levels.
In the second composition, dominated by red and white tones, the red fish symbolize the diversity of endangered species. In most paintings from these series, figures have been included among the fishes. For the artist, they reflect the extent to which our society is dominated by codes which transform people, goods, and living creatures alike into mere numbers.

"The representation of fish and the selection of this theme enable

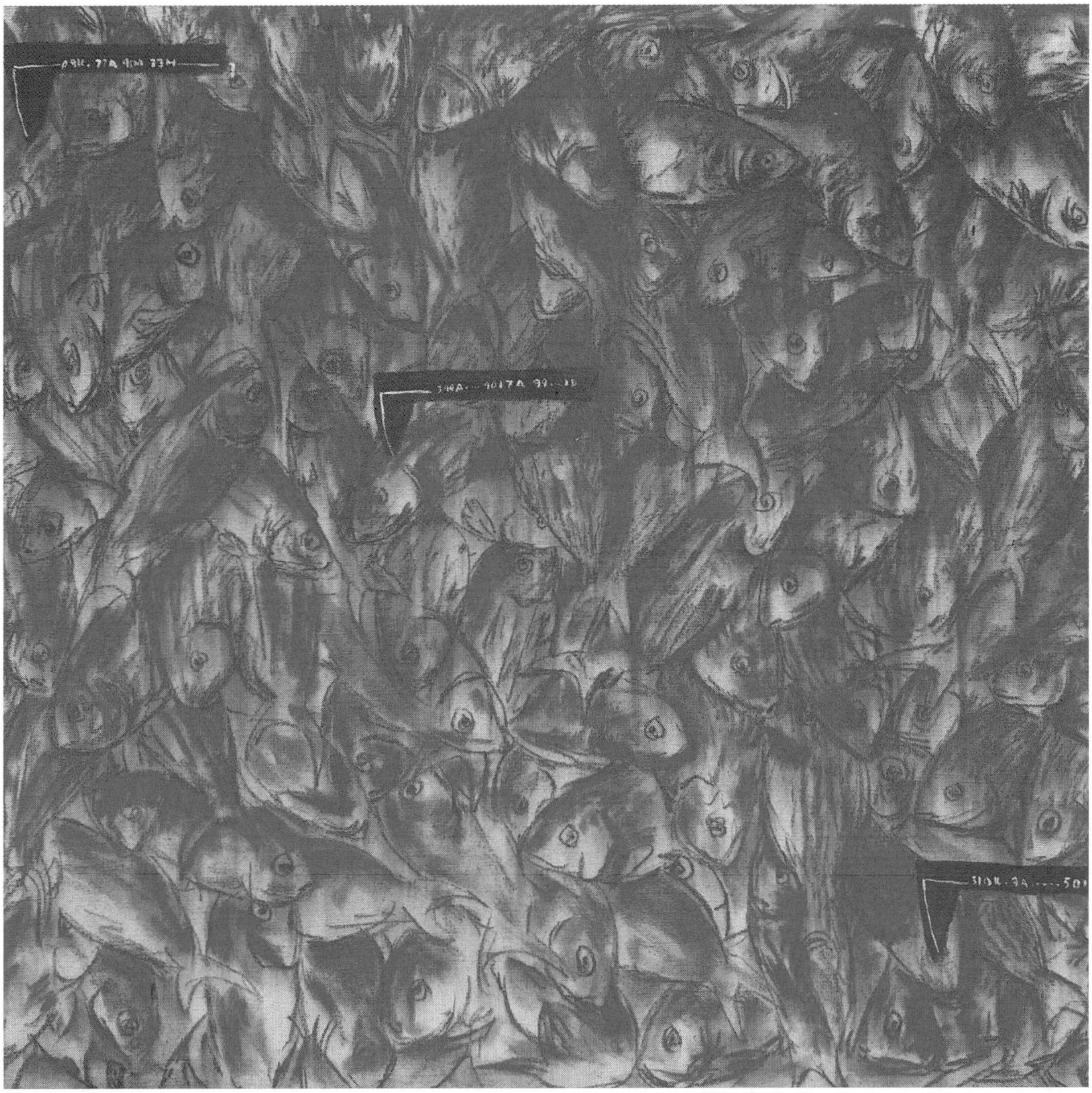

Poisson 2, Poissons series, 2019, Black charcoal, ecru pastel on canvas, 100 × 100 cm

me to expand my color palette. The sea is a vibrant and inspiring universe for any creator. Painting serves as a means to immortalize events and enrich our collective human history, as these works are destined for museums for future generations."

In addition to the issue of overfishing, Cissé addresses the increasing privatization of Senegal's coastline. Consequently, local populations, including himself, find themselves unable to access the sea. They can no longer stroll along the beach, and feel dispossessed of what should be a communal resource. For Cissé, while art can raise awareness about these critical issues, immediate political action is essential to safeguard this shared heritage and natural wealth.

Ale de la Puente

Born in Mexico in 1968

Ale de la Puente was raised in Mexico City and she retains a vivid memory of her first encounter with the ocean. This experience was simultaneously daunting and captivating for her, evoking a sense of the ocean as a terrifying labyrinth where one can never be certain of one's location. These feelings have persisted and continued to infuse her work and research. After learning to scuba dive and collaborating with divers who research the cenotes in the Yucatan Peninsula, and thanks to a transformative university experience sailing with a professor, she recognized the necessity of also learning how to read the sea's surface and the winds. How does one find one's way when at sea, with no land in sight? The sensation of being lost resurfaced, and through techniques and tools such as the sextant, the sea began to reveal itself to her. She soon thereafter enrolled in a class to learn boat-building, driven by a desire to construct her own vessel.

In parallel, de la Puente developed a multidisciplinary artistic practice, grounded in her diverse passions for research, navigation, astronomy, physics, and philosophy. In 2021, she embarked on the research vessel RV Falkor, operated by the Smith Ocean Institute, which was investigating the Pescadero Basin, one of several small ocean basins in the tectonically active Gulf of California. The scientists on board were mapping the seafloor and studying hydrothermal vents, and she was granted permission to use the camera during the periods when their remotely-operated vehicle descended to the sea floor. Her sixty-six minute video work *Nieve Marina* (2023) is the outcome of this experience and her encounters with the marine creatures inhabiting these depths.

The film is conceived as an immersive experience and is projected on a screen larger than human scale. Its extended duration allows viewers to identify with the perspective of the artist who is slowly descending deeper and deeper, meticulously scrutinizing the vast blue expanse in search of marine organisms. As we plunge downward alongside

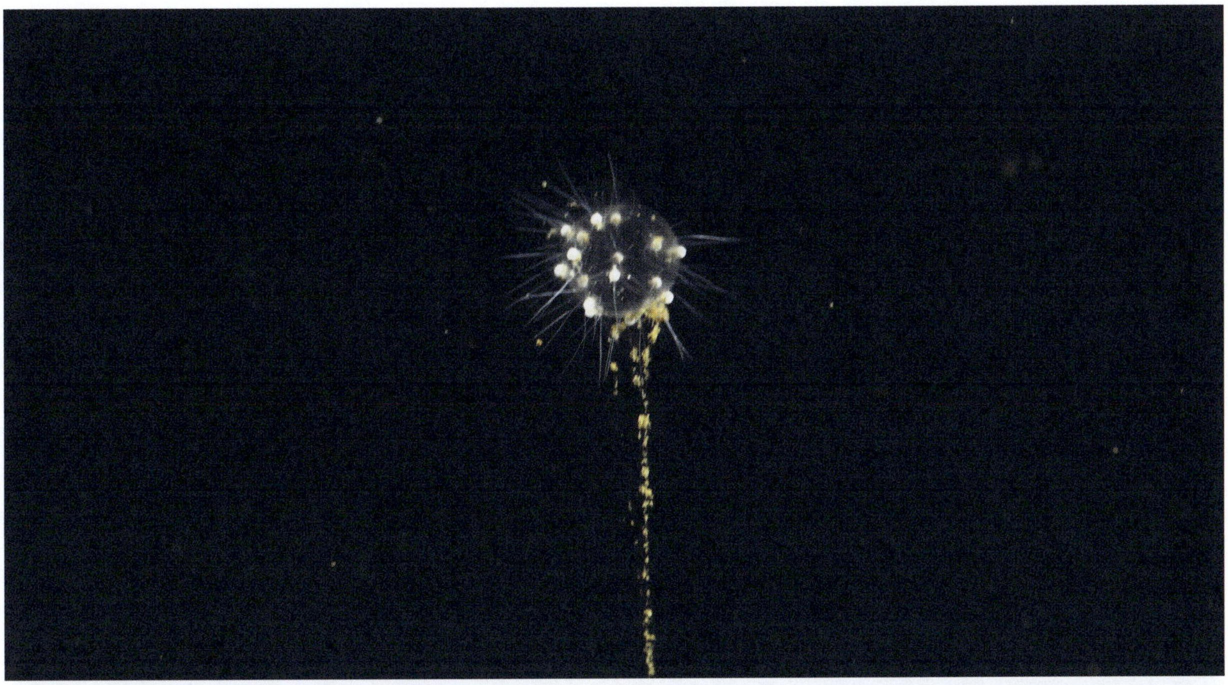
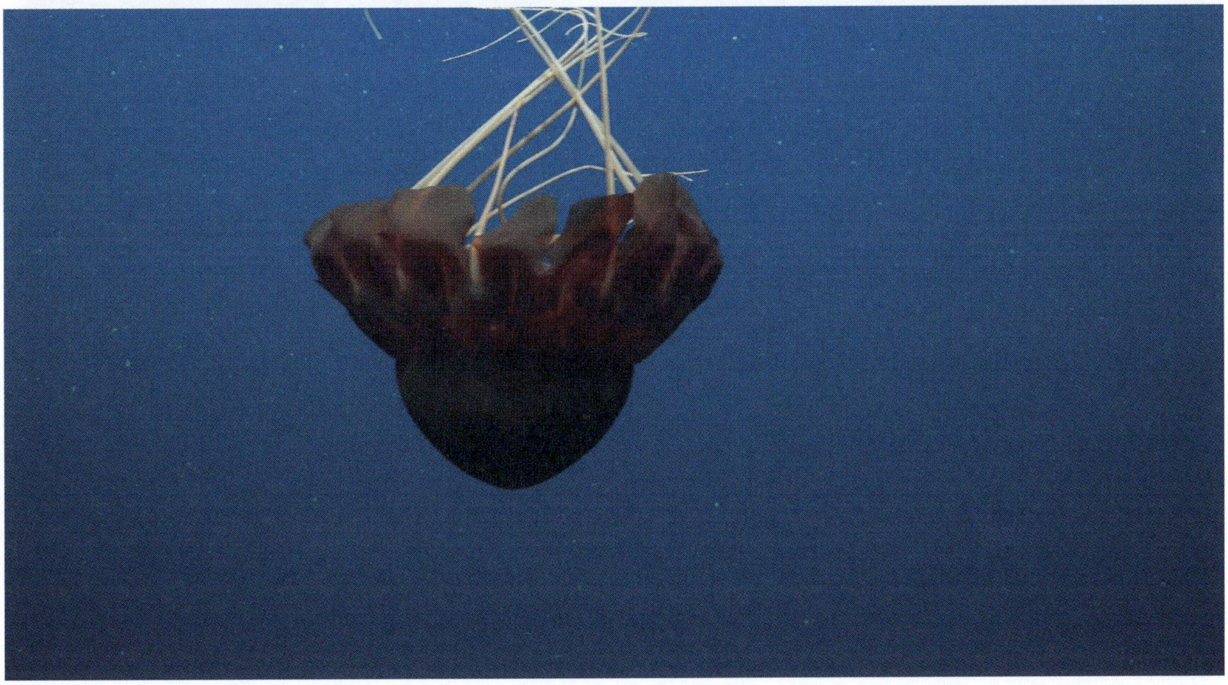

Nieve Marina, 2023, Video, 63 min

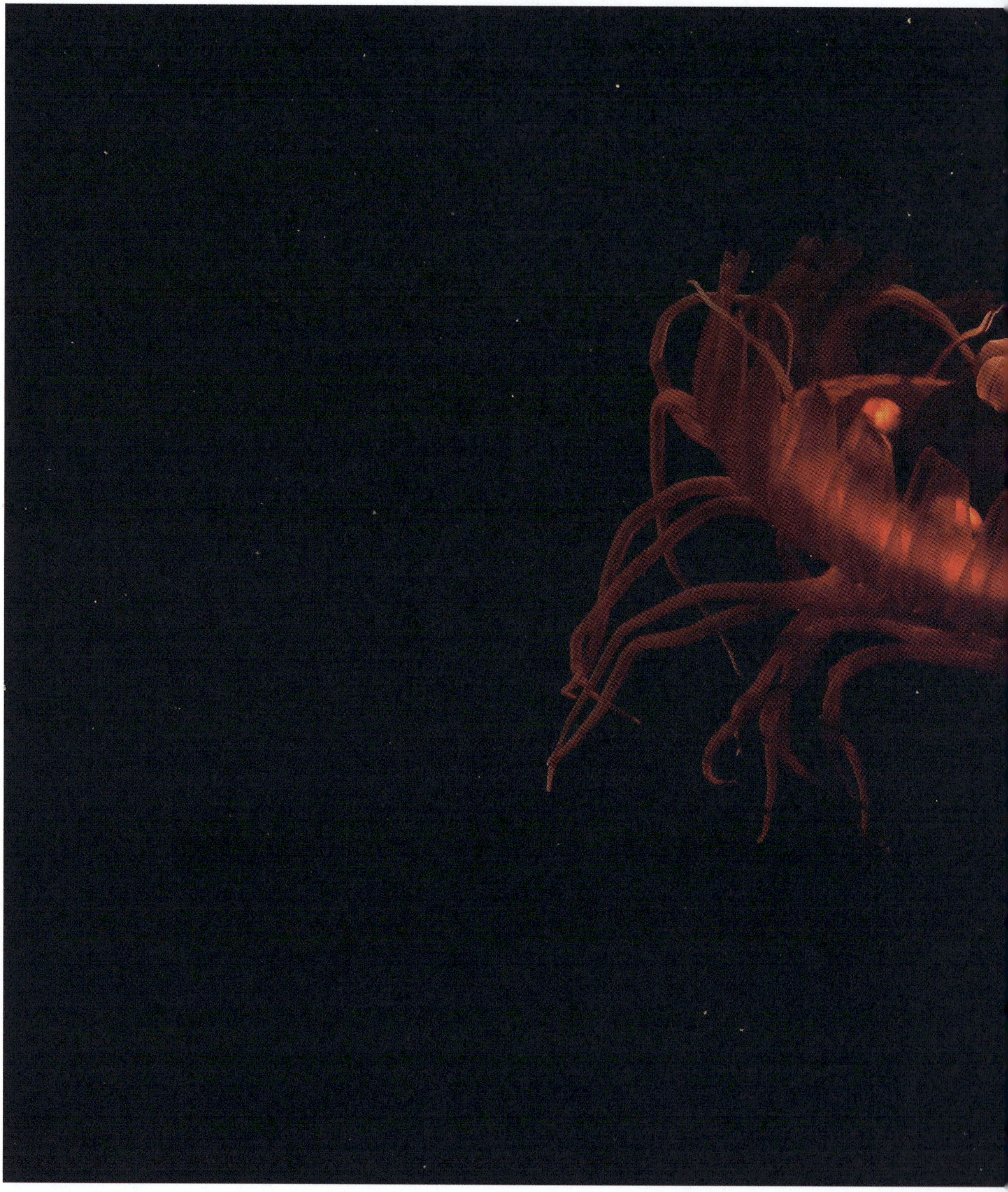

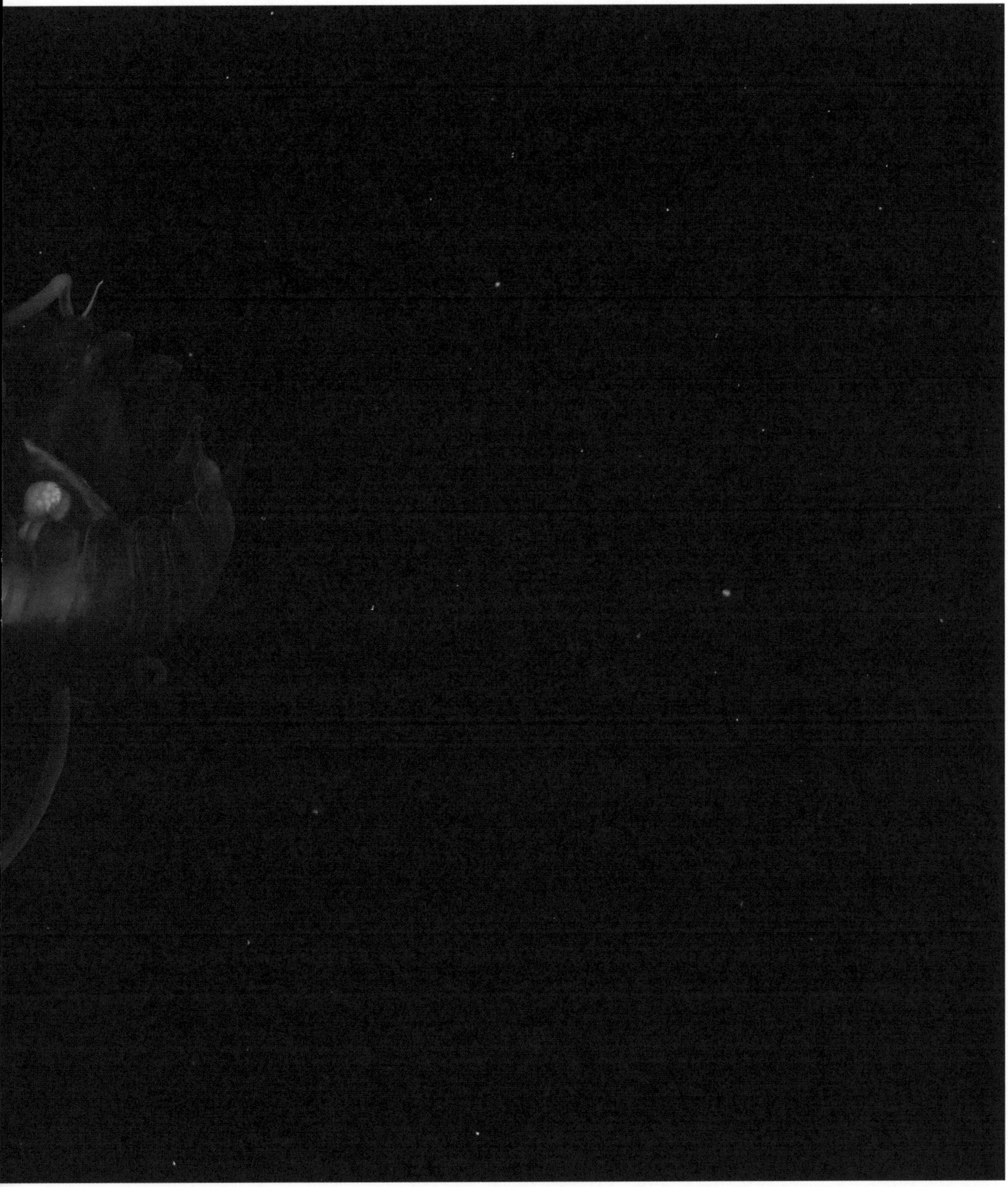

Nieve Marina, 2023, Video, 63 min

her, submerged by what resemble drifting snowflakes, we suddenly encounter these captivating species floating around us, dancing or simply unfurling. Their shapes, movements, and colors are so extraordinary that it is difficult to believe they are real. However, because they are occasionally out of focus, we truly feel that we are with the scientists observing them. We lose sight of them momentarily, then rediscover them. The experience is akin to observing animals in the wild. According to de la Puente, many of these species were unknown to the scientists onboard. The middle layers of the ocean remain a mystery to most, as their investigation requires more advanced technology. In this instance, the descent reached a depth of 4,800 meters, a journey that took approximately two hours to complete.

De la Puente's paramount concerns are beauty and a sense of wonder, yet she did not wish to produce a *National Geographic*-style film. In these mediatized images, what we contemplate is sometimes so incredible that it remains distant from us. In contrast, in her film, she aimed for viewers to feel a greater proximity to what they are observing in order to forge a deeper connection with these species. She emphasizes that these marine creatures belong to our Earth; they simply inhabit the depths. We must recognize this. "I wonder why people usually say that marine creatures seem like spaceships or beings from another planet. We should rather say that we imagine spaceships and beings from other planets to be like these marine creatures who also inhabit our planet."

Deciphering the sea and our environment is one of the central preoccupations of the artist. In September 2017, during the autumnal equinox, she sailed along the equatorial line to reach the precise coordinates where the sun would be exactly ninety degrees perpendicular to the Earth on that line. It is reported that, during these moments, all winds subside and marine organisms come to the surface. With a sense of awe, she observed the phenomenon that she subsequently captured in a satellite image entitled *... encontrando el centro*, (2017) and in other photographs and videos.

Her artistic and poetic practice functions as an invitation for us to dream and transcend our existing knowledge frameworks in order to uncover the laws and richness of the marine world that are still unknown yet within our reach.

Faris Ridzwan

Born in Malaysia in 1993

Faris Ridzwan hails from a family of artists, and his artistic journey is deeply rooted in the Malaysian batik tradition—a technique he learned from his mother. His works are heavily influenced by the lively colors and organic motifs characteristic of this traditional textile art. Large, dynamic flowers, leaves, and geometric patterns often unfold across his canvases, evoking the rich visual language of batik. Recently, Ridzwan introduced a new element into his work: marine life. This artistic evolution brings a fresh twist to the batik genre, which traditionally omits animals and sea creatures, pushing the boundaries of the form while still honoring its heritage.

The artist has explored numerous dive sites around the world, including iconic locations like the Maldives, Mabul, Tenggol Island, and several other remarkable destinations. The *Atolls* exhibition offers a glimpse into his entire journey as a diver, weaving together these underwater encounters. His very first dive took place at Tenggol Island, a small location off the coast of Malaysia. For Ridzwan, the site has personal significance, and this experience left a profound impact on his artistic practice. For him, diving was not only a visual revelation but also an immersive physical experience that connected him to the sea in new ways. Reflecting on this, he remarked, "Diving has been one of the most profound experiences of my life. It's like stepping into an entirely different world, where everything feels surreal."

During his dives, he often reflected on how to share the beauty of these underwater scenes with others. On one of his trips, he noticed that batik items were popular souvenirs in island stores. This sparked the idea of using the batik style to portray these seascapes. For him, adopting batik colors and composition is a way to pay homage to this traditional art form and to create a tangible connection between his experiences and that of his viewers.

This new series brings the underwater world to life on canvas, with corals, seashells, and seaweed spiraling and intertwining in an almost choreographed manner. The artist uses thick layers of paint to create depth and movement, layering colors and textures to evoke the density and fluidity of the ocean. His brushstrokes, at once vivid and delicate, produce a dynamic sense of motion, creating swirling, immersive seascapes. True to his batik-inspired roots, Ridzwan avoids depicting human figures, instead adopting the perspective of a diver. This invites his audience to experience the underwater world as he does—immersed, unencumbered, and free from the constraints of conventional representation. The viewer is left without any clear sense of orientation—there is no up or down, no above or beneath. There is just the vast, boundless ocean.

Shimmer Formation (2024), for example, features a group of manta rays gliding above a coral garden. →

The creatures seem to generate one another, flowing together in an unbroken dance. His framing allows viewers to feel as though they are swimming through the depths alongside the marine creatures he depicts, while his use of vibrant hues stays true to the influence of batik.

Through his seascapes, Ridzwan captures the immensity and mystery of the sea, evoking both its beauty and its unfathomable depths In his paintings, the sea is not just a subject, but a living, breathing entity—as well as a source of joy.

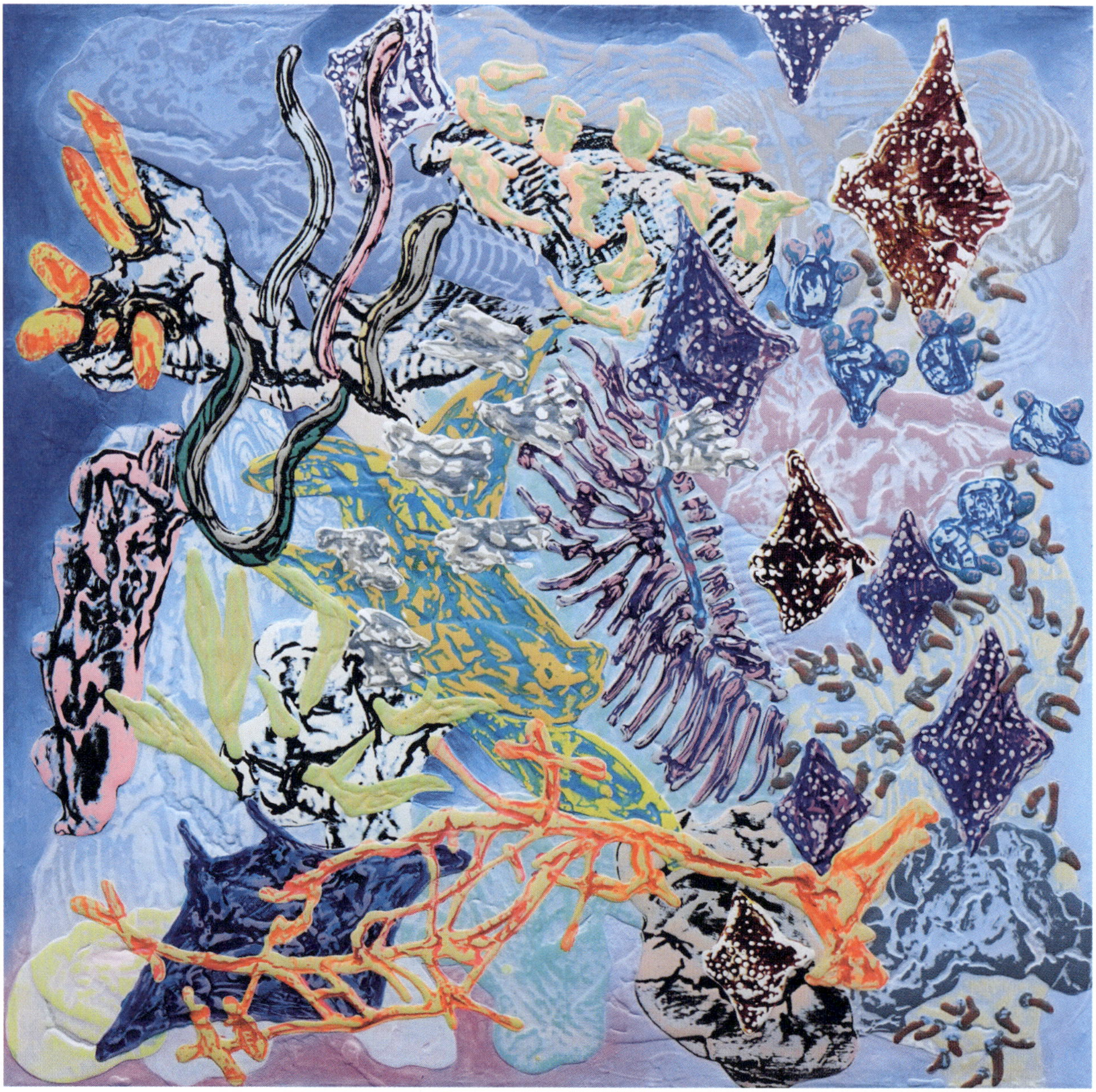

Shimmer Formation, 2024, Acrylic on canvas, 122 × 122 cm

Rebecca Rutstein

Born in the United Sates of America in 1971

In the nineteenth century, artists ventured onto the ocean to capture its natural beauty and wonders. Aboard the HMS Challenger (1872-1876), an artist accompanied the scientific mission, providing artistic perspectives to complement the scientific work. Rebecca Rutstein's practice is rooted in this tradition. As a visual artist, she collaborates with scientists, joining their underwater expeditions and transforming the images and data she collects into artistic experiences. Her work makes accessible what is typically beyond the reach of the general public, bringing visibility to the hidden underwater world and to microscopic creatures that are unfamiliar to most people.

Rutstein believes that art can forge deeper connections between people and the natural world. She often cites the adage, "You only protect what you love." Through her art, the artist aims to spark interest in the marine environment, fostering a sense of concern and connection to this largely unknown domain. Today, only twenty-five percent of the seabed has been mapped, leaving much to be discovered. She finds joy in sharing these discoveries, presenting scientific research from her expeditions in a more accessible and engaging manner.

Rutstein's interest in the ocean stems from her curiosity about geology, which led her to discover a sonar map of the ocean floor underneath the Hawaiian Islands. Her experience diving with the Alvin, a deep-ocean research submersible built in the 1960s, was transformative. The descent takes about two hours. The submersible descends using ballast tanks and drops weights as it approaches the seafloor to maintain neutral buoyancy. At depths of 1,000 meters, total darkness envelops the submersible, revealing the spectacle of bioluminescence, which the artist describes as the most beautiful sight she has ever witnessed. This experience, akin to entering another world, instills a profound sense of insignificance in the observer.

Upon reaching the research site, the submersible's lights are turned on to conduct studies. Inside the sphere, conditions are harsh; it is freezing, and due to the risk of fire, synthetic clothing is prohibited. Despite these challenges and the strict protocols, Rutstein feels privileged to have had the opportunity to dive in the Alvin three times.

Rutstein always collaborates with scientists during her fieldwork, a collaboration that is central to her practice. Beyond the creation of artworks, she values the constructive exchanges she nurtures with researchers, believing that these interactions are mutually beneficial. In the field, scientists tend to focus on their areas of study, while the artist synthesizes their research to provide a broader perspective of the expedition. She continually learns from them and hopes to offer them a different viewpoint on their work as well. She wishes to capture different facets of their research, and, through her

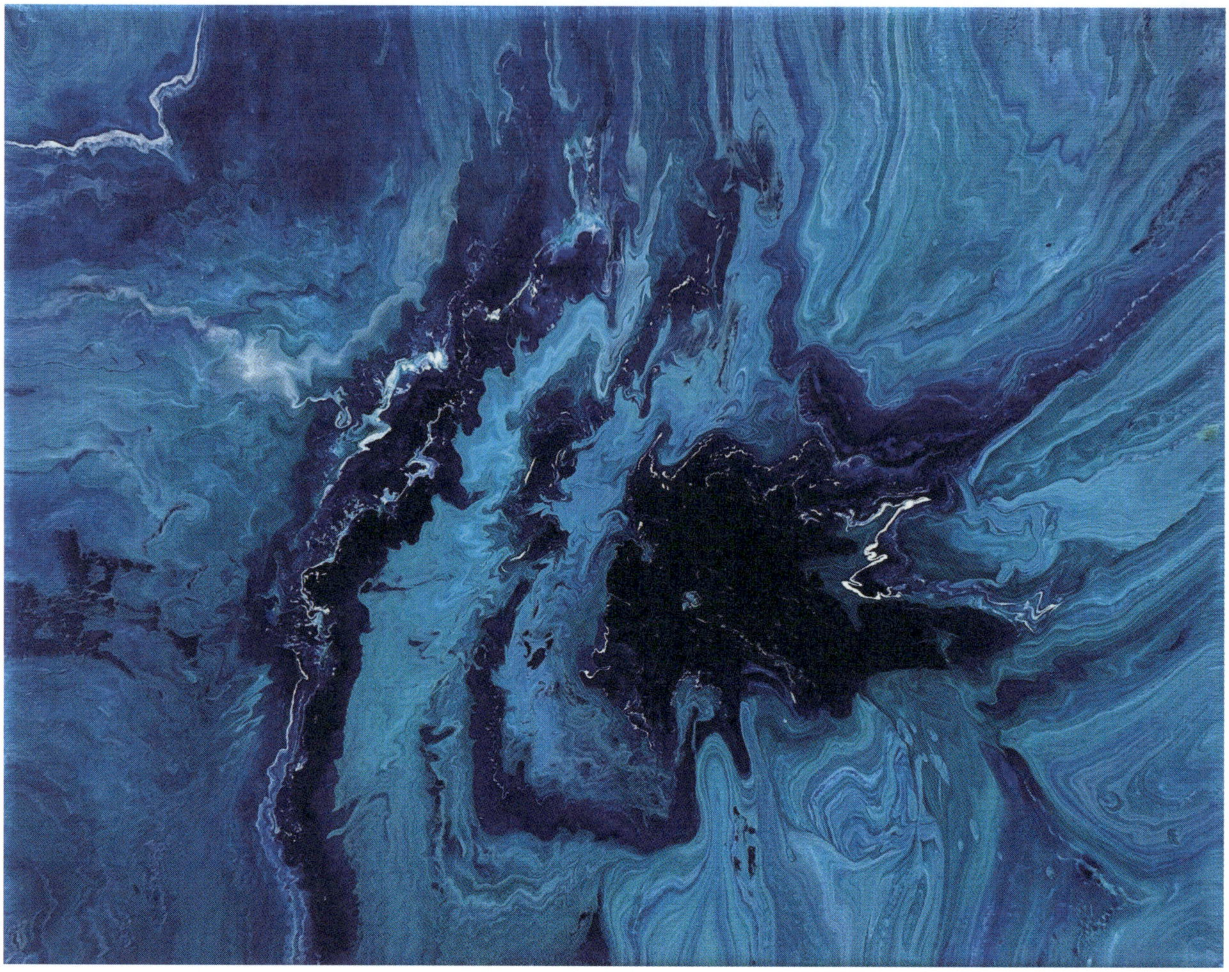

Day 6: Flow, Thirty Days: Chronicling a Voyage at Sea Series, 2024, Acrylic on canvas, 28 × 35 cm

abstract pictorial vocabulary, she seeks to find the visual patterns that connect different scales and areas of research.

When not in a submersible, Rutstein often joins expeditions on research vessels. Aboard these vessels, she paints, as it is her primary medium and her most immediate way of responding to what she observes. Upon returning to her studio, she develops larger installations.

For example, in *Out of the Darkness*, she created an interactive LED sculpture that responds to visitors' behavior. The lights mimic underwater bioluminescence and emphasize the complex communication among microorganisms. Our perception of the underwater world is limited, yet it is known that microbes, viruses, and bacteria communicate in intricate ways. Some creatures increase their luminescence to confuse predators,

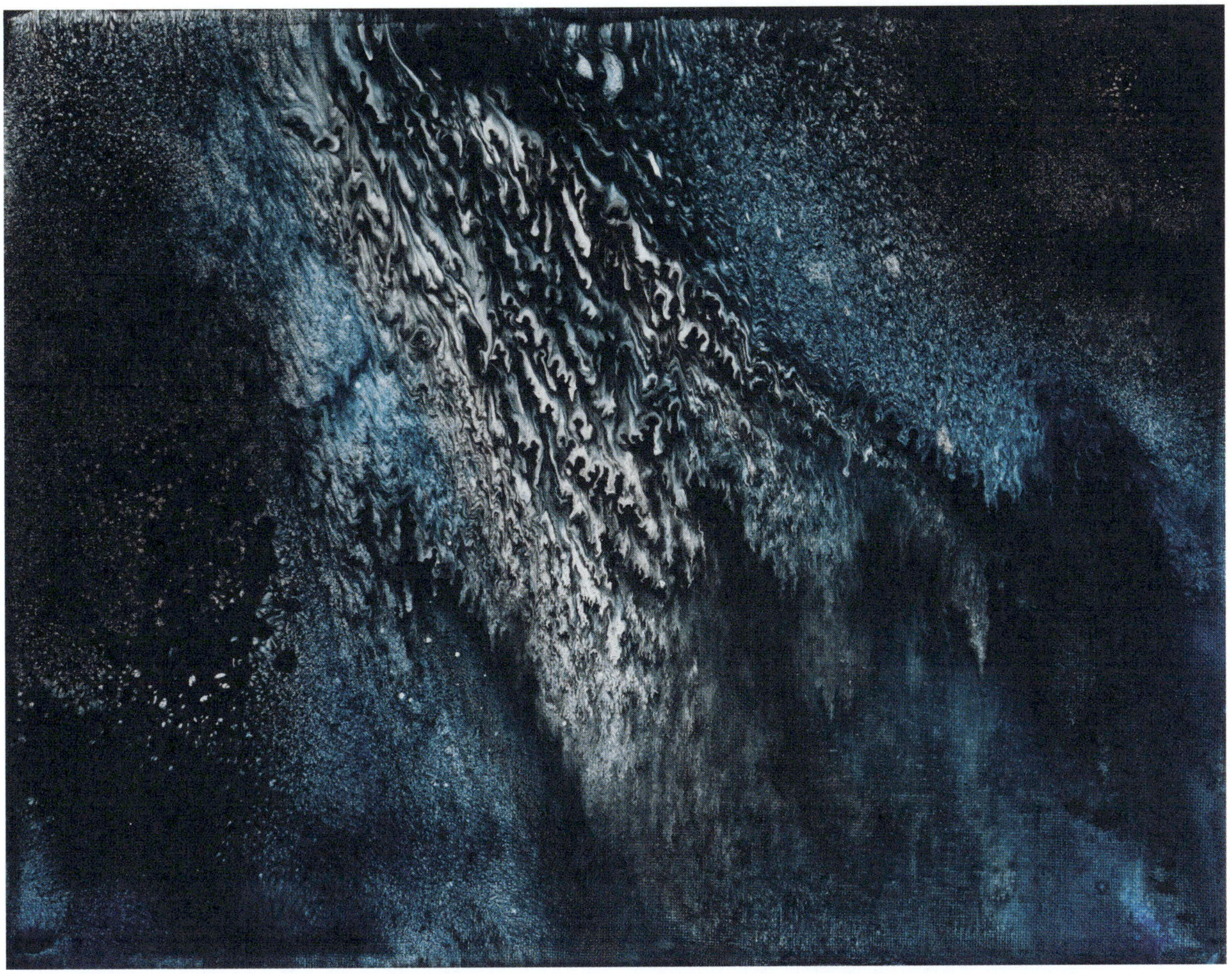

Day 1: Plume, Thirty Days: Chronicling a Voyage at Sea Series, 2024, Acrylic on canvas, 28 × 35 cm

a concept reflected in her installation, which suggests that human presence might represent a form of predation, thus constantly activating the structure.

Rutstein's work encompasses both macro and micro perspectives. Landscapes have always fascinated her, and mapping processes are integral to her practice. Concurrently, she explores patterns at the micro level, such as carbon chains, genes, and cell networks.

She searches for the appropriate language to embody the knowledge she gains from the deep sea. Consequently, most of her works appear abstract at first glance, though they replicate observed patterns or reference data collected by researchers. Her aim is not realism but rather animating and transforming this data into a living experience.

The artist hopes that viewers will engage with her artwork without prior knowledge of her research. She believes that her sensory language might spark their curiosity and interest, ultimately fostering respect for the marine environment. She observes that people, in general, are disconnected from the natural world. They often fail to realize how dependent human beings are on the health of the ocean. For instance, many are unaware of the significant impact marine microbes have on their lives.

In her film *Blue Dream*, Rutstein depicts an effect recently observed by scientists at the sea's bottom: superheated fluids rising from the seafloor, resembling an underwater geyser and creating a mirror-like illusion called "mirrored flange." The vibrant colors, ranging from orange to purple, mimic flames in the film. These venting mineral towers and volcanic chimney ledges are rich in biodiversity, minerals, and metals. She combines video footage and computer modeling to portray the resilience and growth of these microbial networks, our planet's smallest yet most vital living systems in the deep ocean.

Currently, Rutstein is focusing on deep-sea corals, which are less well-known than their shallow-water counterparts. Researchers have discovered how these corals obtain the energy and nutrients they need by trapping tiny organisms in passing currents. Portions of the modular structures of this interactive installation on display will become an artificial reef to help rehabilitate deep sea coral communities decimated by the 2010 Deepwater Horizon Oil Spill in the Gulf of Mexico.

Kim Yip Tong

Born in Germany in 1991

Kim Yip Tong's passion for the ocean was nurtured by the privilege of growing up in a house situated directly on the beach in Mauritius. The stretch of coastline where her family resides is characterized by volcanic rocks, which harbor a remarkable biodiversity. Throughout her childhood, she spent countless hours playing, fishing, and exploring these rocks with her siblings. Yip Tong expresses a deep connection to the local ecosystem, considering the crabs, fish, and sea shells as part of her extended family.

"This incredible lifestyle has fueled my imagination and my custodian feeling of responsibility. Today, my practice is driven by contrasted feelings of love and anger," she says. Over time, the once-small village where she grew up has evolved significantly. What was once a sparsely populated area where she and her sister often played alone is now a highly developed region.

Her close relationship with these ecosystems, which she feels few people

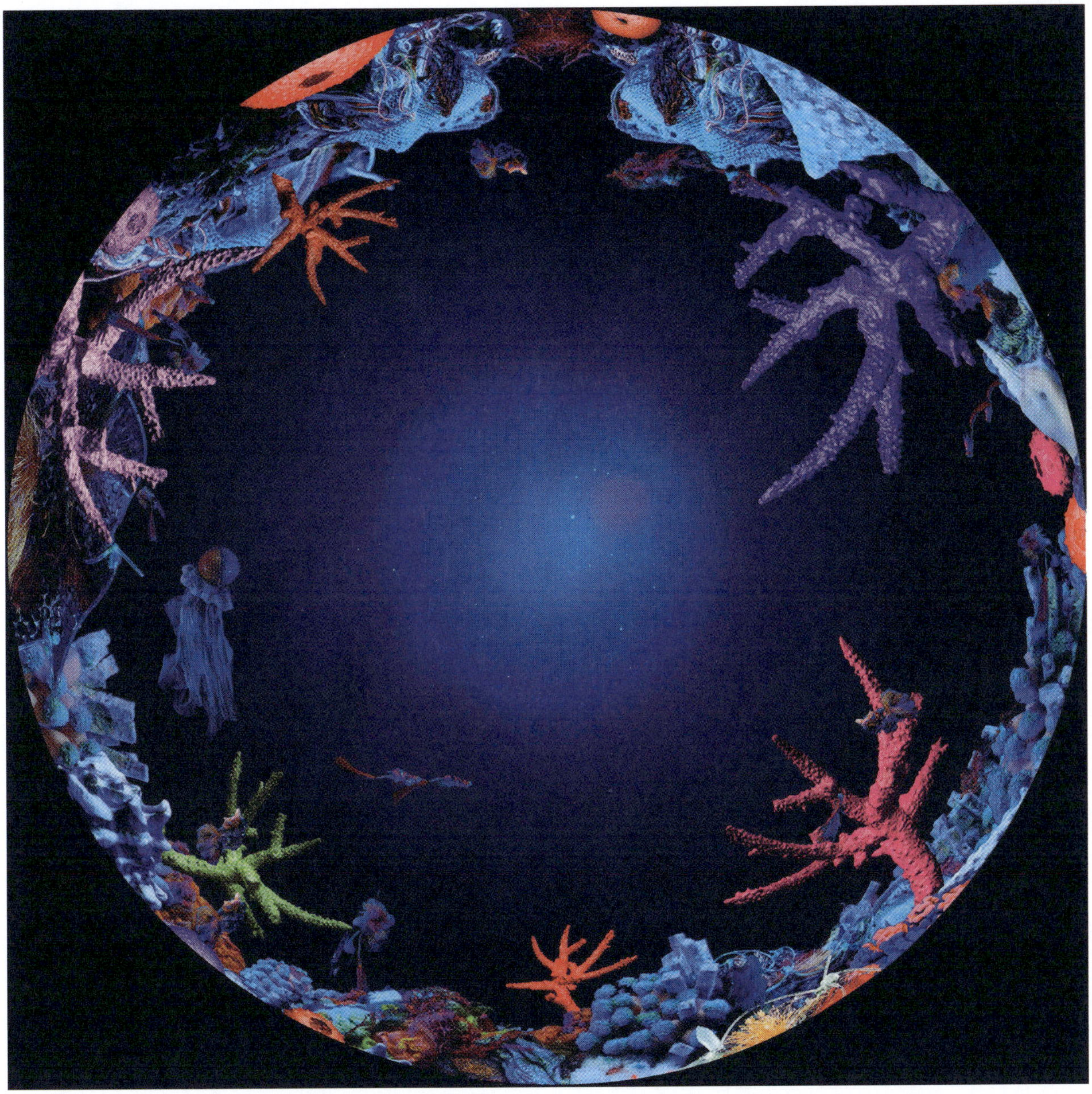

Anthozoa, 2017, Video 4D, 2 min 09

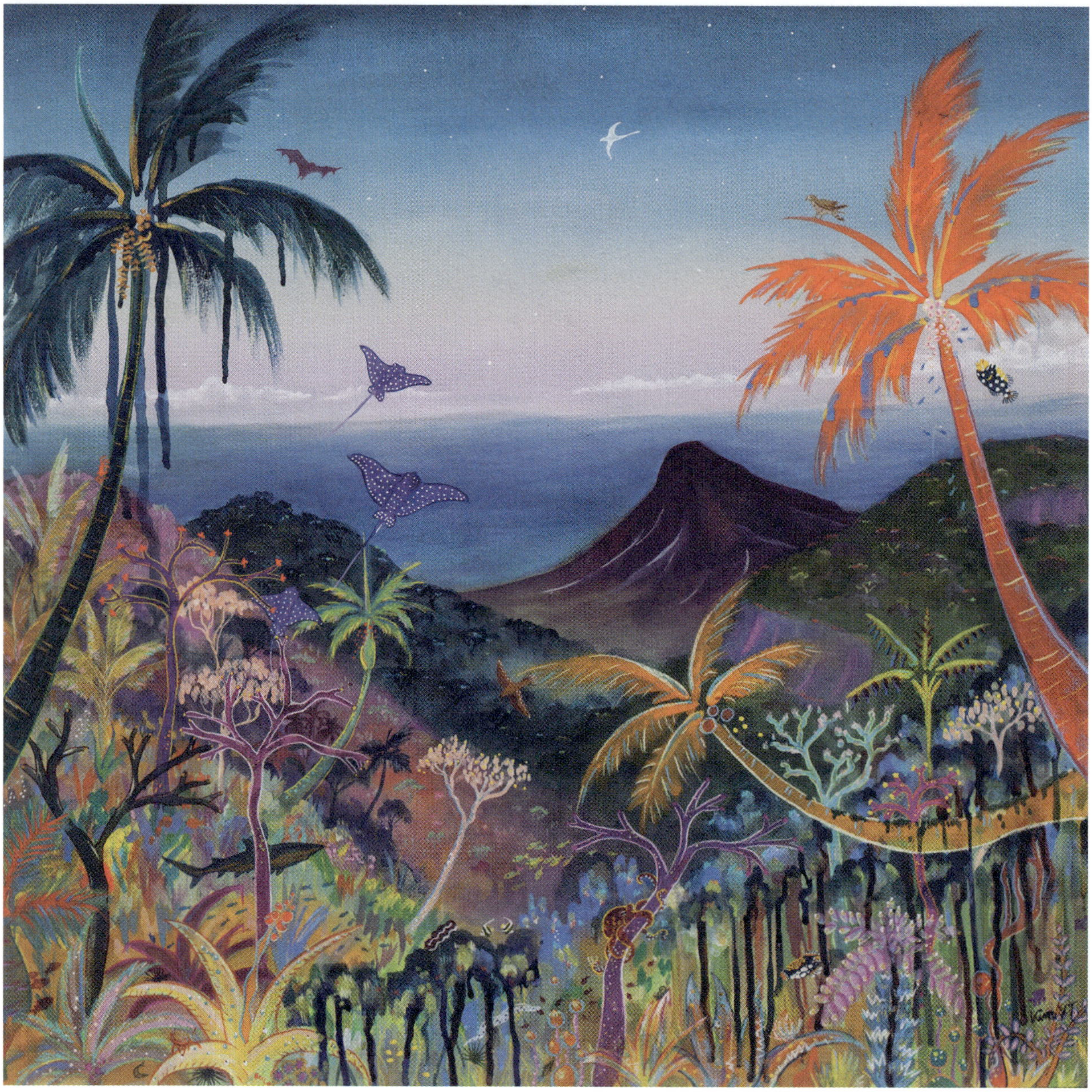

Lafore Palmie, 2021, Mixed Watercolours on paper, 60 × 60 cm

take the time to truly engage with, has allowed her to witness firsthand the gradual and often detrimental changes occurring over the years.

"I'm not a scientist so I wouldn't risk myself in quoting facts and numbers etc. But I can see in front of our home the extent of the damages. Despite occasional sightings of turtles, rays, and sharks, the signs of ecological distress are clear to even a semi-trained eye. The ocean is ill and it's really heart-breaking to witness."

Most of Yip Tong's artworks celebrate the magnificence of the underwater world, perhaps as a means of demonstrating its resilience and opposing the detrimental effects of climate change and human pollution. Her series of paintings, for example, portray the landscapes of Mauritius and its marine species in a naïve, almost childlike manner, as if to preserve the viewer's capacity to marvel at their beauty. *Anthozoa* (2017), an animated film, features resilient coral reefs that grow and move against a backdrop of recorded submarine city sounds. While many of the depicted reefs are undergoing bleaching, some manage to recover and continue to thrive, symbolizing the possibility of regeneration amidst environmental degradation.

Another notable work, *Pie dan lo* (2024), is an animated documentary that recounts the 2020 oil spill that occurred in Mauritius. Rather than focusing solely on the ecological disaster itself, Yip Tong highlights the collective response of the local population. In an extraordinary act of solidarity, the island's residents attempted to mitigate the damage by constructing floating barriers made from sugarcane leaves and cut hair. Through this piece, the artist underscores the resilience of both human communities and the natural world, emphasizing the strength that emerges in times of crisis.

"This oil spill is just the tip of the iceberg of our deeply dysfunctional society. In many ways, we are all both victims and complicit in the state of the ocean . . . I find it very difficult to articulate these sensitive issues that are also political, which is why I channel these conflicting emotions into my projects as effectively as I can. When I reflect on the situation deeply, all I want to do is scream and cry, as it is challenging to maintain hope in the face of such realities. Yet, I return to snorkeling or surfing and feel immense gratitude for these moments of bliss. The ocean is so wonderful!"

Juria Toramae

Born in Morocco in 1983

Juria Toramae was born in Morocco, raised in Egypt and Thailand, and educated in Malaysia. In 2010 she moved to Singapore where she began investigating the local coastline and marine life. For her, the sea represents a form of stability, a way to connect all these different territories and to think beyond maps.

In *Temporality* (2013), Toramae used photographic archives of Singapore's seascape to reconstruct a collective memory of the place. The shoreline of the country has drastically changed over the last decades, creating a sensation of loss and confusion for those who refer to the landscape as memory artifacts. This series questions how the sea is affecting the life of people in the recollection of their personal memory and sense of belonging.

This project marks the beginning of her photographic archive from Singapore. Eager to explore the local seashore, she joined a group of marine conservationists to record marine species from the shallow waters. For years, she has photographed them, creating a very significant data base and learning about the local ecosystems of the reef. While this group often focused on endangered species, she started questioning their scientific taxonomy and work methodology. Her concern was to learn about all kinds of marine life, and she did not agree with any form of hierarchy or selected mode of seeing.

In her video installation *Pelagic Dream* (2021), she used an algorithm to create a flow of images featuring imaginary marine species based on her photographic archives. Thanks to artificial intelligence, the forms continuously recompose and transform, without singularizing any specific creature. This dreamlike flow highlights the multiple varieties of marine creatures and also hints at the sea's still unknown species. Toramae imagines that the ocean is dreaming and that, perhaps, it continues to generate new creatures.

By injecting hybridity—thus fiction—into the archive, Toramae also plays with the idea of "alien creatures." "What does it mean to be alien?" she asks. "Sea creatures existed long before humans, so they are not alien. They only look alien because we are not used to seeing them."

In Singapore, where most people tend to think of the local seashore as a polluted place, empty of life, showing these beautiful images is a way of starting discussions about the thriving local marine life and its preservation. In parallel, the artist is investigating the impact of land reclamation on the seashore, one of the pressing issues in Singapore. Because of intensive activity, almost ninety percent of the local corals has undergone bleaching.

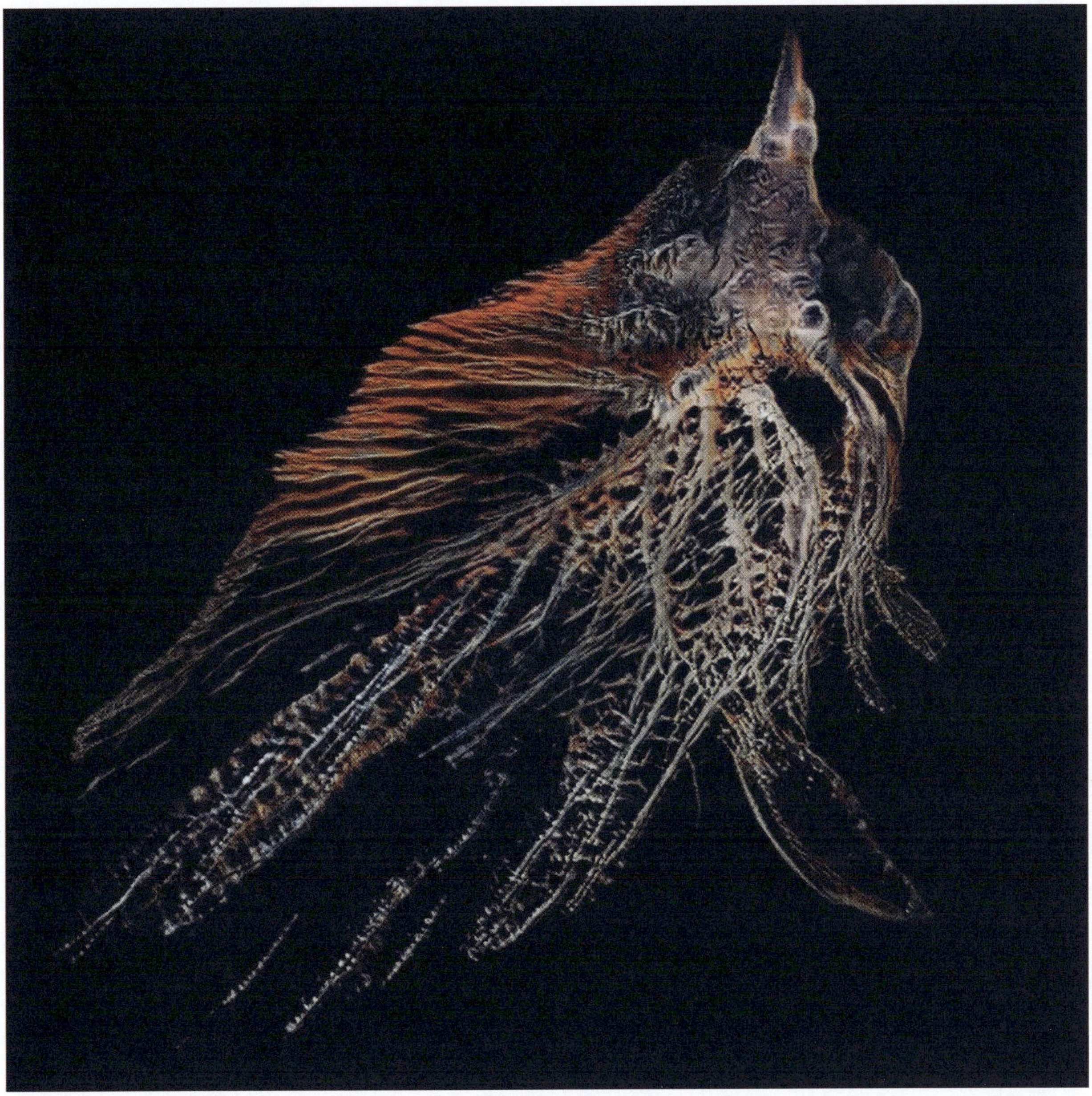

Pelagic Dreams, 2021, Video, 2 min 35

Learning From the Ocean

I need the sea because it teaches me
I don't know if I learn music or awareness,
if it's a single wave or its vast existence,
or only its harsh voice or its shining
suggestion of fishes and ships.
The fact is that until I fall asleep,
in some magnetic way I move in
the university of the waves.

Pablo Neruda

1 Ravi Agarwal
Born in India in 1958

2 Martha Atienza
Born in the Philippines in 1981

3 Nicolas Floc'h
Born in France in 1970

4 Kaisen Jane Jin
Born in Korea (Jeju Island) in 1980

5 Mario Macilau
Born in Mozambique in 1984

6 Manel Ndoye
Born in Senegal in 1986

7 Temandrota
Born in Madagascar in 1975

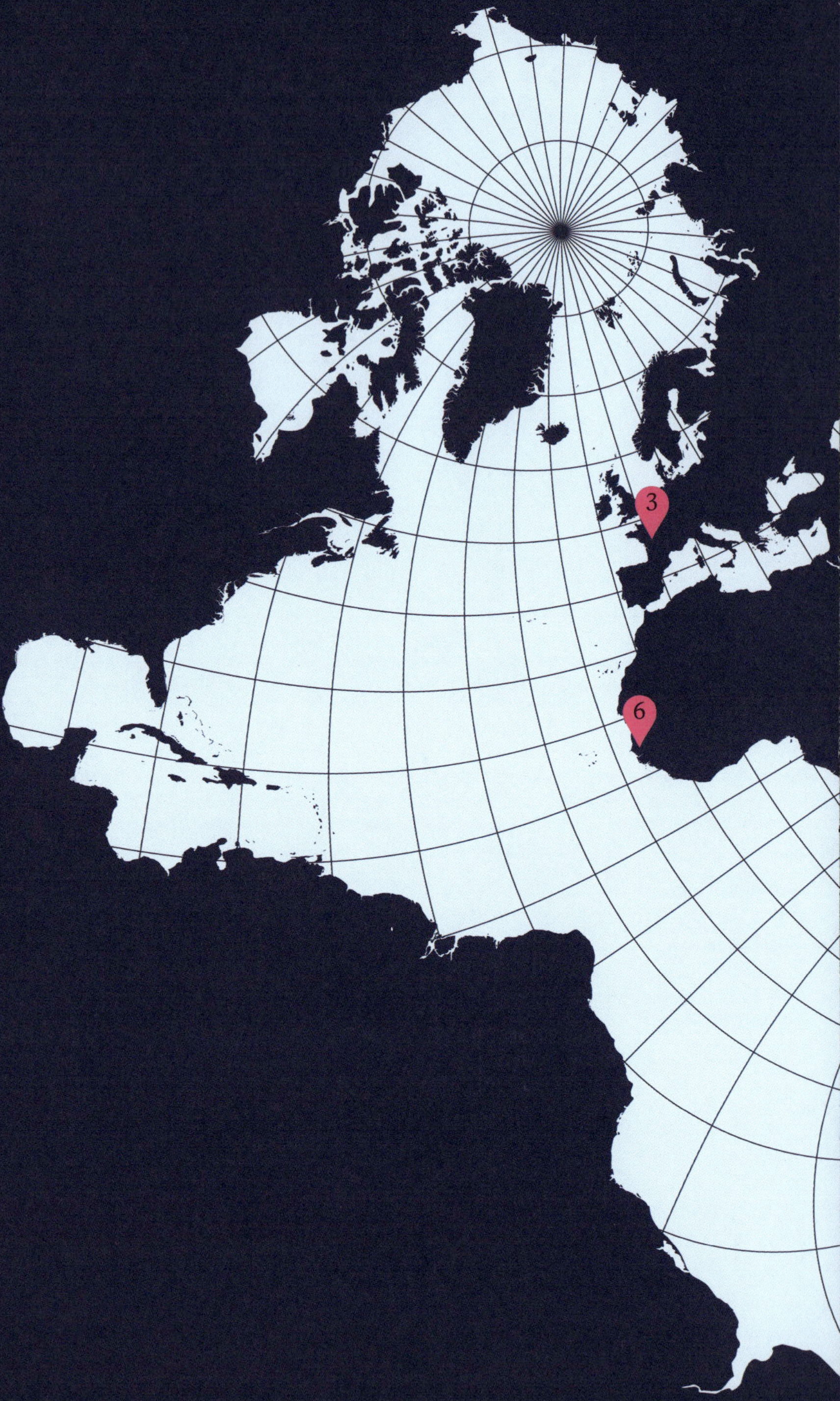

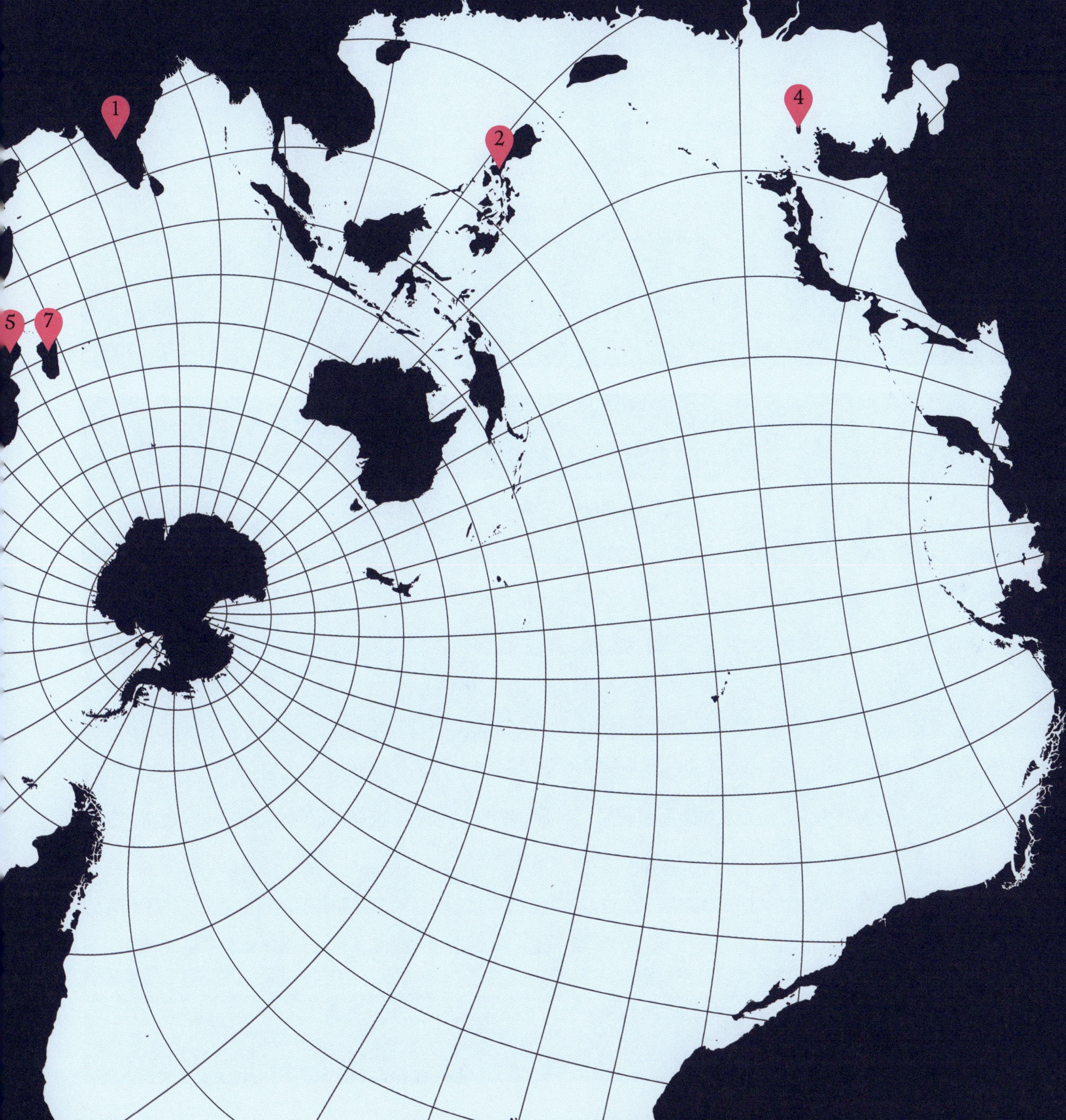
1
2
4
5
7

1.

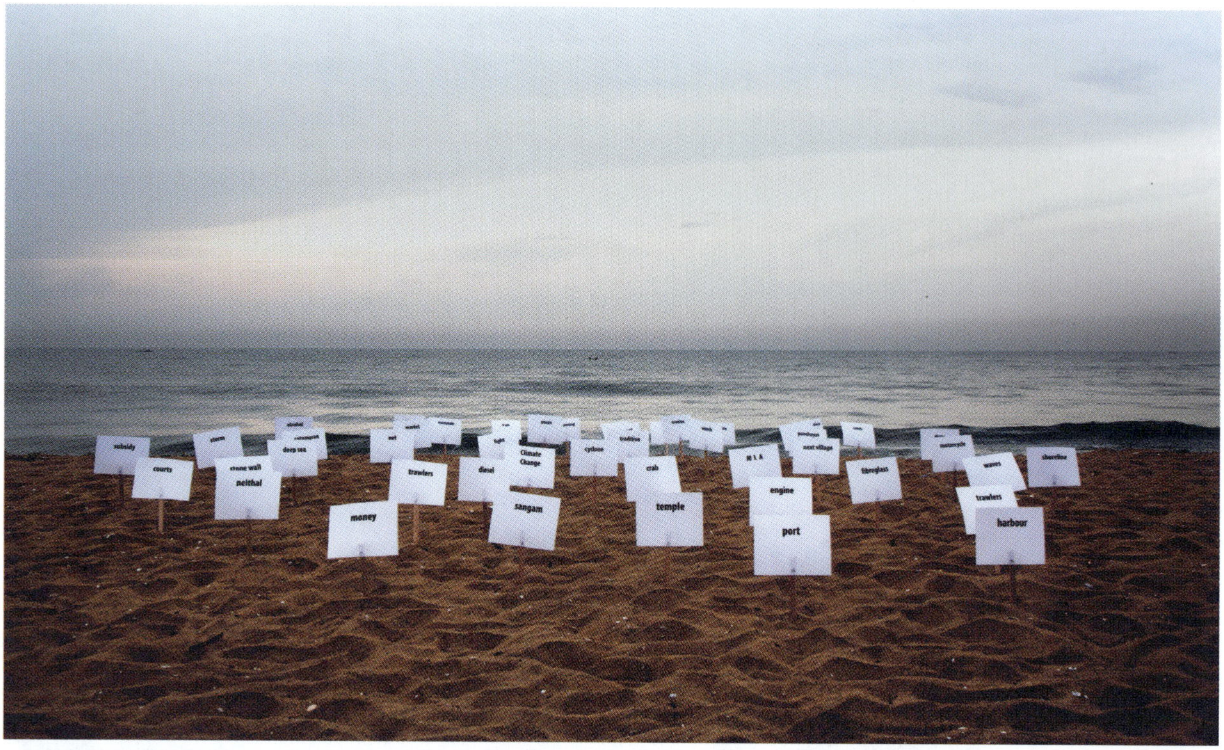

2.

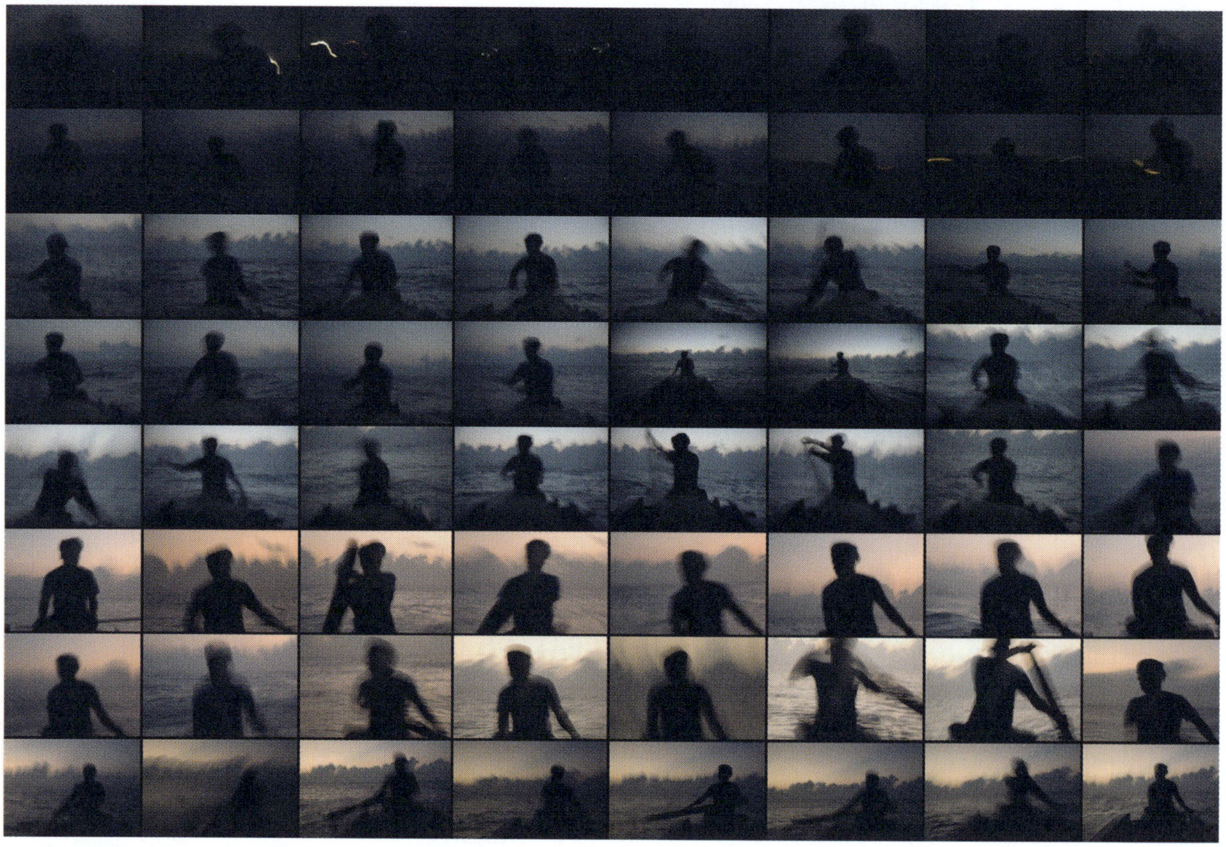

1. *Rhizome*, 2015, Archival photographic print, 80 × 120 cm
2. *4 am*, 2016, Archival photographic print, 66 × 95 cm

Ravi Agarwal

Born in India in 1958

"Artists through embedding their practice in the ecological discourse and collaborating with other practitioners can foster fresh interdisciplinary understandings of the entangled questions of ecology." (*Routledge companion to contemporary art, visual culture, and climate change*. Ed. T.J. Demos. New York: Routledge, 2021)

Ravi Agarwal is an artist, a documentary photographer, and an environmental activist. His pluri-disciplinary practices overlap and complement each other, and aim at addressing today's ecological crisis and environmental injustice.

Agarwal grew up on the banks of the Yamuna River, a 1,300km-long river that crosses the megalopolis of Delhi. With his camera—he received his first at the age of thirteen—he observed the changes of its waters, its ecosystem and environment, recording its serious deterioration due to climate change and intensive urbanization. For decades, he documented its flora and fauna, the rise and fall of the coal-fired power plants on its banks, and the life of the population living from and around it.

Before 2013, the sea was an alien element to the artist. Then he decided to study the impact of climate change on the coast and on the lives of fishermen from a small fishing community in Tamil Nadu, in the south-eastern part of India. Tamil Nadu is one of the largest fish producing states in India. It also offers a very rich marine ecosystem with great biodiversity. However, recently the state has been facing many challenges and traditional fishing activities may well be on the verge of disappearing.

New techniques of fishing are putting pressure on the local ecosystem and on the natural stock in a national context of overfishing. Besides, mechanized boats compete inequitably with smaller traditional boats. Despite recent regulations, there are frequent conflicts between artisanal fishermen and those involved in mechanized fishing operations. With increasing tourism, the price of land also soared, and many fishermen sold their places. In addition, the neighbouring town built a harbour that impacted the ocean currents, resulting in sand erosion of the beach where the fishermen used to land.

Else All Will Be Still (2013-2015) is an ensemble of works, mainly photographs and videos, derived from the artist's long-term fieldwork among the villagers. The series offers a multitude of perspectives from which to grasp the complex reality of this community whose traditional way of life is bound to disappear. The work focuses on the fishermen's work techniques and daily gestures and emphasizes their specific relationship with the local ecosystem. At the same time, Agarwal proposes a deeper reflection on the unfathomable nature of the sea and shares his own experience and attempts to capture its inexhaustible features.

Art offers Agarwal the means to reflect on this multiplicity: according to him, activists, in order to be heard, need to deliver clear messages, and these obviously lack nuance. Similarly, academic researchers must express themselves with clarity. They are usually asked to develop relevant arguments. In contrast, as an artist, Agarwal does not need to comply with any rule and feels free to propose something different, more open, perhaps more confused as well but, he hopes, closer to life.

Else All Will Be Still is an attempt to seize reality with its complexities and ambiguities. The artist's questions are numerous: what is nature and what is the sea? How can we engage in nature differently and move away from the Western conception of nature as a separate entity? How are these fishermen coping with environmental deterioration, how can we learn from them? In his diary, which is part of the series, he asks: "What are the boundaries of 'ecology,' of 'nature,' of 'knowing?"

Agarwal emphasizes how much the life of the fishing community is entangled with the sea and notes that, perhaps, "in such entanglements could lie hidden secrets of other ways of being and re-learning how to coexist with the planet." (Agarwal Ravi, "Introduction" in *Weather Report – The Crisis of Climate Change*, Omita India International Centre, Winter 2019-Spring 2020)

Such a symbiosis inspired Agarwal to rethink the idea of nature no longer seen as a mere resource or as an object that can be exploited and commercialized. Just as the fishermen do, he wanted to learn how to live "on, off and in" nature and the sea, an endeavor which is, in fact, more complicated than it seems. Fishermen's lives cannot be romanticized and, besides, the artist was confronted with the open sea for the first time. He acknowledges that the sea he saw was clearly not the sea they saw and reflected on his own vulnerability and inability to grasp its essence (Conversations with the artist to prepare the exhibition *Boundless Sea*, Guangzhou Academy of Fine Arts, 2020). For him, the issues raised by nature are still not understood today. In India, the question is seldom addressed, and these small-scale fishermen have little agency regarding the decisions that are made about the development of their own land and economic activities.

As an invitation to penetrate their daily lives, to see their work techniques, and to address some of their concerns, *Else All Will Be Still* gives these fishermen a voice and emphasizes their culture. The series does not produce new sets of dogma, but initiates dialogues rendered even more necessary by the indeterminacy it creates. The series could be conceived as an initiation or an invitation to penetrate the world of these fishermen and to think anew, with them, about our daily interactions with the sea and nature from a more collaborative, ethical and mutually respectful perspective.

Martha Atienza

Born in the Philippines in 1981

"You cannot address marine environmental issues without talking about the specific social and economic situation of our place."

Martha Atienza's relationships with the sea is complex; it is both physical and emotional. She spent most of her childhood on Bantayan island, a group of islands located in the Visayan Sea in the Philippines. Her grandfather was a lighthouse keeper, her father was a sea captain and her mother worked for a while on the ships of an American cargo line. Later on, her father set up a shipping company in which most of their relatives got involved. Consequently, Atienza grew up surrounded by seafaring people, in addition to the local fishermen from the island.

The artist had a romantic idea of the ocean and of life on board because navigating on the sea had been a way, for her father, to discover the world. In 2010, she began documenting the lives of these men and set off on a cargo ship carrying coal. Her film, *My Navel is Buried in the Sea* (2011), explores the relationships between seafarers, fishermen, and the sea, based on her experience. For her, it was an eye opener. The film aims to initiate a dialogue between these communities and to discuss the socio-economic situation of the local fishermen: on the island, most of them cannot survive anymore on fishing. The depletion of fish and new fishing techniques are pushing them to become seafarers instead.

From then on, Atienza followed different crews and trading routes, delving further into history, trade flows, and maritime power relations. She is eager to trace the connections that link the various archipelagos of the Pacific and to highlight the communities' common history, knowledge, and navigation skills. As she collected stories from the local communities, she was also drawn to the study of the ecosystems, and began documenting the changing of the coastline under the effects of climate change. She observes that the rising sea level issue has become almost normalized in daily discourse. It is taking place so quickly that people witness its effects and are beginning to worry about the future.

Our Islands 11°16'58.4"N 123°45'07.0"E (2017) is an artistic representation of a yearly event that takes place in the artist's hometown. The film captures the essence of the Ati-Atihan festival in Madridejos, where men participate in the festivities wearing creative hand-crafted and often humorous costumes. This festival derives its name from the Aeta people, who are believed to be among the earliest inhabitants of the Philippines, predating the Austronesian migrations by around 30,000 years. For the work, the artist asked participants to reproduce the parade underwater on the floor of the Visayan Sea. The film showcases some of the men gracefully floating past the cameras, while others inadvertently

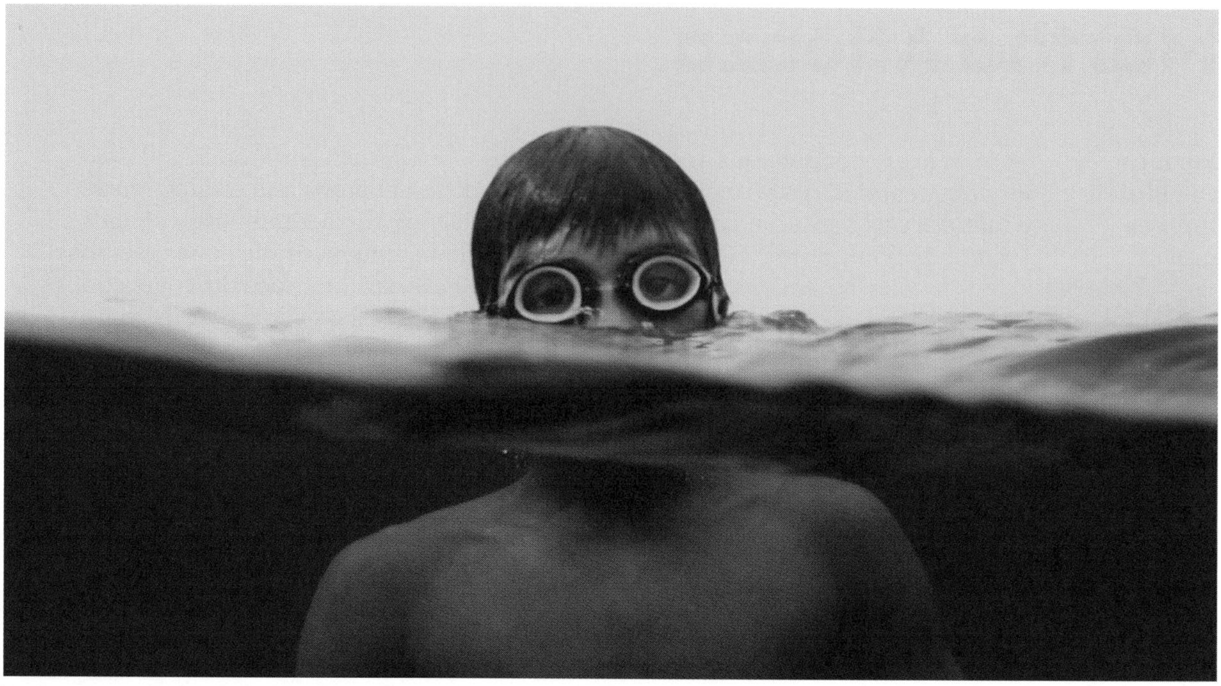

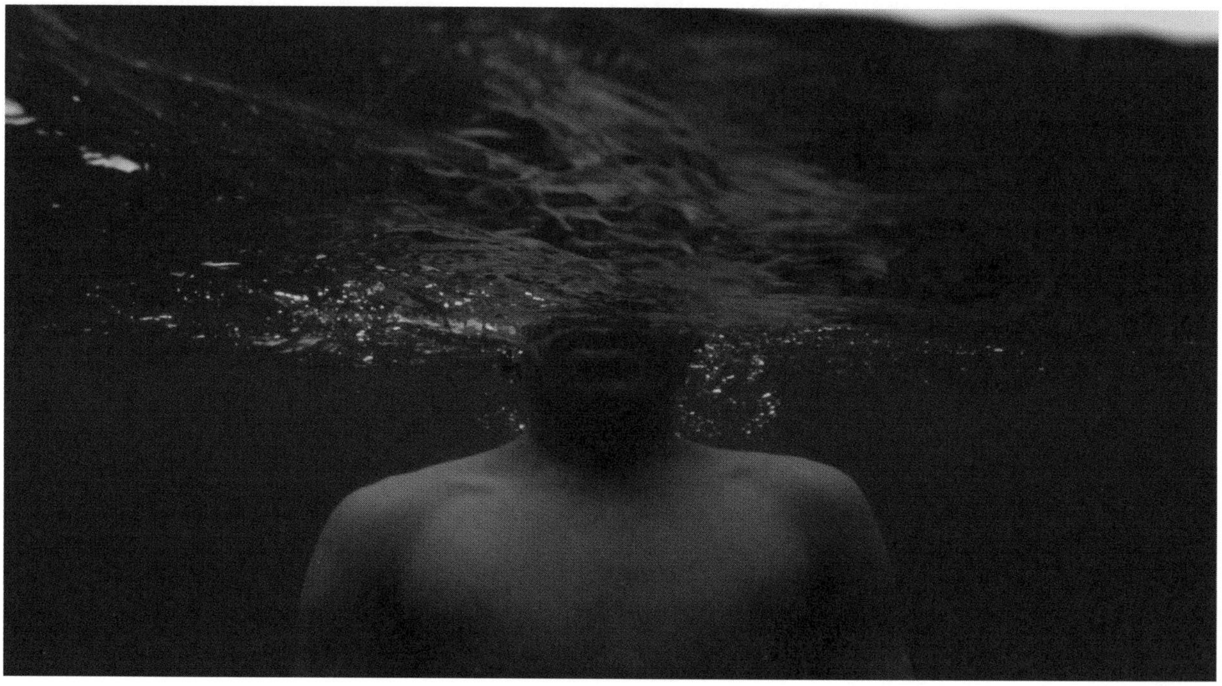

Tarong 11°16'12.0"N 123°45'23.4"E 2019-08-06 Tue 2:27 PM PST 1.50 Meters High Tide,
Kaongkod 11°16'12.0"N 123°45'23.4"E 2019-08-03 Sat 12:14 PM PST 2.03 Meters High Tide,
2013, 3-channel HD video installation, 79 min

crush dying coral. Their movements are slow, almost comical, and the procession resembles a dream. But Atienza is highlighting here the pressing issue of the rising sea level and the profound impact of human activities on the local marine ecosystem.

"We need to think more locally and give agency back to the local communities. For too long, fishermen have suffered discrimination, their knowledge has not been taken seriously. Now we have to listen to them and help them understand that they are key players in this unprecedented situation."

For Atienza, art can play a vital role in bringing about social change and, even more, to mobilize people. What is at stake is not only communicating a specific message but also about starting complicated conversations between various groups of people and favoring collaboration.

The Bantayan archipelago has been a protected area since 1981, when it was given the status of "Wilderness Area." However, increasingly, legislations are being lifted, in favor of commercial deals and real estate companies. "Who owns the land? Who owns the sea?" asks the artist. For the Istanbul Biennale in 2022, she created *Fisherfolk Day*, a video dealing with the issue of protecting areas and marine regulations. Above all, she wanted to establish a day of celebration and to use the work as a platform to reunite fisherfolk, NGO leaders, the government, and all others concerned by marine issues. The film follows the parade of fishermen, all navigating their traditional boats decorated with flags for the occasion. It is in slow motion, as if to root its action in a far more extended time period, into the island collective heritage. Because fishermen tend's to be dispossessed of their land, and of their rights, it is paramount for the artist to revive this memory as a way of resisting the disappearance of their living conditions.

"Climate change has a strong impact here, on the people, and on the coastline. Right now, we are facing a severe drought, and farmers cannot cope. As an alternative to fishing, people have developed seaweed farming, but when the weather is too hot, the seaweed is affected by microbes and production diminishes. People are feeling all these effects simply because it results in hunger. We have no choice but to discuss solutions. It is imperative for companies and institutions to take action, but, through art and collaborative processes, we can strive to find our own local solutions, because we cannot just wait for others to do something."

Fisherfolk's Day has continued to be celebrated since its foundation, providing an opportunity to share concerns and knowledge about the future. With Bantayan islanders, the artist founded GOODLand, an organization aiming at preserving the local environment and the protected marine areas, among other objectives. "This is a life-long task, says Atienza, but it is only through collective thinking and action that we will change the system."

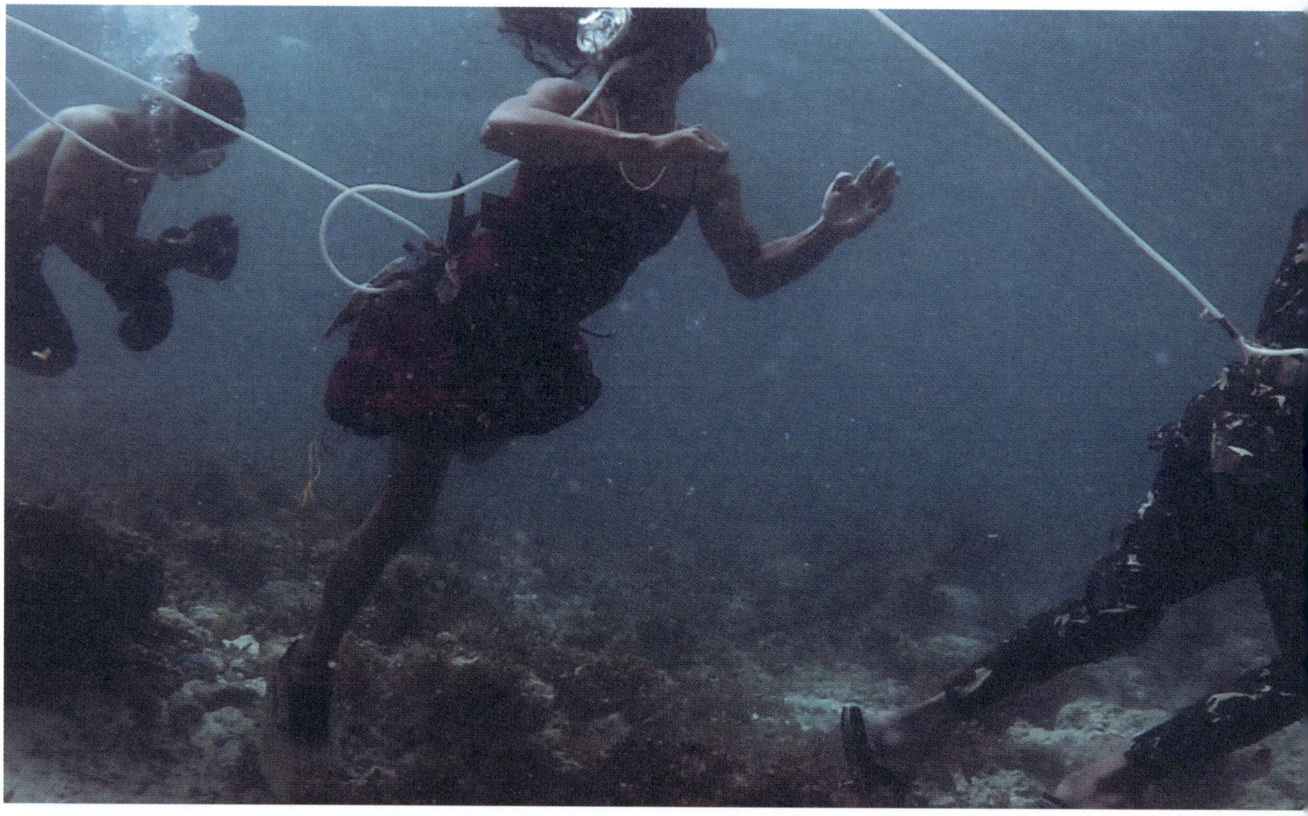

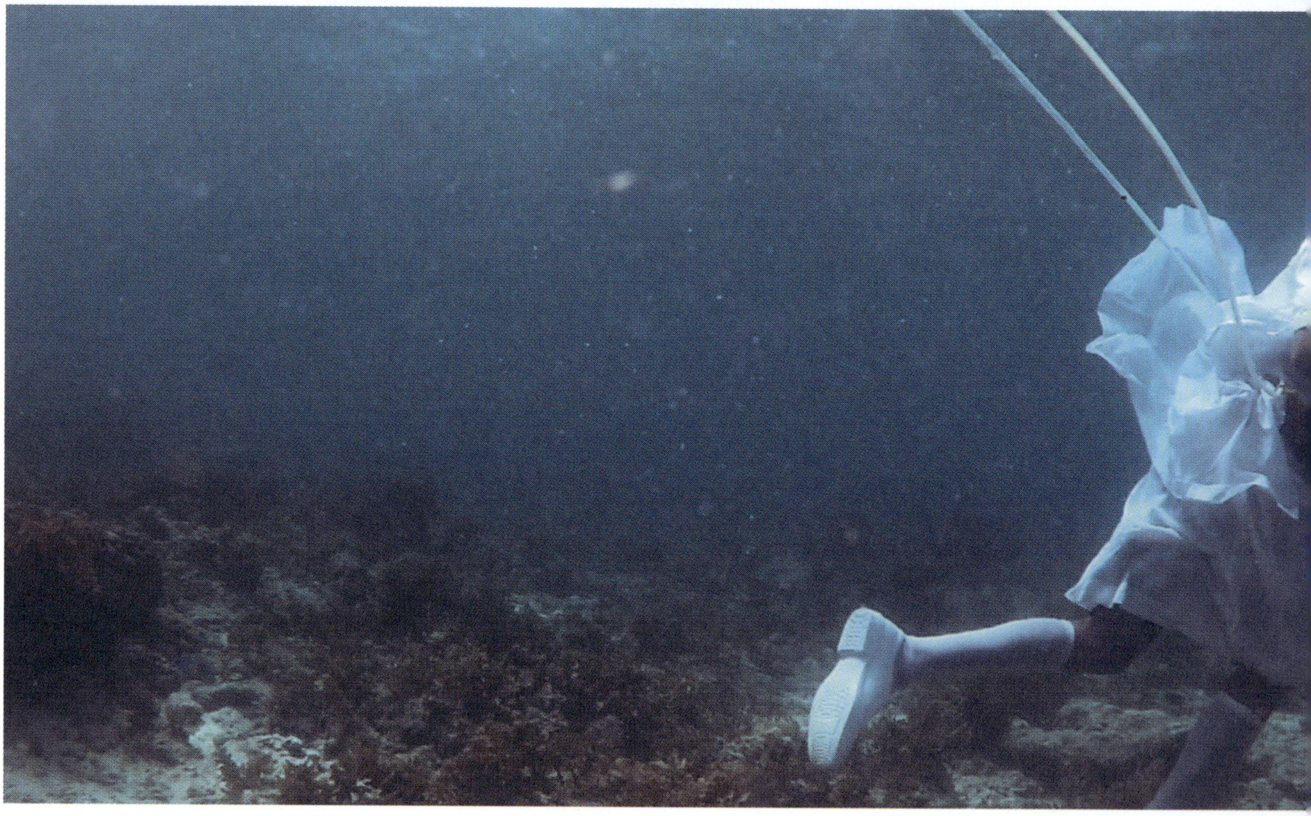

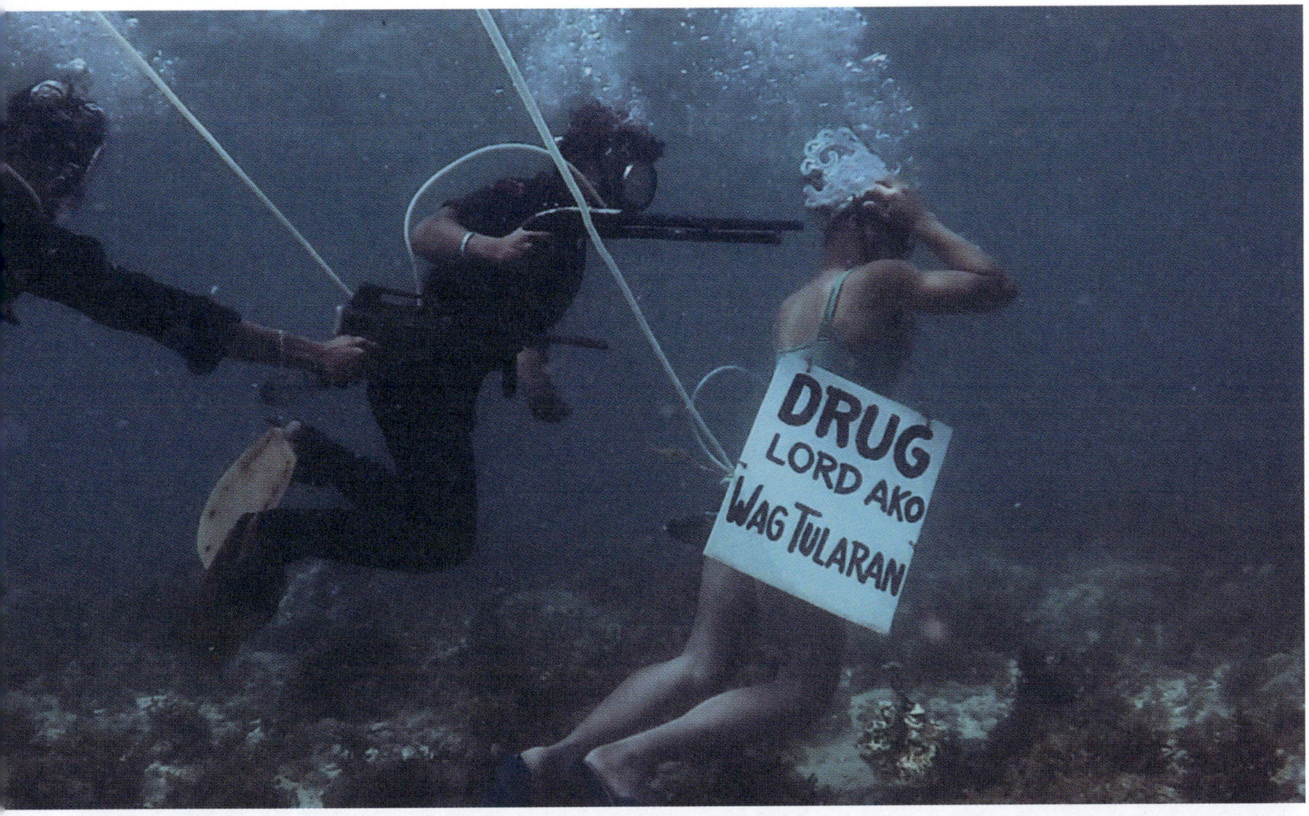

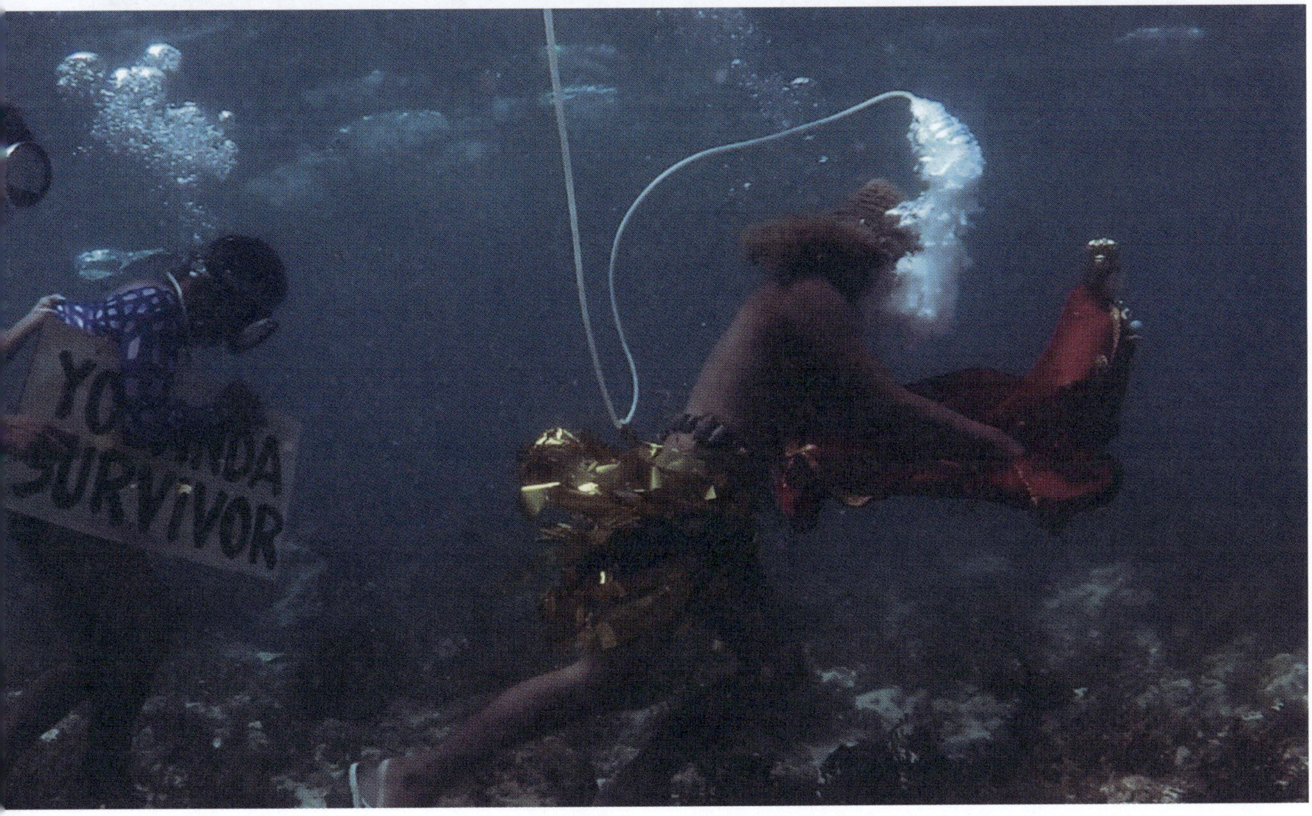

Our Islands 11°16'58.4"N 123°45'07.0"E, 2017, Video, 72 min, loop

Nicolas Floc'h

Born in France in 1970

"How to document a landscape that has not been documented yet? How do we look at the underwater world, and what do we look at?" It is with such questions that visual artist Nicolas Floc'h began what he calls his "artistic exploration," a practice that combines the poetry of art with the rigor of science. Over the past ten years, he has been trying to establish a typology of underwater spaces by exploring and comparing seascape territories.

Nicolas Floc'h grew up in Brittany, a peninsula in France. As a regular diver, he has extensive experience of the sea as well as of rivers. Over the years, he has observed the transformative impact of climate change on the local landscape of his native region, particularly at the estuaries of the Loire and Vilaine rivers, which flow into the Atlantic Ocean. It was during this time that he realized the lack of historical references to compare the current state of the environment with earlier periods, highlighting the global lack of archival images depicting seascapes. In contrast, there are ancient archives showcasing glaciers, enabling us to compare their past and present conditions, often spanning over a century. This comparative perspective is absent when it comes to the underwater world.

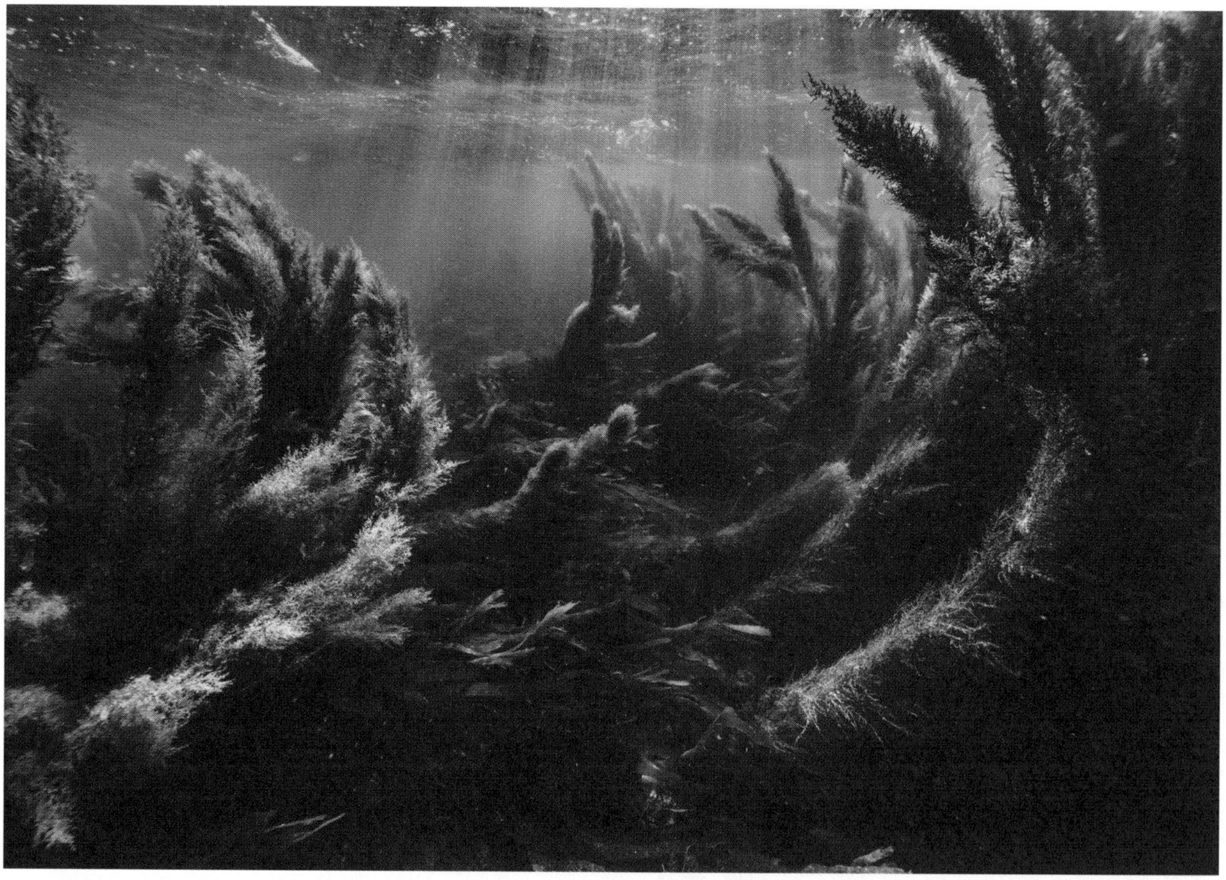

Paysages productifs, Initium Maris, Sargassum muticum (sargasses), –3 m, Hœdic, 2019, Piezo print on fine art paper, 110 × 154 cm

Marine landscapes have indeed too often been neglected and are usually not considered as an object of study in themselves. Seascapes are commonly regarded merely as backdrops to underwater photography, featuring animals or divers, which accentuates their anthropocentric nature. Furthermore, many individuals tend to associate seascapes exclusively with tropical and picturesque settings, often overlooking their own coastlines and nearby underwater territories. However, seascapes encompass a remarkable diversity, and it is precisely this richness that Floc'h endeavors to document. His primary focus lies in representing the various typologies of maritime landscapes, free of narrative elements, with a combination of poetic expression and precision. He thus aspires to foster a fresh perspective on maritime space, encouraging individuals to reconsider their own perceptions of, and relationships to it. These underwater territories, which house diverse and essential ecosystems, deserve greater attention, especially today with the acceleration of the marine effects of climate change.

Floc'h's approach to the sea and to photography is very rigorous as he always follows a given protocol. Underwater, he consistently employs a wide-angle lens and takes his photographs using natural light, subsequently printing them on carbon. However, this systematic approach does not imply an attempt to achieve objectivity. Floc'h does

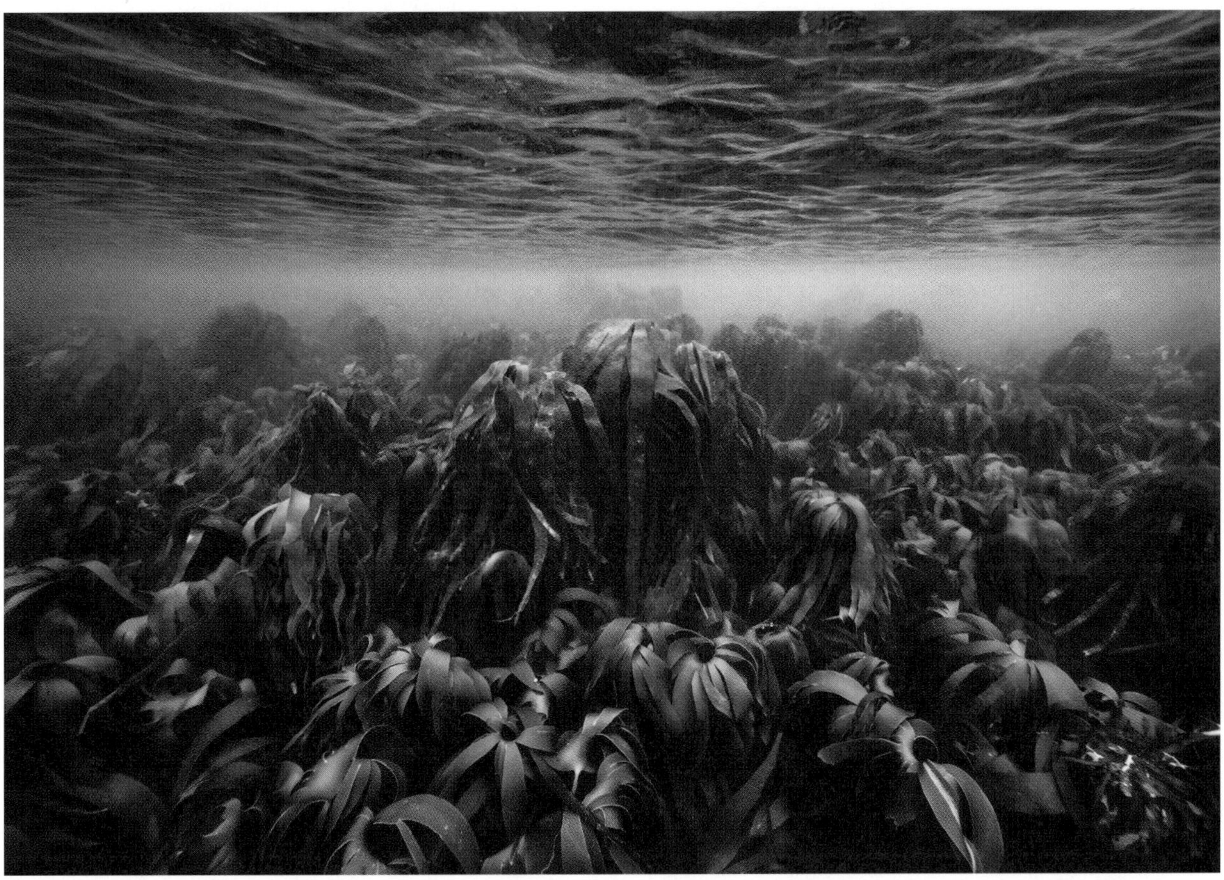

Paysages productifs, Initium Maris, forêt de laminaires, –6m, Île de Molène, 2019, Piezo print on fine art paper, 110 × 154 cm

not seek to follow the Düsseldorf School of photography and willingly embraces aesthetic choices. Through his images, his intention is to portray the dynamic nature and diversity of the marine environment, while consciously avoiding the pitfalls of stereotypes. Much of life underwater remains invisible to the human eye. Deciphering the sea is a challenging task that requires time and patience to master. The majority of biomass exists in microscopic form. While plastic pollution is readily observable, phenomena like water warming are more difficult to perceive. However, by adopting a different perspective beneath the sea's surface, one can learn to interpret the water's subtleties–its texture, color, and density. Even seemingly invisible occurrences, such as the impact of ocean acidification on marine ecosystems, can be discerned by observing the landscape's metamorphoses.

The artist's ongoing long-term project, *Initium Maris* (2015-), comprises a collection of black and white photographs depicting the seabed along the coasts of France. The choice of black and white photography serves to situate his practice within the broader historical context of exploration and discoveries. By including himself in this lineage, which originated in black and white imagery, he aims to include maritime territories in this existing archive. Additionally, the absence of color allows him to eschew exoticism and stereotypes, while maintaining a visual uniformity across his images, emphasizing contrasts and monochromatic tones.

In his series *Deep Sea* (2021), the artist ventures into the depths of the ocean (between 700 and 1800 meters), attaching his camera to a robot to film the abyssal seabed. The approach and resulting images bear resemblances to space exploration, evoking vast unknown spaces brimming with mysteries and promises. The main difference, he likes to emphasize, is that there is no identified life in space, while the marine environment is full of life.

In contrast, the *Water Color Project* (2015-2019) comprises a series of vibrant and almost abstract images. Here, the artist focuses on the color of water, exploring its nuanced variations that serve as indicators of the oceans' health. Each color becomes a distinct signature, reflecting the specific characteristics of a particular territory–such as the quality and quantity of sediment, the surrounding rocks, and even the impact of human activities, including the use of fertilizers and other pollution. When studying an ocean, Floc'h also examines its tributary rivers and coastlines, recognizing the inseparable connection between water and land. Viewing the sea entails observing an interconnected flow that transcends boundaries. In his series dedicated to the mouth of the Tagus River in Lisbon–a mural composed of 408 images–he documents the shifting water colors as the river merges with the Atlantic Ocean. From red to dark blue, beneath their visual beauty, the work offers a direct portrayal of the deterioration of water quality, the impact on local biodiversity, and the state of the seabed.

Throughout his journey, Floc'h collaborates with scientists

and researchers, although he conducts his projects independently. He has participated in various residencies focused on his maritime research, including his works, *Arcus* in Japan (2001-2002) and *Tara Pacific* (2017). These encounters have played a pivotal role in the development of his artistic practice. Through these collaborations, he has gained scientific knowledge and a deeper understanding of the phenomena he has observed over the course of decades. Working with experts has not only allowed him to connect his local observations with broader issues but has also highlighted the potential for a complementary relationship between the work of scientists and that of artists. For example, while scientists may not typically record the color of water visually, focusing instead on data and diagrams, Floc'h believes that there can be a mutually beneficial convergence between these different approaches.

Currently, the artist's photographic archive consists of approximately 30,000 images, representing a body of work that can be considered

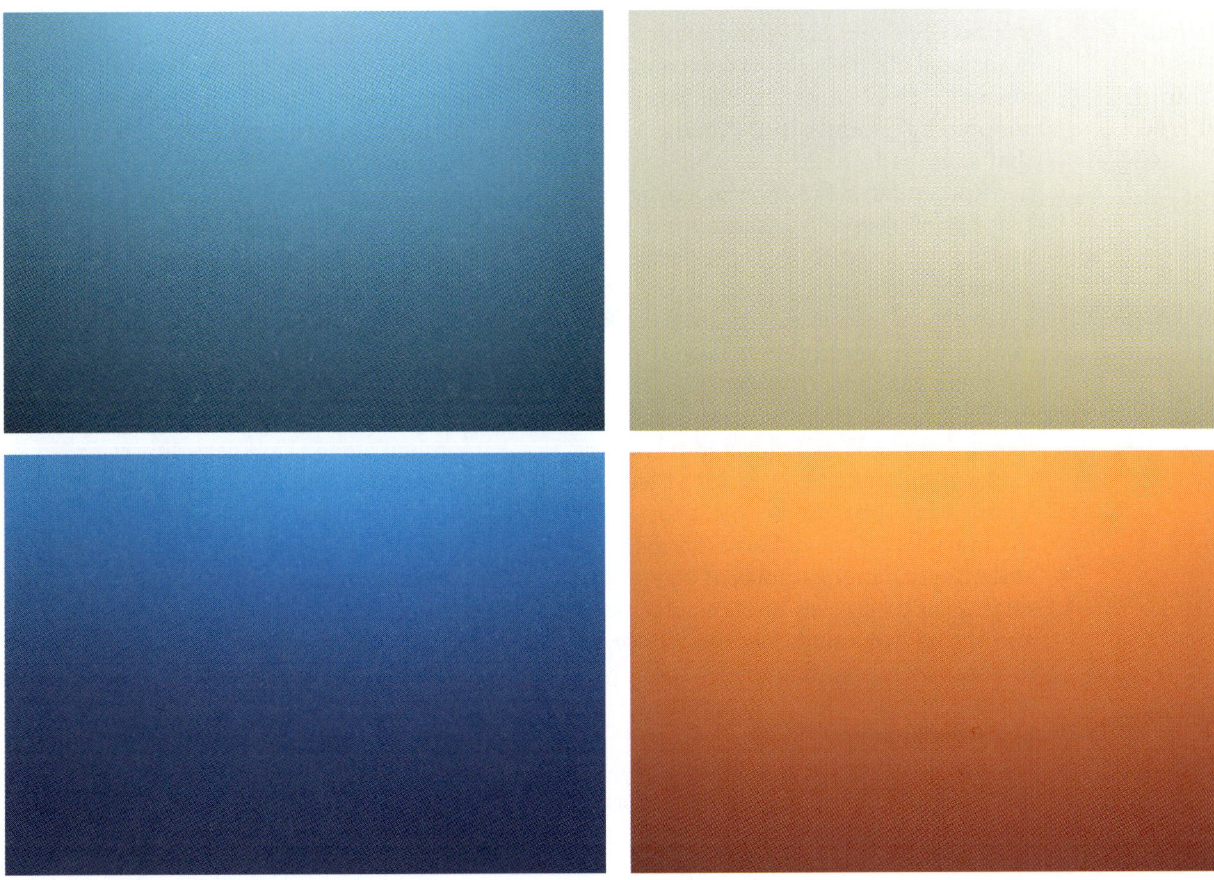

La Couleur de l'eau Series, Baie du Stiff, Ouessant, 2021 / White River, Mississipi, 2022 / Porquerolles, 2020 / La Loire, Paimboeuf, 2021
Pigment prints on fine art paper, 110 × 154 cm (each)

part of a collective visual memory. Floc'h does not primarily consider his practice as pedagogical, nor does he define himself solely as an activist. He identifies first and foremost as an artist. Nevertheless, he hopes that his art can contribute to a better understanding of the world we inhabit. In addition, he publishes books for those who wish to delve deeper into his work, aiming to foster a collective vision of the ocean that evolves and transforms over time.

Jane Jin Kaisen

Born in Korea (Jeju Island) in 1980

Jane Jin Kaisen was born in 1980 to a family from Jeju Island, which is situated off the southern extremity of the Korean Peninsula. She was adopted by a family in Denmark that same year. After reuniting with her Korean birth family, she learned that her mother and grandmother used to live as *haenyeo*, those "sea women" who engage in free diving on a daily basis to harvest shellfish, such as abalone or sea urchins, as well as seaweed. Now inscribed on the UNESCO list of the intangible cultural heritage of humanity, the knowledge and traditions of these ancient communities of women demonstrate profound respect for the marine and coastal ecosystems, and offer salient examples of a symbiotic relationship between human activities and the environment. Kaisen, whose research-driven and multidisciplinary artistic practice also interrogates the history and identity of the Korean island, has observed this distinctive matriarchal culture, which has informed the conceptual foundation of some of her films.

In her film *Of the Sea* (2013), we follow Kaisen as she walks along the volcanic coastline near the hometown of her grandparents. She carries fishnets, diving tools, and a book written by her grandfather, a local teacher and historian, which chronicles the Jeju Haenyeo Anti-Japanese Resistance movement that occurred during 1931-1932, a period of Japanese colonial occupation. Initially, this was a labor uprising. According to the artist, the sea divers were the public face of the resistance movement as they marched along these very shores to protest the economic exploitation of their labor. Here, the artist interweaves her personal narrative with history and cultural heritage to reflect on the pivotal role played by women in shaping local culture. The work suggests that it requires the same courage to oppose a colonial power as it does to brave the sea on a daily basis. The film's soundtrack, a song composed during the resistance movement, poetically captures the harshness of the daily routine of the *haenyo*. Written by a social activist and leader of the movement

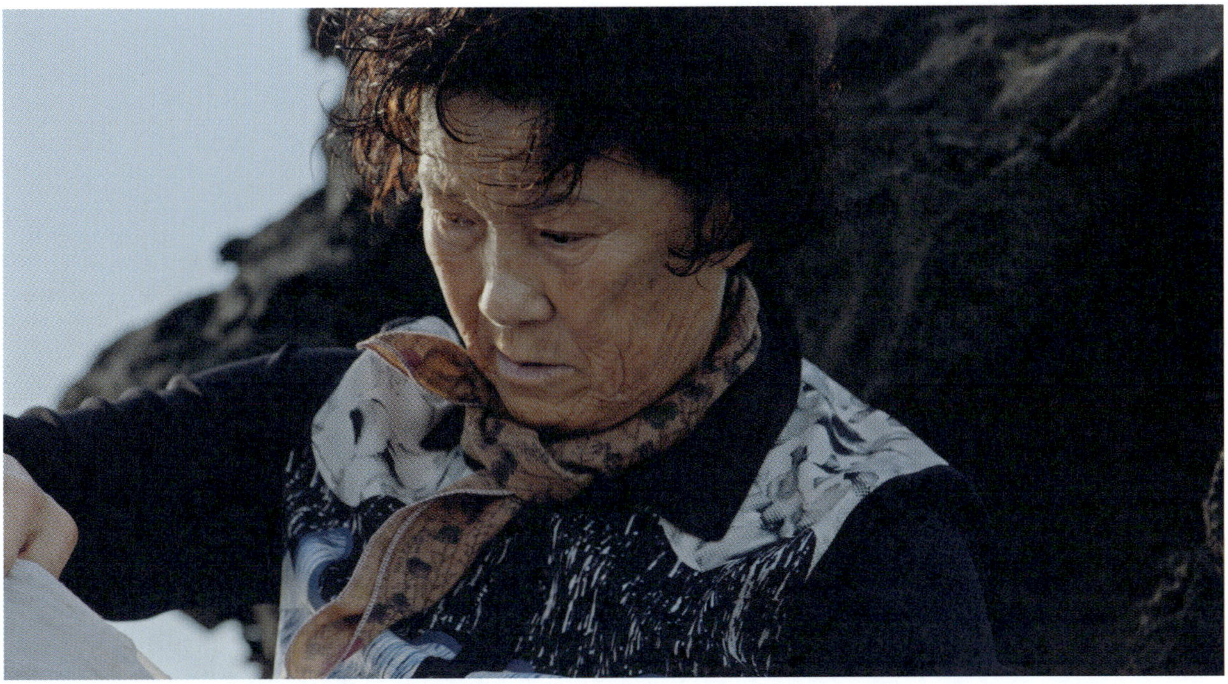

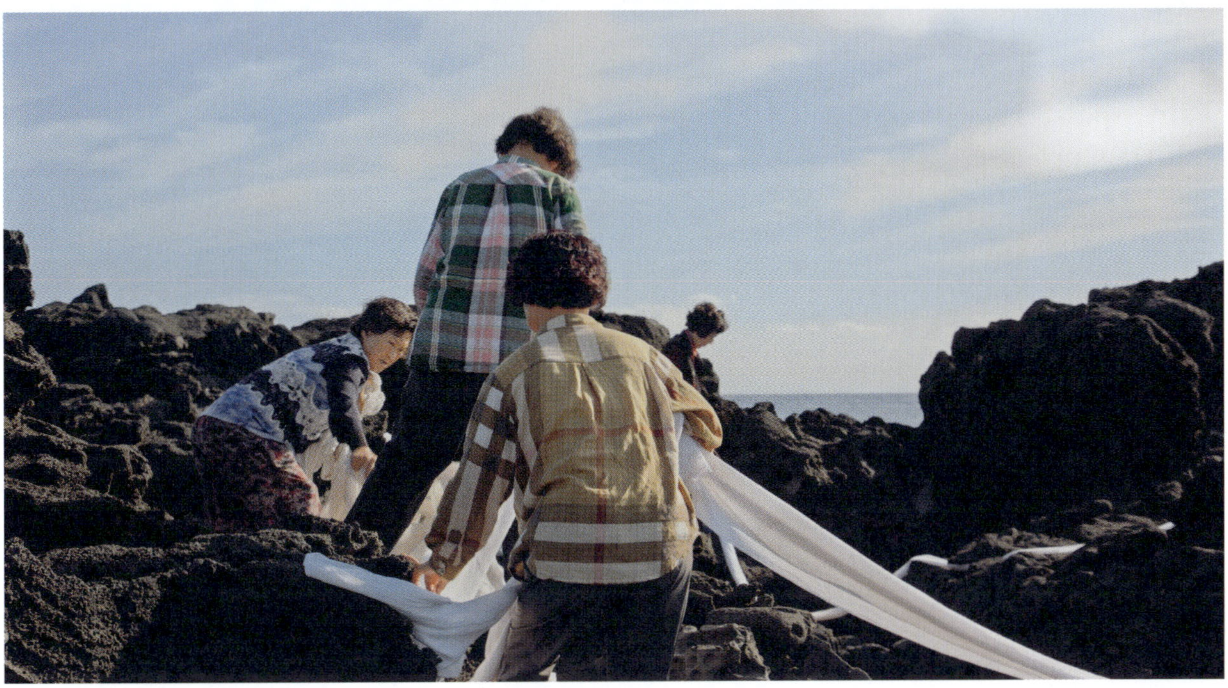

Halmang, 2023, 4K Video, 12 min

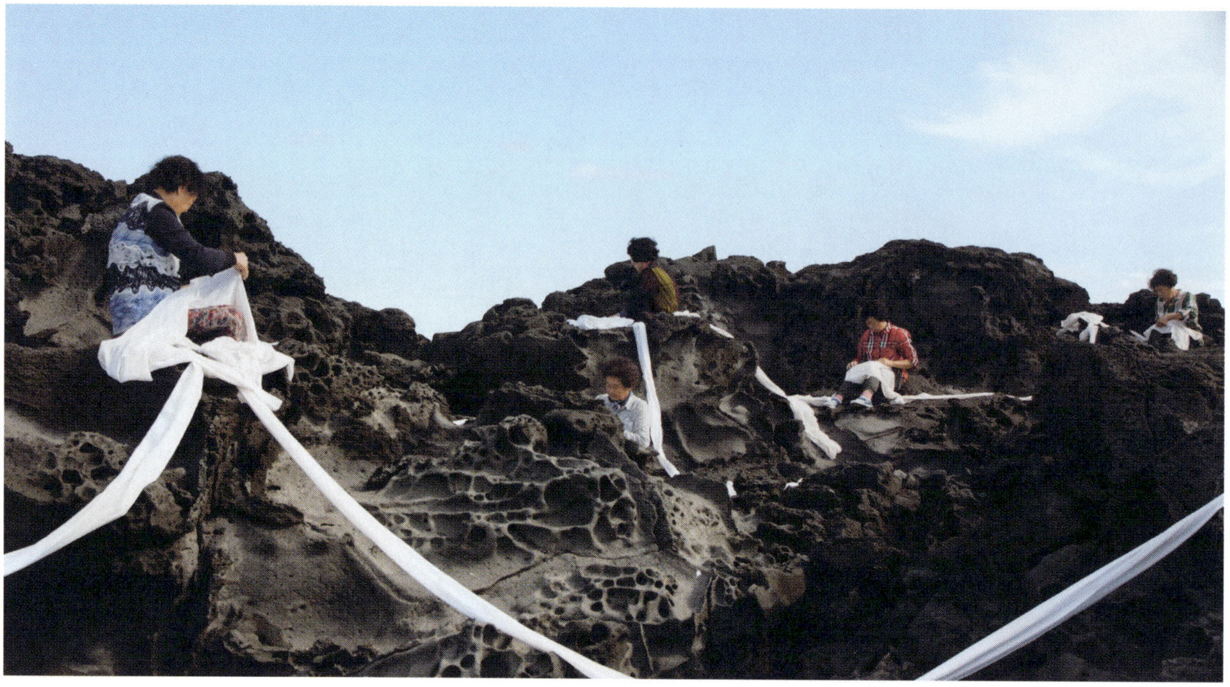

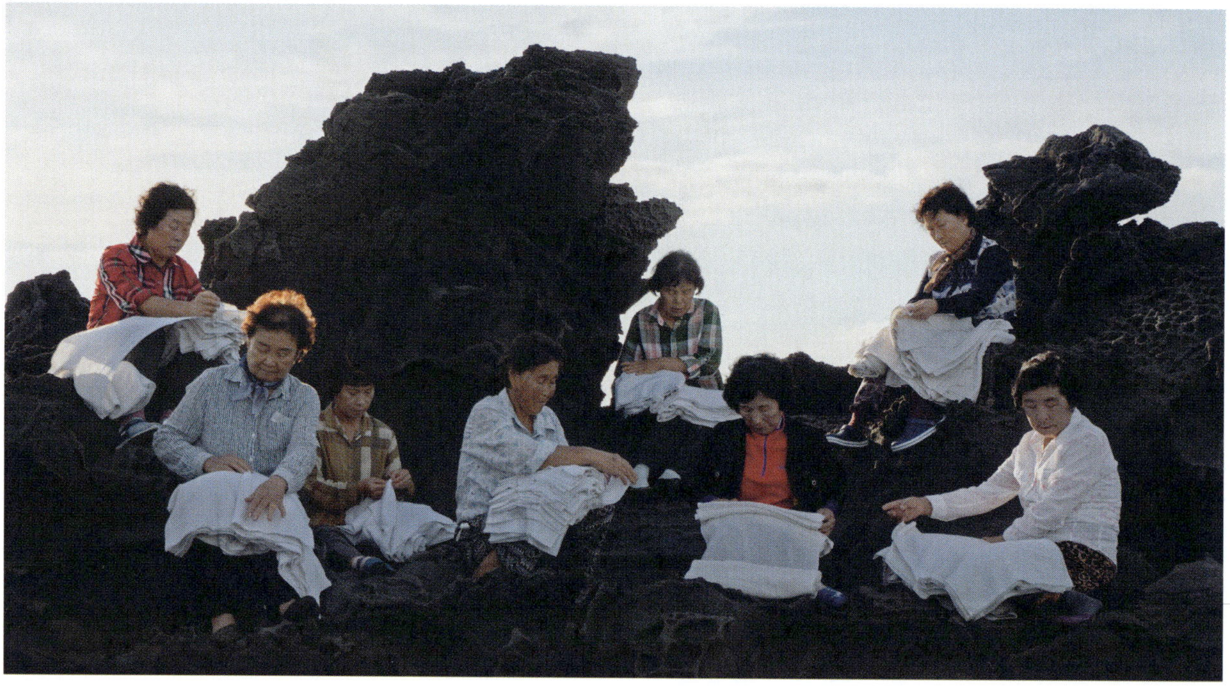

Halmang, 2023, 4K Video, 12 min

while he was imprisoned, the song was sung by *haenyeo*, although it was banned.

Free-diving as a sea-based vocation is more than just an occupation; it is a way of life that demands an unwavering commitment to the sea and the women-centric community with whom they dive. Many divers have perished underwater, yet others continue to plunge into the depths even at an advanced age. Due to the harsh nature of their work and the modernization of society, this mode of living is on the verge of disappearing, along with the ancestral, indigenous knowledge linked to the sea and its shores. The artist poses the question: how can the essence of sea diving be preserved? The island has become a prime tourist destination for Koreans and resorts continue to proliferate along the coastline, threatening the indigenous traditional way of life and transforming the landscape. The folklorization of the sea divers is on the rise, but little is being done to limit the island's development and protect its coastal ecosystems.

Over the past twenty years, Kaisen has kept returning to Jeju Island and has been closely observing the changes that are altering the coastline and transforming the local culture in the hometown of her grandparents. Most of the divers are now in their seventies and eighties. In her film *Halmang* (2023), she documents a group of them as they reverentially fold a long, traditional white cotton cloth, known as *sochang*. This fabric is pivotal to their daily life, used for carrying their babies as well as performing shamanic rituals and rites of passage. For the artist, it represents the cyclical nature of life and death. The film takes place on the same lava rock coastline as *Of the Sea*, where the artist's mother and grandmother once dove. However, Kaisen opts not to depict them diving, as she wishes to avoid perpetuating stereotypical imagery of the free-divers. Instead, the viewer contemplates the women's silent communication and collaborative work. This solidarity, which is vital when they dive, is also central to their relationship with the sea, as they form a unique, interconnected entity, linking the sea, the land, and humankind. The film's deliberate slow pacing allows the artist to focus on the women's gestures and movements, a poetic approach that situates their ritual within a longer, cross-generational timeframe and invites meditation. The steady rhythm of the ebbing and flowing waves further anchors the work in a distinct spatio-temporal dimension. At a certain point, a traditional *haenyeo* song accompanies their measured movements. Progressively, the elderly women unfold the cloth and display it along the rocks, forming a long, white spiral that starkly contrasts with the darkness of the lava rocks. This resembles an offering that echoes the multitude of rituals the sea divers perform before descending into the waters.

The work was filmed near a lava rock islet that used to serve as a shrine for the wind goddess Yongdeung Halmang. Shamanic goddesses in Jeju are referred to as *halmang*, and this is also the Jeju term for "grandmother" and a respectful way of addressing women. Today, the sea women continue to engage in shamanic rituals to pray for protection and plentiful catches. After their dives, they make offerings to the sea as a gesture of respect and humility. In the artist's opinion, we still have much to learn from these sea women, who possess a profound knowledge of the sea and the local coastal ecosystem. Their daily diving activities also allow them to record environmental changes and observe the effects of sea warming on regional fish and mollusk populations.

Mario Macilau

Born in Mozambique in 1984

"Photography is a very powerful tool, and I don't really judge the still images for its aesthetic qualities only, but also see it as a tool that can be used for positive influences and to bring hidden identities to the world." ("On life, photography and faith," *Rituals, Spirits and Sea*, interview with José Eduardo Agualusa, Jackson Lawson edition, 2019)

Mario Macilau grew up in southern Mozambique in Maputo, a city by the sea. He started his photographic journey in 2003 and went professional when he traded his mother's cell phone for his first camera in 2007. His photographic practice involves long-term projects that often deal with the ecological and social issues of his country. For a while, he followed street children to bear witness to their living conditions. Recently, he has been focusing on the local seaside of his native area, its landscapes, andthe coastal communities that inhabit the region. His work captures the diverse ways in which people utilize the beach: for prayer, fishing, vacationing, or simply escaping the oppressive heat of the desert to enjoy the refreshing ocean breeze. Concurrently, he is documenting the rapid changes occurring along thecoastline, including new construction projects. These multifaceted perspectives form the core of his long-term project.

The series *Faith* focuses on animism and the rituals performed on the seashore. The local community regards the ocean as a sacred entity, performing rituals to connect with its spirit or deity. Macilau shares a profound connection with the sea as well, believing it to be created by a powerful deity who governs natural elements such as the wind and currents. However, his personal beliefs do not interfere with his investigation into the enduring features of these religious practices, which persist despite radical societal transformations, the effects of global capitalism, and environmental deterioration. His poetic images reveal the contradictions between a changing world and deep-rooted traditions, highlighting the gaps that may form between different socio-economic and cultural systems.

Simultaneously, Macilau's photographs of fishermen at work

capture the ontological link between human beings and the sea. The children and men depicted appear almost symbiotic with the marine environment, empowered by their ancestral knowledge of this ecosystem. Macilau spends extensive hours on the shore, contemplating these scenes and collecting stories from fishermen, with whom he has built strong relationships. "I am not only photographing with my camera but also with my soul." The artist feels an intrinsic part of this community and, through his art, he aims to amplify their voices and provide insights into their perspectives and experiences.

By addressing the impact of climate change, urban development, and a modernity that tends to marginalize significant portions of the population, Macilau's work seeks to shift public perception and question how we might learn from existing traditions that emphasize respect for ecosystems. He wishes to initiate discussions about ecological issues, highlighting the lack of legislation in Mozambique to protect the environment. Currently, there are no designated marine protected areas, and there is inadequate legislation of burgeoning construction projects, land privatization, and ocean pollution. His work underscores the urgent need to respond to climate change while fostering a deeper appreciation for the ocean's cultural and spiritual significance.

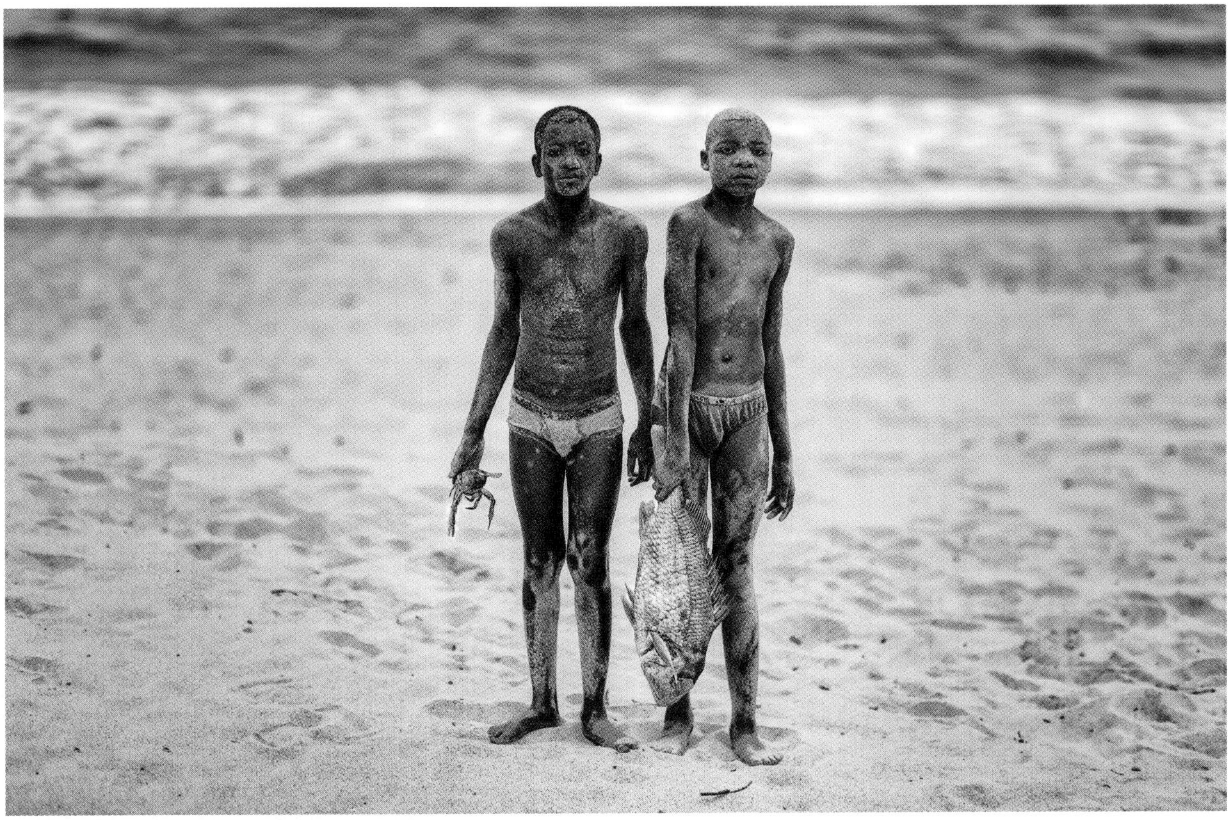

For Sale, Faith Series, 2018, Archival pigment print on cotton rag paper, 60 × 90 cm

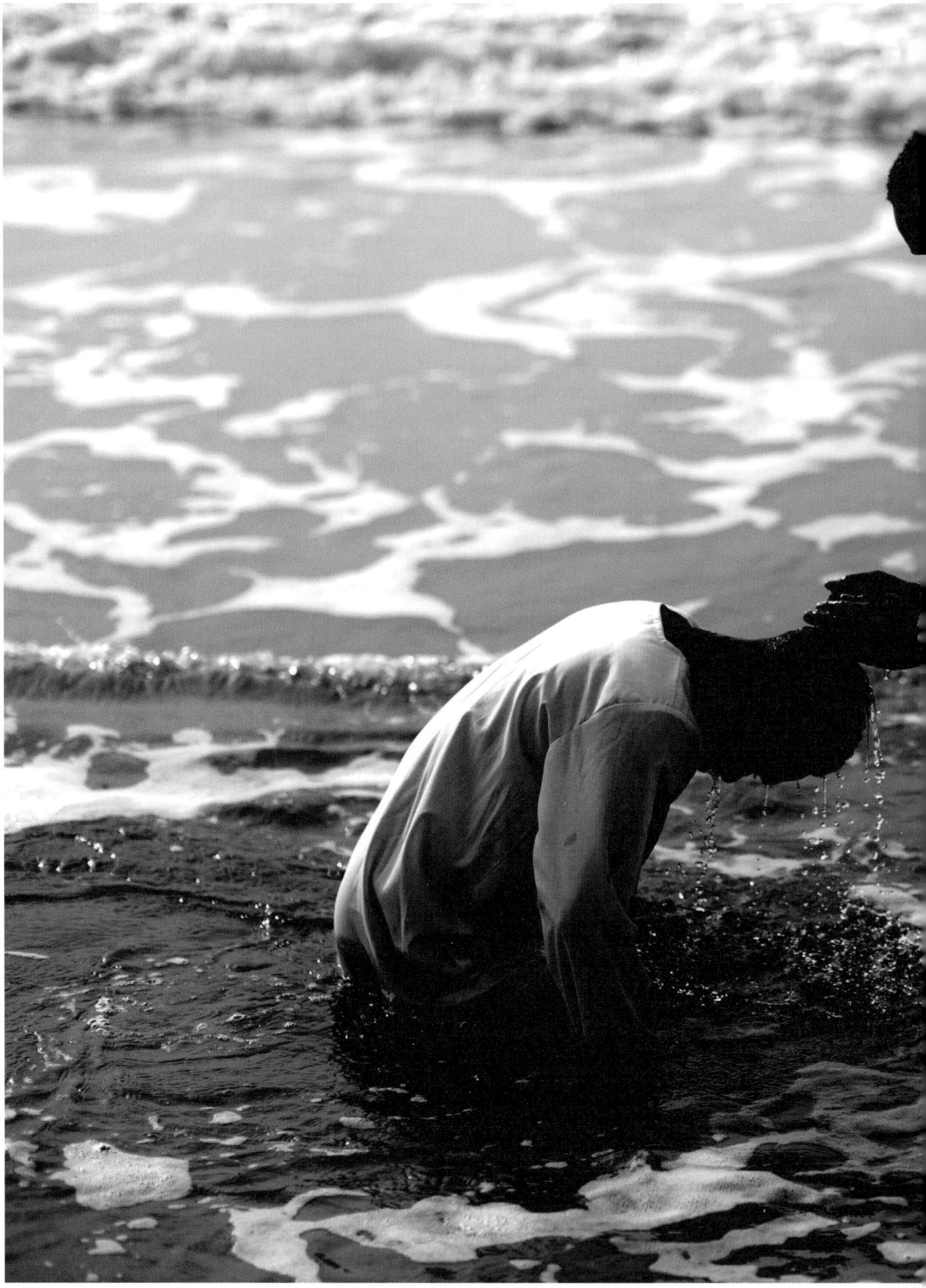

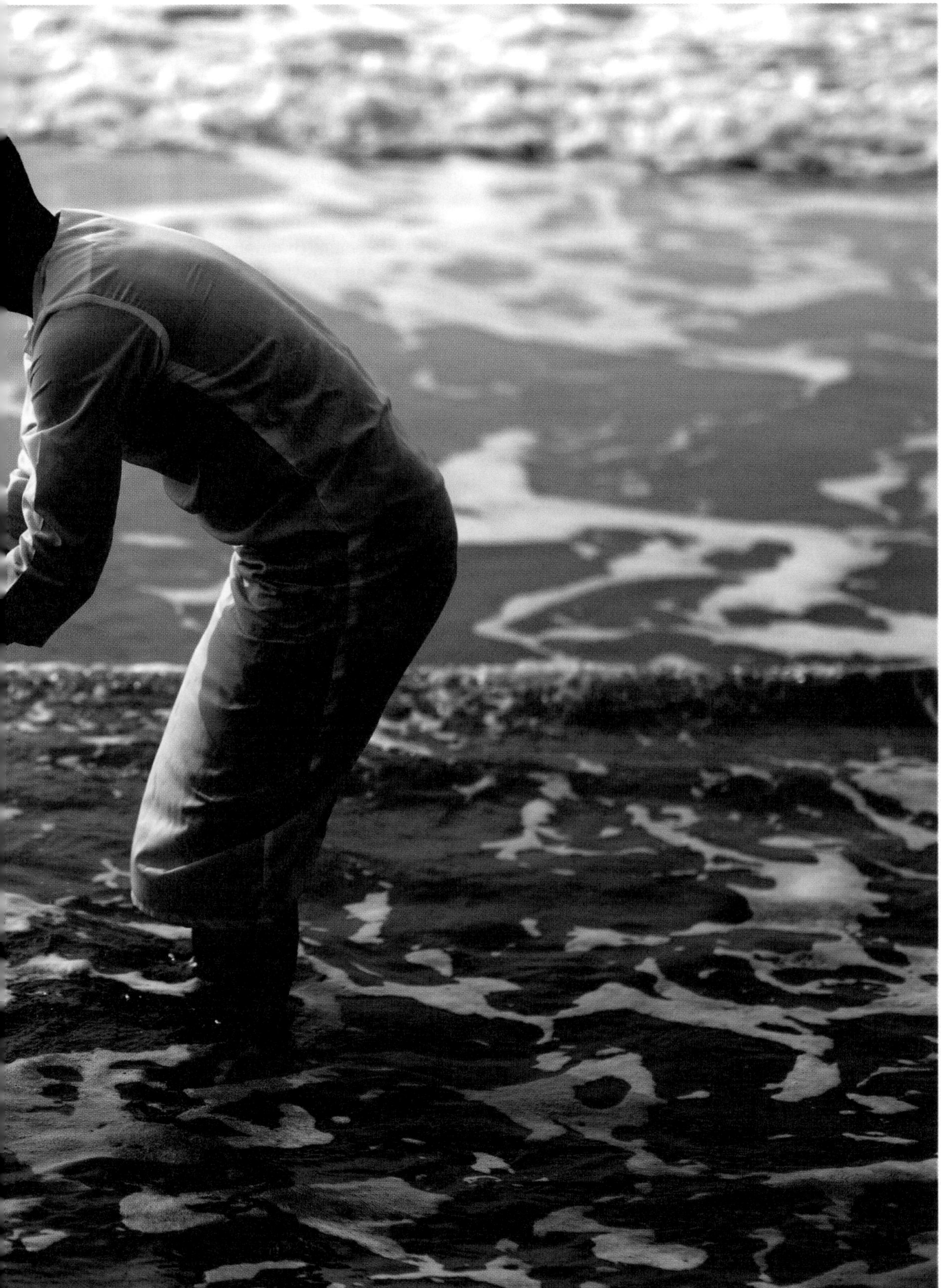

Baptism, Faith Series, 2018, Archival pigment print on cotton rag paper, 60 × 90 cm

Manel Ndoye

Born in Senegal in 1986

Manel Ndoye comes from the Lebu ethnic community, a vast community spread across all of West Africa. Most of the Lebu are fishermen, but Ndoye grew up among farmers. For him, water primarily meant the rain rituals and the associated dances when the villagers sought to ward off drought. When he arrived in Dakar, he discovered the *Ndawrabine*, another traditional dance, one that is closely linked to the ocean. This dance, performed by women, is entirely inspired by the movements of the fishermen: the motion of the paddle, the casting and then pulling of the net, the movement of the boat and bodies, back and forth. For him, it was an

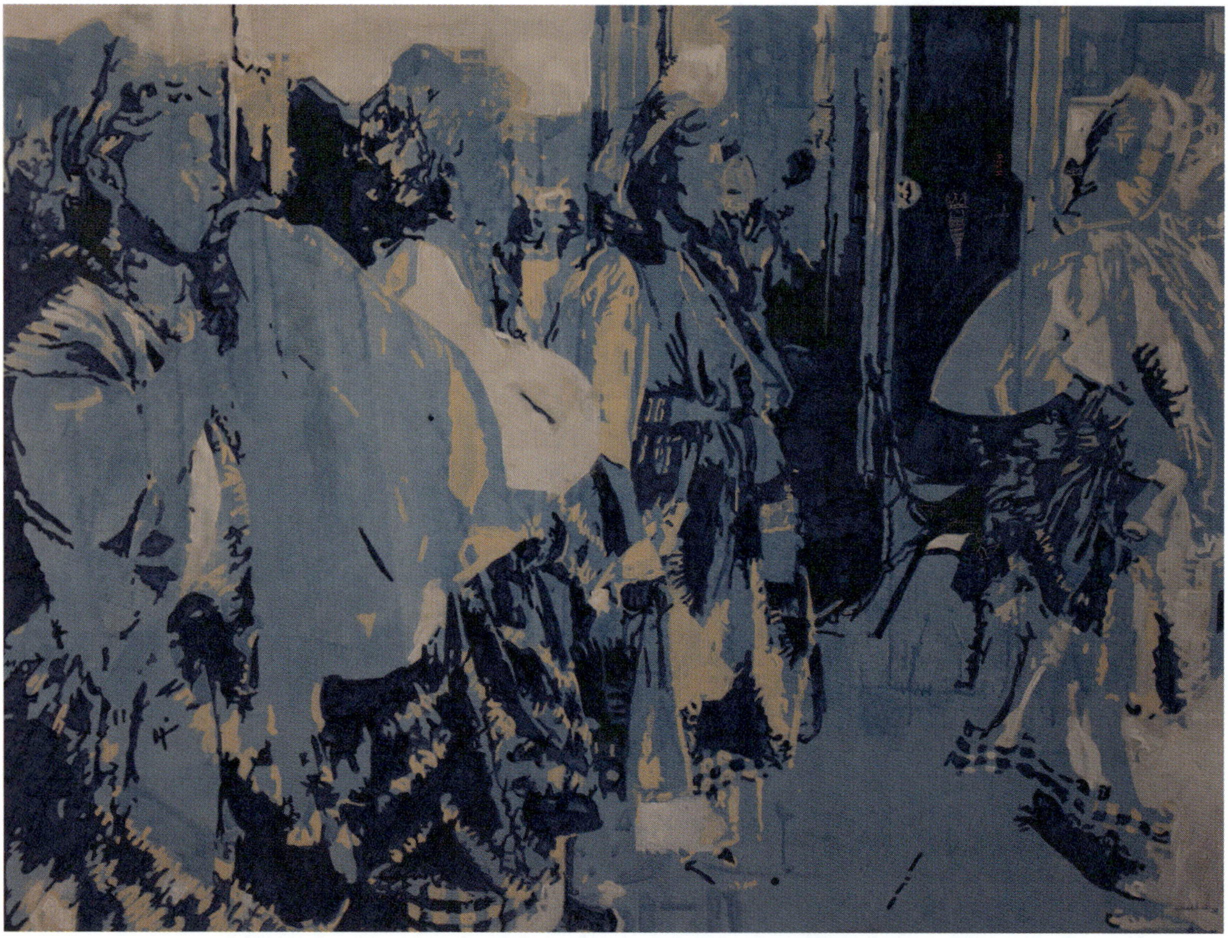

La mer, une inspiration, 2024 , Acrylic on canvas 104 × 134 cm

incredible discovery, and he immediately sought to learn more about this tradition. However, there are few written documents, so he turned to the local populations to study this ceremony. "It may be one of the few dances directly inspired by fishing practices," Ndoye says. On the beach, the women dance to encourage their husbands and sons before they left to go fishing. Ndoye was immediately fascinated by these gestures, by the slowness and precision of the movements, and by the beauty of the women who performed them. "Each gesture embodies not only a way of being in the world but also a way of connecting with the ocean. One can read in it an entire symbolism of the lifestyle of these fishermen, as well as their spirituality."

Ndoye practices painting and tapestry, with a body of work that oscillates between figurative and abstract art. He plays with the shimmer of colors and the vibration of lines to best capture the movements and rhythms of the daily life of coastal communities. He also pays tribute to an entire living ecosystem. Above all, for him, it is essential to raise people's awareness of the beauty of the marine ecosystem and the urgency of respecting this fragile environment to which we are intimately connected.

Today, like in many other coastal communities, life is increasingly difficult. Yet, according to the artist, if everyone respected nature and took only according to their needs, there would be enough fish for everyone. "Everyone, with just one canoe," he says. The artist plays with the symbolism of animals to address these contextual issues: the cat, for example, is a familiar animal from his childhood. Isolated from other animals, it would come to steal the fish before it had been shared. It represents the greed and individualism that the community must fight against. In his paintings, we also encounter the penguin, whose habitat is disappearing due to global warming. Alone, detached from the continent, floating on drifting icebergs, it embodies the current anxieties of the villagers who see the sea level rising and fear seeing their homes disappear. Fish, finally, are omnipresent in Ndoye's work. Together, tightly packed, in schools, their dance echoes the women's dances. "There is a common rhythm," he says. "We all resonate together." In some compositions, the artist focuses on their vibrations: with vertical lines, he expresses their descent into the abyss, the dynamics of the depths. "It is a celebratory dance," he adds.

His works reflect the beauty of these aquatic worlds and are an ode to the sharing of resources and the wonders of nature.

Temandrota

Born in Madagascar in 1975

"I am inspired by the breath of the ocean, by its expansive reach, both physical and mystical." For Temandrota, born Randriahasandratra Razafimandimby, the sea symbolizes nomadism, uncertain futures, and the fragility of life. The ocean is also inevitably linked to the narrative of his native island; humans first approached Madagascar from the sea, and the stories of these early encounters continue to permeate local culture.

Pirogues, the traditional canoes of the region, lie at the heart of Temandrota's recent work. As a child, he observed the transformation of these boats, noting how fishermen adorned their prows with paint and carved sayings that conveyed their personal stories. Before launching the boats, the community would engage in rituals. In this context, the wood becomes a vessel, and the fisherman merges with the sea, knowledgeable of its laws.

The pirogues evoke the traditional lifestyles of these local fishermen, particularly the semi-nomadic coastal communities of southern Madagascar known as the Vezo. Their name translates to "the people who fish," yet they are also described as those who "struggle with the sea." This historical nomadic lifestyle has always been tied to the cycles of drought, but today, it is increasingly challenged. Climate change, the rapid development of port infrastructure, overfishing, and industrial fishing compel the Vezo and other clans to migrate further inland.

For Temandrota, the pirogue serves as a receptacle for their voices. He works with salvaged, abandoned boats, recycling them as a means of reintegration and preservation within the realm of imagination. His sculptures draw inspiration from the symbols of various clans, such as birds and zebus, embodying their ways of life. The forms, colors, and textures provide a visual translation of local songs, evoking images of bivouacs, campfires, dances, and beliefs.

"The canoe is integral to our traditions; it is central to our songs and communal life," he states. He perceives it as a traveling sculpture, one that communicates with the sea and the elements. However, it also represents a form of fragility: small-scale fishermen are increasingly unable to sustain themselves. They depart at dawn, uncertain of their return and whether they will bring back any fish.

His work resonates with the current struggles affecting the island, particularly regarding the extraction of natural resources and the dominance of certain multinational corporations over the territory. In southeastern Madagascar, incidents involving contaminated water are multiplying. "The entire ecosystem is being disrupted," he says. "Industrialization

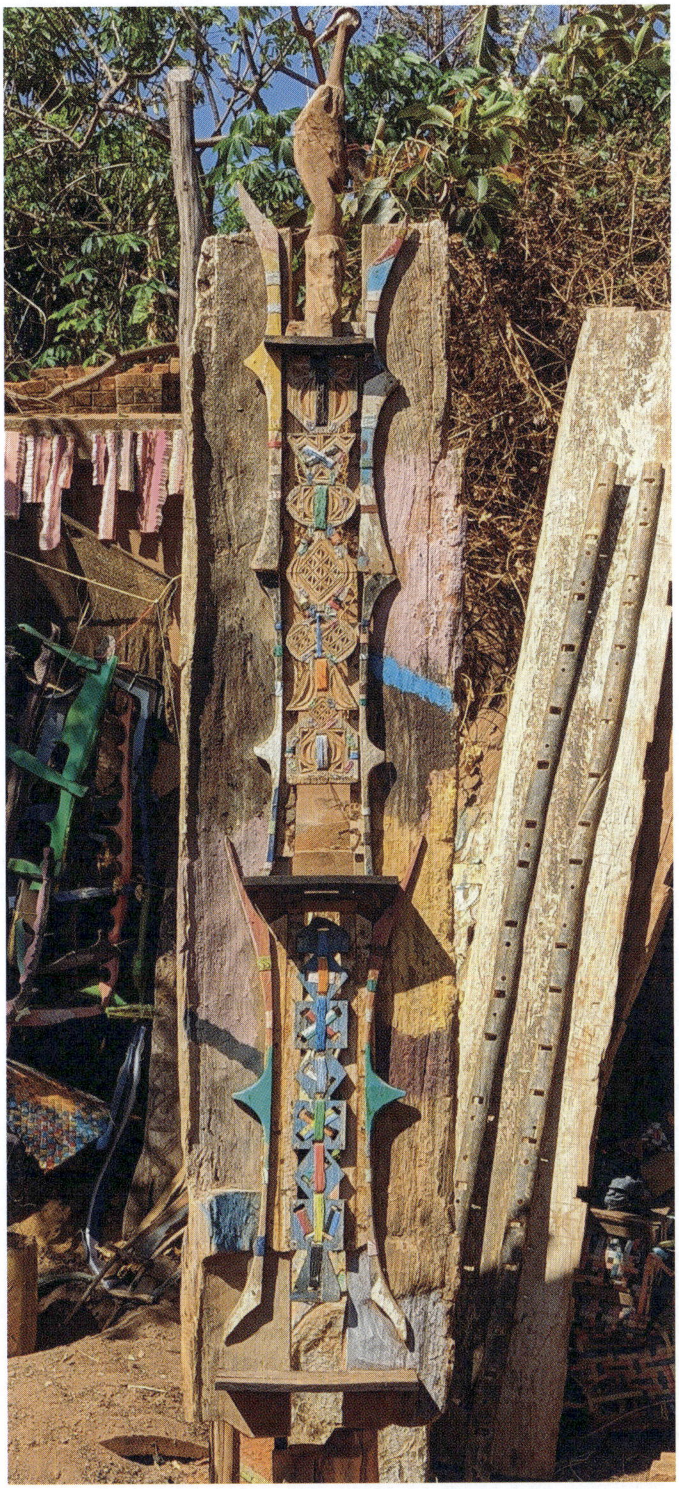

Totem Agnombe Vosy, Clan Ombalahimasy,
undated, Sculpture, 421 × 83 × 13 cm

and the expansion of extraction activities are leading to population displacement and the destruction of biodiversity."

The southwest coast of Madagascar is renowned for its exceptional coral reef ecosystem and rich environment. Temandrota hails from this region and, as a child, endured the famine brought on by drought. This experience has heightened his awareness of ecological issues and the imperative to listen to and adapt to the environment. He identifies as an activist, although his artistic work addresses these concerns in an indirect manner.

"I would like us to seek a common pathway–together, human beings and nature, human beings and the sea, all huddled together at the hollow of this pirogue."

Sylvain Agostini[1]

Researcher and marine biologist at the Shimoda Marine Research Center (Japan)

Shimoda Bay, located at 34° N along Japan's Pacific coast, lies at the crossroads of temperate and subtropical zones, where a warm ocean current flows northward to Tokyo. This temperate zone at the edge of the subtropics hosts a unique mix of temperate and tropical species, including some corals. However, climate change and warming currents are driving rapid ecosystem changes, a phenomenon known as "tropicalization."

Rising sea temperatures have brought an influx of tropical fish, including herbivorous species that graze on kelp, once abundant in the bay. These kelp forests, vital for sheltering marine life and supporting local fisheries, are vanishing within a few years. In their place, new coral species are appearing, but the transformation is not without consequence. The warming waters are also becoming more acidic as CO_2 levels rise, altering the chemical makeup of the ocean and lowering its pH. This acidification poses additional challenges for marine life, particularly for corals, which struggle to thrive in such conditions.

While some researchers view tropicalization as a potential silver lining–arguing that temperate zones might serve as refuges for corals damaged or bleached in tropical regions–this perspective often ignores the broader ecological costs. The disappearance of temperate ecosystems not only threatens biodiversity but also disrupts local livelihoods. For example, the loss of kelp forests has devastated populations of abalone, a key species for *ama* divers and local fisheries.

To study these changes, Sylvain Agostini collaborated with Nicolas Floc'h during a residency aboard the *Tara*[2]. Their work focused on Shikine Island, where volcanic activity naturally releases CO_2, creating a preview of what acidified oceans might look like in the future. What they observed is that high acidity prevents corals from thriving, undermining the idea that temperate zones can serve as sanctuaries for coral colonization.

Agostini found his collaboration with Floc'h to be both enriching and inspiring. Floc'h's artistic lens provided a broader perspective, complementing Agostini's scientific focus. Together, they combined aesthetic imagery with hard data, creating a richer narrative that communicated both the beauty and urgency of marine conservation. Floc'h's photographs from 2017–2018, for instance, now serve as an archive of the bay's ecosystem before the disappearance of the kelp forests. These images are not just scientific references but also emotional touchpoints that Agostini uses to engage students and local communities, highlighting the rapid pace of environmental change.

In response to these challenges, the focus is shifting from mitigation to adaptation. According to Agostini, artificial reefs have become a promising strategy, offering habitats for algae and marine species, creating fishing zones, and alleviating pressure on natural reefs. Local communities are finding ways to adapt as well. Some fishermen, for instance, are transitioning to work as diving instructors, while efforts are under way to manage populations of herbivorous fish that threaten algae farming. Once-thriving industries, such as the cultivation of wakame (a type of algae), have largely disappeared due to overgrazing by tropical fish, but replanting projects using artificial reefs offer a glimmer of hope for restoring this lost resource.

Collaboration with local NGOs and dive instructors also helps raise awareness among tourists, fostering broader engagement with environmental issues. Agostini emphasizes that the ocean is central to our society—not just economically, but socially and culturally. To ensure a sustainable future, we must engage with the ocean responsibly, finding ways to balance productivity with preservation. While the ocean aligns with Goal 14 of the United Nations Sustainable Development Goals, he stresses that eradicating poverty and ensuring education—Goals 1 and 4—are prerequisites for addressing environmental challenges effectively.

---

1. Based on an interview with the scientist, May 2024.

2. Since 2003, the Tara Ocean Foundation, the first public interest foundation in France dedicated to the Ocean, has been conducting scientific expeditions to study marine biodiversity as well as observe and anticipate the impacts of climate change and pollution.

Displacing our Gaze

We are the sea, we are the ocean

Epeli Hau' ofa

1. Ursula Biemann
Born in Switzerland in 1955

2. Nadia Huggins
Born in Trinidad and Tobago in 1984

3. Kristina Pulejkova
Born in Macedonia in 1988

4. Robertina Šebjanič
Born in Slovenia in 1974

5. Shakil Solanki
Born in South Africa in 1997

6. Lara Tabet
Born in Lebanon in 1983

7. Alberta Whittle
Born in Barbados in 1980

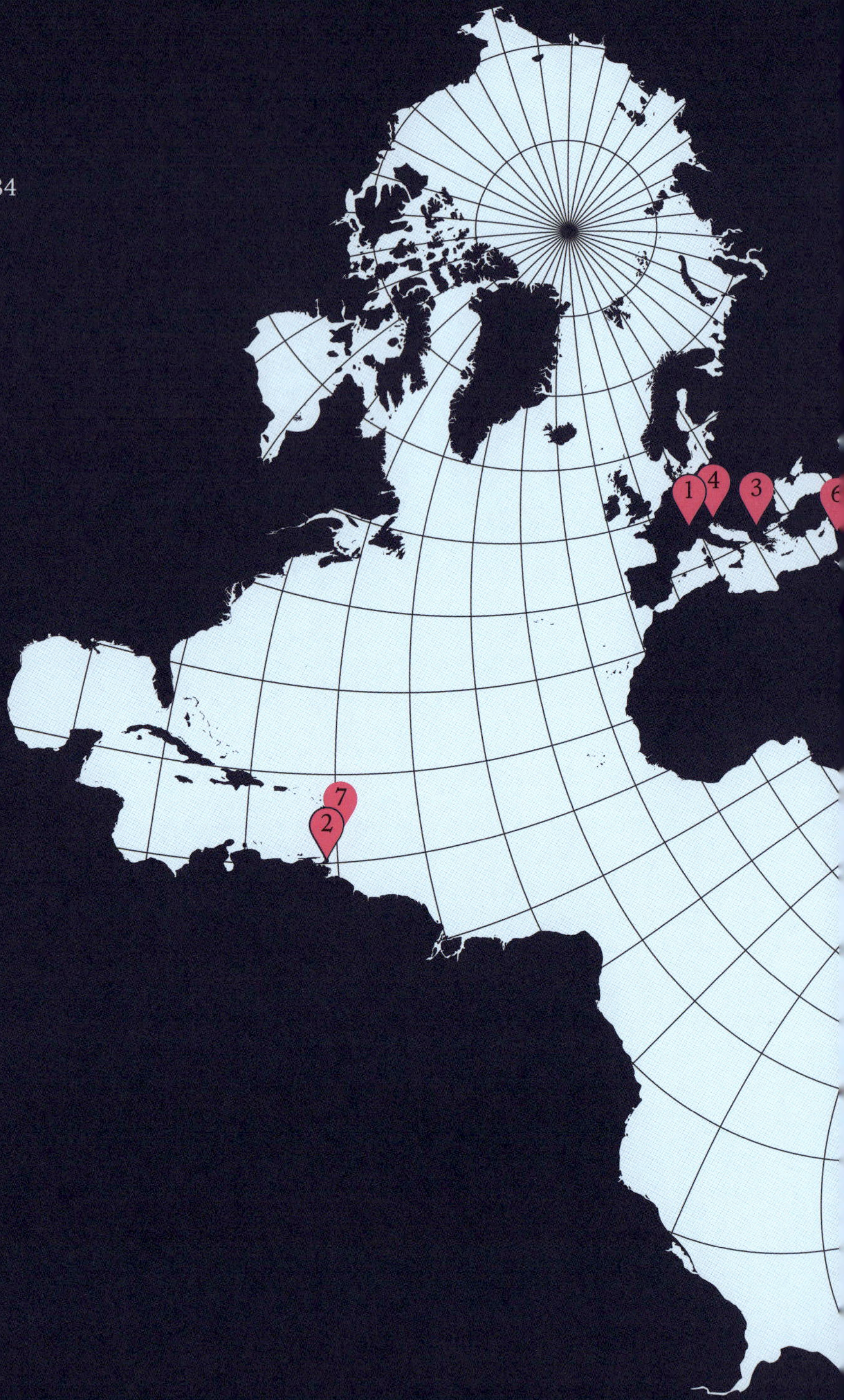

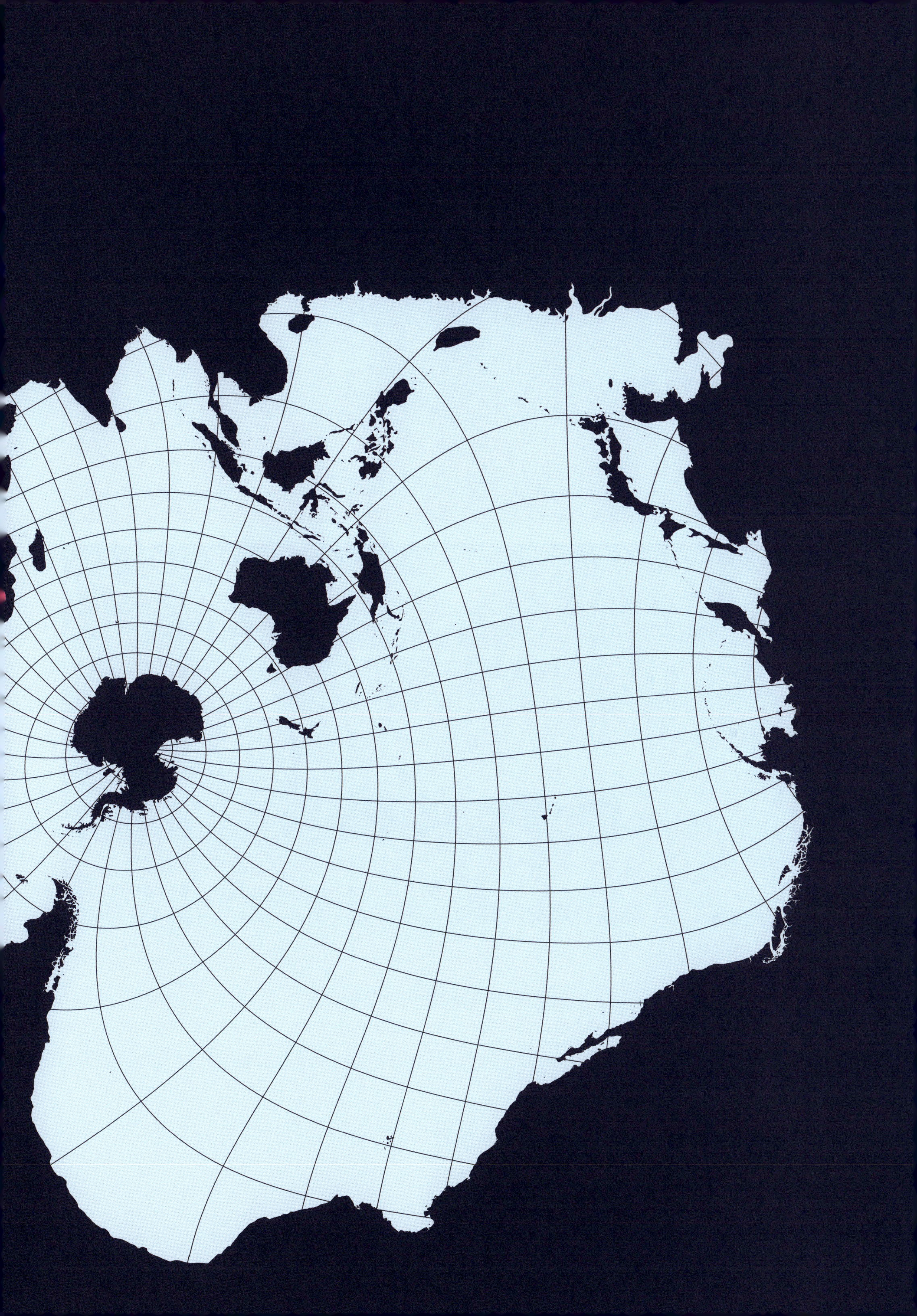

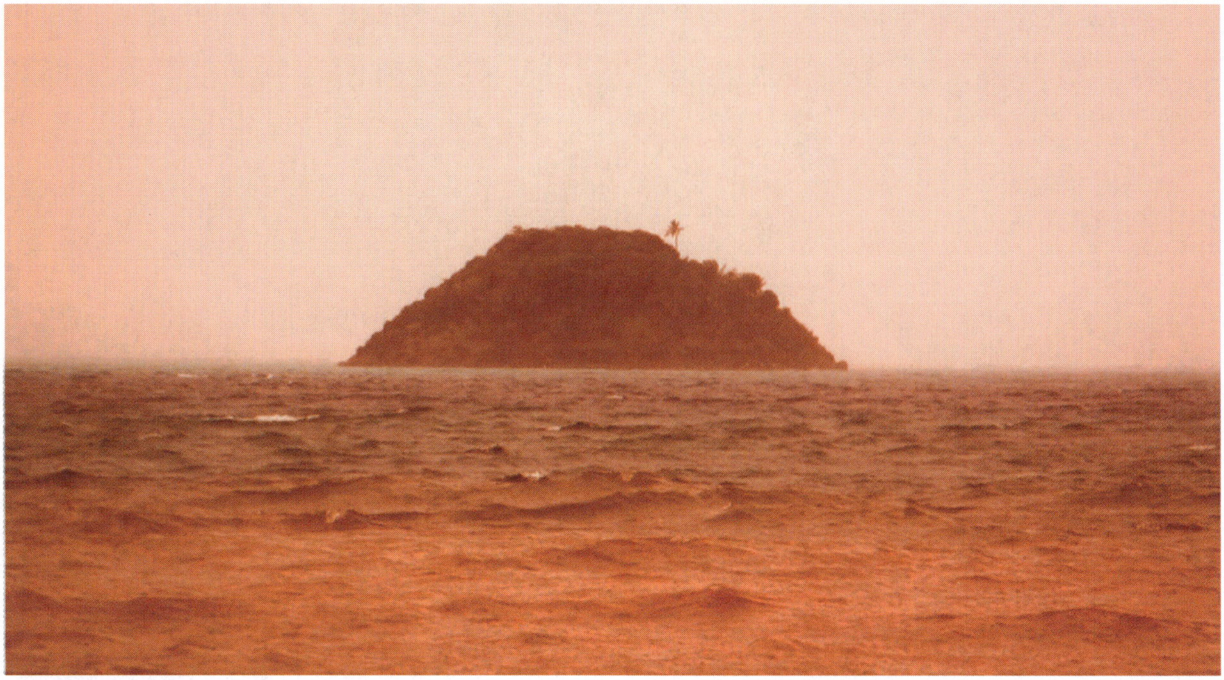

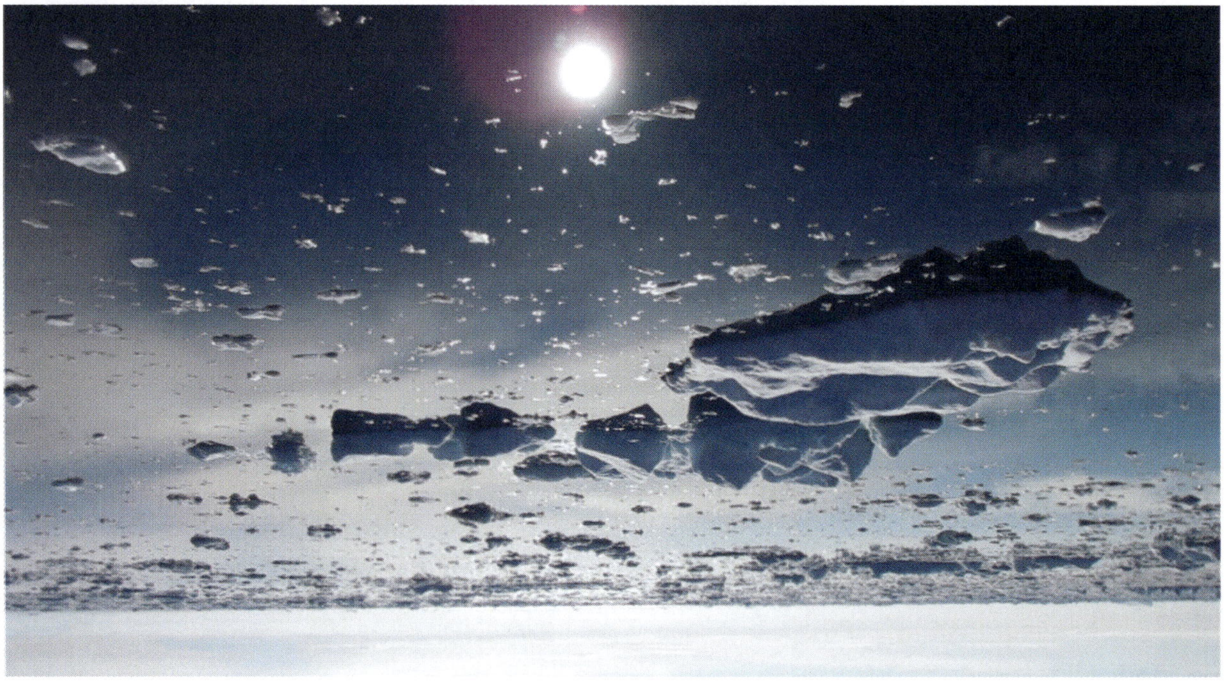

Subatlantic, 2015, Video essay, 11 min

Ursula Biemann

Born in Switzer-land in 1955

"It is fairly easy to access vast amounts of scientific data about the climate and environment, but rational thinking alone cannot make us understand the magnitude of the change we humans are in for." ("Poetics and Politics of Worlding," becomingearth.unal.edu.co)

For over a decade, Ursula Biemann explored the topics of migration and transnational borders, creating quasi-ethnographic videos and artworks that address human mobility and the politics of containment. After this long-term project, she directed her attention to planetary ecologies, opening up to cosmopolitical and biocentric perspectives. With *Deep Weather* (2013), she tackles the issue of water for the first time. The film presents two interconnected narratives: oil extraction in northeastern Canada, known for its severe environmental impacts, and scenes of villagers in the Bangladesh delta struggling to contain rising sea levels. At that time, awareness of climate change as a planetary issue was just beginning, and Biemann sought to highlight the connection between these two locations and their contrasting situations. On one hand, there were devastating industrialized open pits; on the other, people manually building embankments with mud bags to prevent flooding.

Following a maritime expedition in the Shetlands, Biemann continued her exploration of watery connections, focusing on the ocean not merely as a surface and route but as a living, three-dimensional space. Her film *Subatlantic* (2015) is a spatio-temporal dive into the sea, telling the story of a scientist measuring landscape changes over approximately 2,500 years. From the outset, viewers are immersed in a semi-fictional world, following the narrative of an immortal woman who studies glacier melting and rising seas–though she remains unseen. The term "Subatlantic" refers to the last phase of the Holocene epoch, inviting viewers to reconsider oceanic landscapes as carriers of our history. Like *Deep Weather*, the film underscores the deep solidarity linking distant territories, past and present. The weather, streams, currents, winds, and migratory patterns of fish and birds are what connect the Shetland Islands to Greenland's Disco Bay and the Caribbean Islands. "Underwater passages accumulate impersonal intelligence," narrates the voice-over, and the poetic flux of images evokes these hidden interconnections that link and impact all living things.

At one point in the film, the camera is inverted, making icebergs resemble clouds in the sky. The title itself appears twice at the beginning, as if reflected on a mirror. This inversion of perspective encourages viewers to de-center and rethink their relationship with the ocean and with the natural world at large. What might the view from the bottom of the sea be? How could we address the perspectives of non-human species? The artist aims to foreground what has long been neglected and relegated

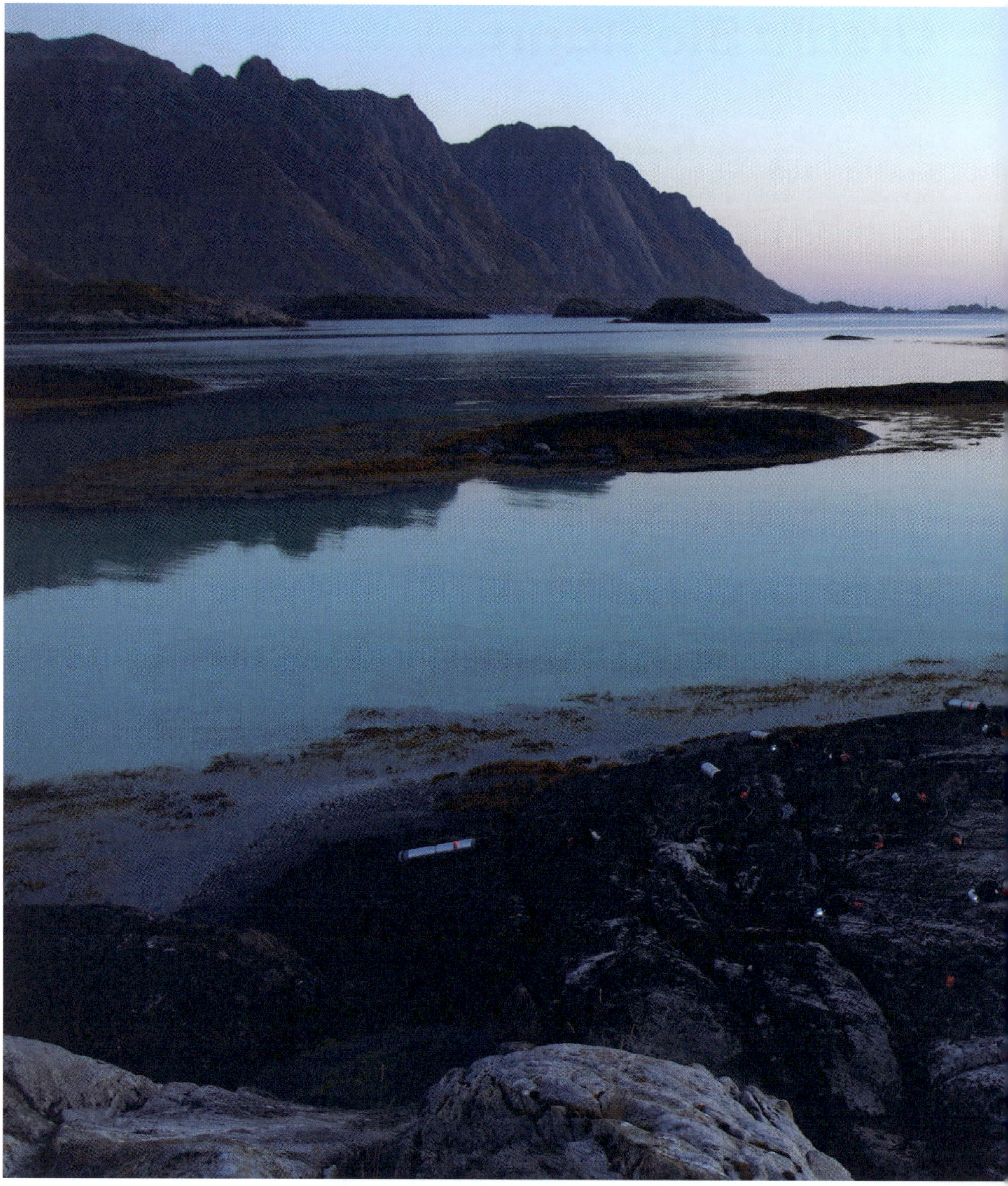

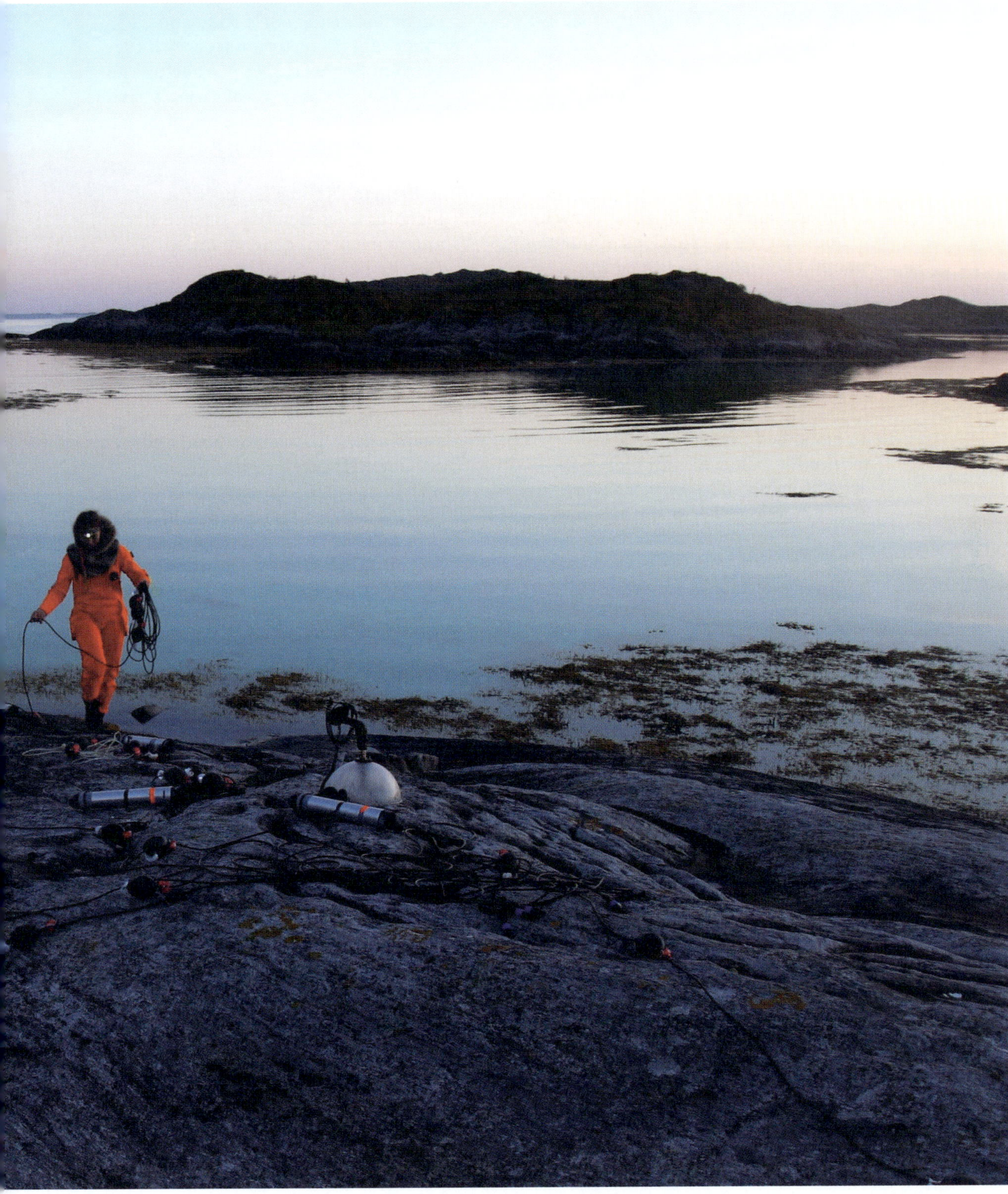

Acoustic Ocean, 2018, Video installation, 18 min

to the background in our approach to the world.

Ursula Biemann's work is deeply influenced by her extensive experience with various indigenous communities across the globe. *Subatlantic* reflects a sensory and empirical connection to the marine world, inviting viewers to question the foundations of our shared knowledge, our category of thoughts and the tools we use to understand our environment. This theme is further explored in *Acoustic Ocean* (2018), which can be seen as an extension of *Subatlantic*. In this film, Biemann introduces an embodied fictional character for the first time, a Sami biologist-diver from northern Scandinavia. The film follows her scientific work as she records underwater sounds using a hydrophone, a device originally developed for military purposes during the Cold War to detect submarines. For this scientist, however, "her hydrophones function as external organs, enabling her to deeply immerse herself in the aquatic habitat." Here, the technology becomes an extension of the researcher's body, allowing her to listen to sounds imperceptible to the human ear.

Biemann raises fundamental questions about our ability to approach the natural world with existing paradigms and tools. While she does not dismiss Western science, she highlights its limitations in rethinking our relationships with non-human species and the underwater world. As the voice-over states, "Geometry's power has lost its grip on the mind." In contrast, indigenous modes of knowledge encourage using one's body and inner sensations to connect with the surrounding world. Experiencing the world subjectively, rather than striving for an objective distance might be a key to better understanding it.

Biemann's films are poetic, blending documentary, science fiction, and moments of contemplation. Despite being rooted in long-term research, they do not require intellectual understanding; instead, viewers are invited to let go and follow the flow of images, sensations, and thoughts, feeling the fluidity and connections within the oceanic symbiosis. A particularly moving sequence features a Sami woman walking towards the camera and sharing her personal story, which is in fact the story of the performer. She discusses climate change, how her grandmother observed its effects, and its impact on her community, particularly the endangerment of reindeer due to their inability to access food. "The reindeer is our guardian, as we are his guardians," she says. There is no judgment in the expression of this fact, and there is never any moral dimension or dramatization in Biemann's practice.

"My work remains very suggestive. It triggers unsettling emotions, but never in a scary way. It is not didactic either. Today, information is already available everywhere. My job is to create worlds and not to represent reality. My artworks need to be conceived as open proposals."

Biemann's recent work is less directly critical; she expresses fatigue from pointing out the flaws in our global system and its destructive effects. She feels disenchanted by this critical tradition and doubts that art can change people's thinking. However, she believes that art can influence our collective perception of the world. It is a powerful medium to convey complex issues and nurture our collective imagination. Ultimately, since everything we perceive is filtered through these collective interpretations, art can alter our vision of reality. Through her fictional characters and situations, Biemann encourages viewers to imagine their place in the world: where we are, how we position ourselves, and how we can reimagine the future and expand our vision of the world.

Nadia Huggins

Born in Trinidad and Tobago in 1984

Nadia Huggins grew up in St. Vincent and the Grenadines, surrounded by seashores. As a Caribbean person, she feels that this this connection to the sea is culturally rooted in her. Since 2015, she has been documenting the seascape of Indian Bay Beach, a popular beach where she swims daily. Gradually, through her photographic practice, she has been collecting an important archive from this specific underwater landscape, shooting the local corals and various marine species. Initially, it was a way to describe her "home" and to witness the changes she could observe every day. Gradually, her practice became more conceptual. "There are street photographers who capture what they see in urban areas. My practice is not so different, except that I shoot marine landscapes and the creatures living underwater."

According to the artist, among the population of Caribbean islands, many people do not know how to swim. And when they do, they do not venture too far and stay where the water is shallow, where they can still see through the clear water. In Saint Vincent, although, most people still rely on coastal and marine resources to make a living, they do not perceive the sea as a safe place. Researchers have tried to explain this cultural pattern, which may have many causes, including a history of trauma, the absence of compulsory swimming lessons, and a collective fear that is transmitted unconsciously from generation to generation (cf. *Immersion*, Artaud Hélène, Paris: La Découverte, 2023).

Huggins was lucky enough to learn how to swim from an early age. For her, the sea is a safe place where she can be herself and think anew about her identity. What remains from a human being plunged into the ocean? When diving, she needs to hold her breath, and experiences every limitation of her body. She also explores what

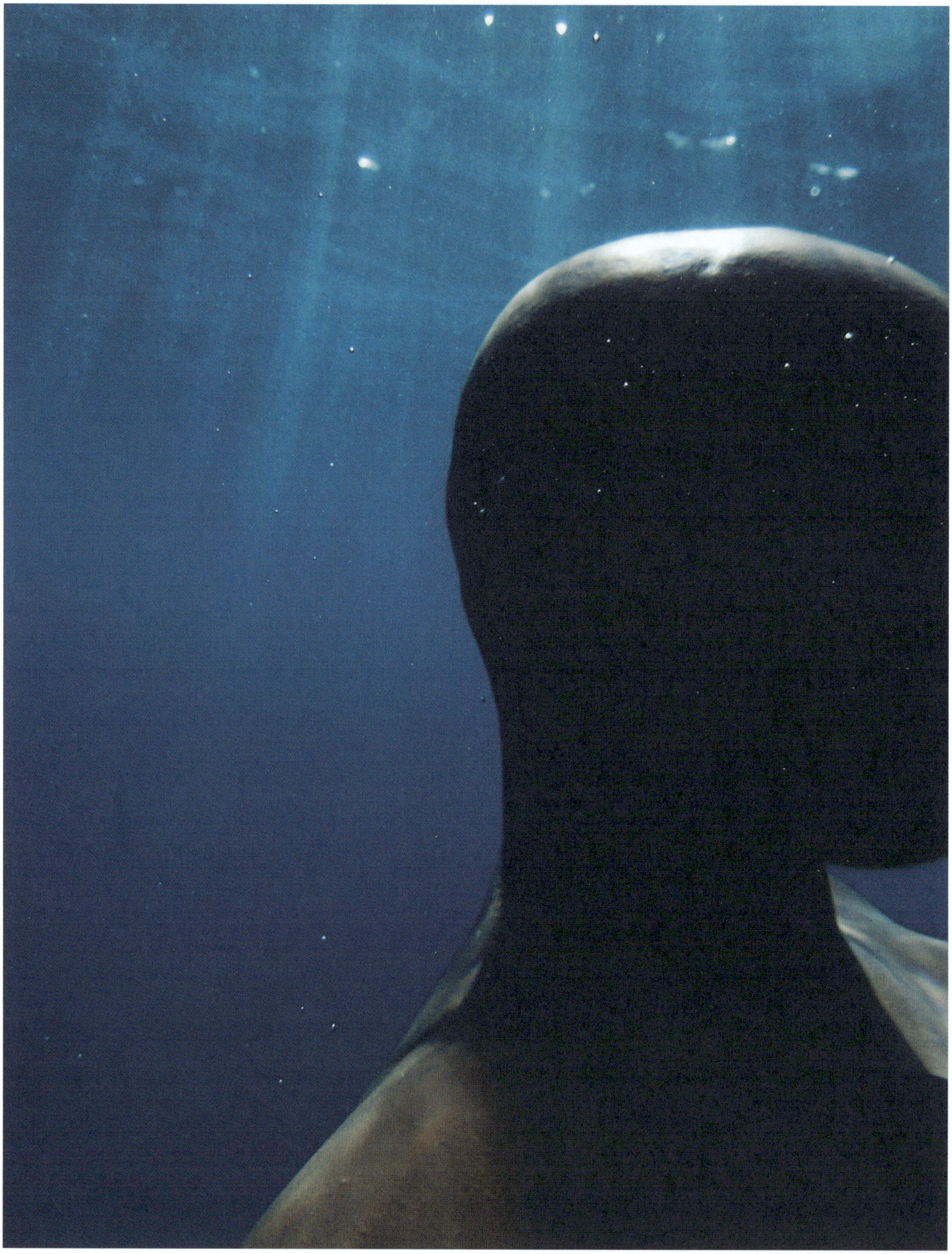

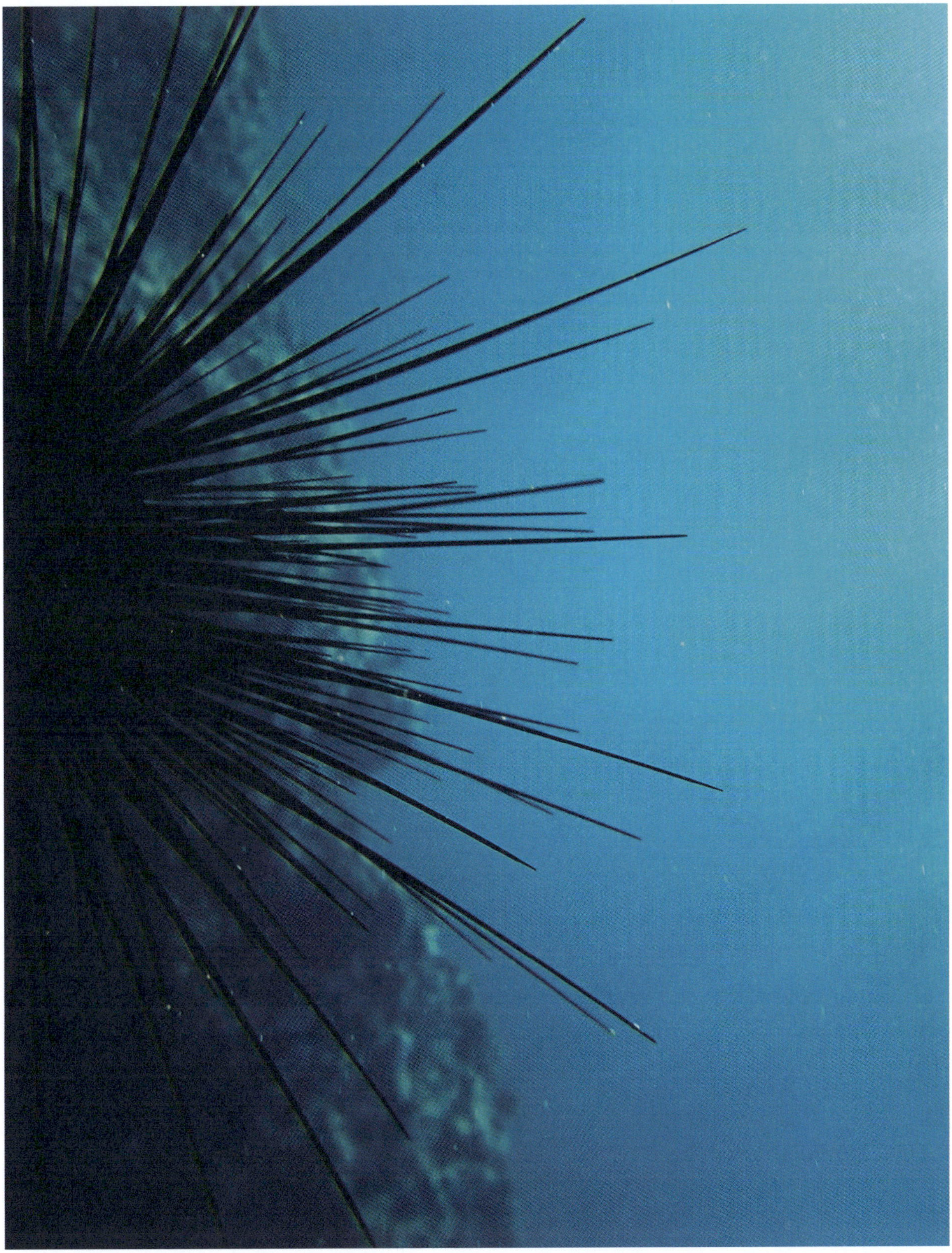

Transformations No. 1, Fighting the currents Series, 2015, Digital photograph, 76 × 118 cm

makes her human. "How can I exist without the social constructs that dictate identity on land?" she asks. While submerged within the water, what remain of someone might be merely his or her physiological features. The ocean dissolves any form of identity.

The beach from which she dives is also a historical site: it is the place where the first Indian indentured laborers are believed to have landed. These workers were transported from India to work in European colonies following the 1833 Slavery Abolition Act, replacing enslaved Africans on plantations. Although this significant part of history is not explicitly addressed by the artist, her exploration of identity is partially rooted in this context. The ocean serves as a space where she can reconceptualize identity, traditionally viewed as a rigid cultural marker.

Huggins approaches the sea from within, yet she does not fully embrace recent hydro-theories, where scholars and art practitioners emphasize the watery composition of the human body to imagine a complete merging with water. When she swims, she only briefly becomes one with nature. This transitory experience is why she considers photography the optimal medium: it allows her to capture and freeze these fleeting moments, making her image the point of connection.

Transformation (2014-2015) stems from her daily practice of photographing her body and the reef. The series forms unique diptychs, featuring her face, hand, or shoulder extended by a marine animal. The artist draws upon the grammar of carnival performances, where identities can flip. The series suggests a fluid transition from one form to another, an idea reinforced by the title *Transformation*. For Huggins, the notion of belonging within any space—natural or man-made—is perpetually shifting. The two parts of the diptychs never fully merge, though, separated by the frame. It feels more like a direction, a desire, each creature aiming at meeting the other. Above all, the series dismantles any hierarchy between the human and non-human worlds. What, then, is transformed? If not the body itself, at least our conception and perception of underwater life, which is often neglected, exploited, and destroyed by human activities.

The ocean, for the artist, is a place where all kinds of bodies can find a space for freedom. Her ongoing series *Disappearing People* (2018-) features traces of bodies in the water formed by a multitude of bubbles. From the image, we can imagine their movements and guess their trajectories, but it is impossible to define their identity. Something, or someone, was there, interacting with the water, yet the sea has erased their identity. Thus the ocean offers us a new space to think about ourselves.

In her seascape archival portraits, Huggins tries not to humanize her subjects, yet she acknowledges that this is still the best way to connect with the marine creatures she is photographing. Who will pay attention to black sea urchins? Usually, in paradisiacal Caribbean postcards, they do not feature in the landscape. When she takes their portraits, she wishes to create a feeling of similarity so that viewers change their perspective. Finding the right lens is always challenging.

One of the most pressing ecological issues she has been confronted with is coral bleaching. The island is very vulnerable to the effects of climate change, the absence of legislation, and increasing development projects. Many coastal areas have been sold by the government and privatized for hotel construction. As a result, local populations are displaced and fishing communities struggle to make a living.

Beautiful corals, pristine beaches, and white sand are part of the constructed image of the archipelago. Recently, a black sand volcanic beach was entirely replaced by white sand in one of the bays to better match tourist expectations. Once again, ecological issues coalesce with culture and politics. Eurocentric visions of history are replaced by a commodification of what is perceived as "exotic." In the context of the Caribbean islands, Huggins' quest for identity reflects a more general desire to recover agency and to decentralize our gaze.

Kristina Pulejkova

Born in Macedonia in 1988

Grounded in scientific research, Kristina Pulejkova's interdisciplinary practice explores how the use of technology might lead to greater forms of sustainability in relationships between humans and nature. Combining moving image, sound, and immersive installation, the Macedonian artist proposes fictional narratives that invite viewers to change their personal, human-centered perspective and embrace different ways of seeing and feeling. *Wedding Journey* (2019), an immersive installation consisting of 360 VR video, holographic projection on water screens and sound. It is an exploration of the mysterious life cycle and migration of the critically endangered European eel.

"Although I spent every summer of my childhood and adolescence at Lake Ohrid in North Macedonia, where European eels are part of the cuisine and folklore, I only discovered that they are trapped there by dams built for hydroelectric power stations in 2014. This discovery then led me on a path to grasp the life and mysterious journey of these incredible animals that connect the oceans with the rivers and lakes of Europe."

The European eel has a life cycle marked by long migrations. Adult eels spawn in the Sargasso Sea in the Caribbean, and their larvae migrate to European shores following the Gulf Stream current. Eels typically live for five to twenty years in freshwater and brackish water, such as rivers, coastal lagoons, and lakes, before returning to the sea to spawn once for their "wedding journey," and then die. In the past, European eels could be found in various European water. However, over the past twenty years, there has been a significant decrease in the number of eels reaching European river systems, resulting in the European eel being listed as critically endangered. This is notably due to water pollution, overfishing, and the construction of dams across rivers.

"I wanted to shed light on the story of this amazing creature, and ultimately raise public awareness on the topic of the eel as people know very little about its plight.

"Whenever I would talk about eels, people would often show disgust at their appearance. Although technically a fish, eels are very snake-like, they're slippery and slimy so not many people can relate to them and connect with their story. Eels have a human lifespan –they can live for up to eighty years. I was fascinated by their determination to get to the Sargasso Sea, where they originate, and how, driven by instinct, they return there. I find it quite mysterious. They have to travel about 2 thousand miles. During the journey, they eat nothing, concentrating only on making their way through the cold ocean water. In some way, this reminded me of the human striving for self-fulfillment–going above and beyond one's strengths to reach a moment of self-actualization.

"When drawing parallels, there is the risk of becoming too anthropocentric, but I wanted to take that risk, as I could see too many similarities between our species."

With the VR video artwork *A Calling, Deeply* (2019), Pulejkova invites the viewer to feel and move like an eel embarking on its migration journey. Similarly, the installation *Do you think about home?* (2019) offers a sensual experience, where the viewer might identify with the dancer whose body moves through the water screen. The dance is both fluid and limited, free and constrained by the frame, creating a sense of wonder and anxiety. The audio track, with the sound of human breathing that accelerates and increases with time, strengthens these feelings.

"For me, dance was a way to bring grace into the work. I watched many videos of eels moving under water and I wanted to bring this to the human body, to make the dancers look like and embody these eels. I collaborated with choreographer Georgia Tegou, who really took this idea on board. There were moments when for me the dancers became eels, one moment showing the exhaustion after the long migration, the other the frenzy of the upcoming act of love."

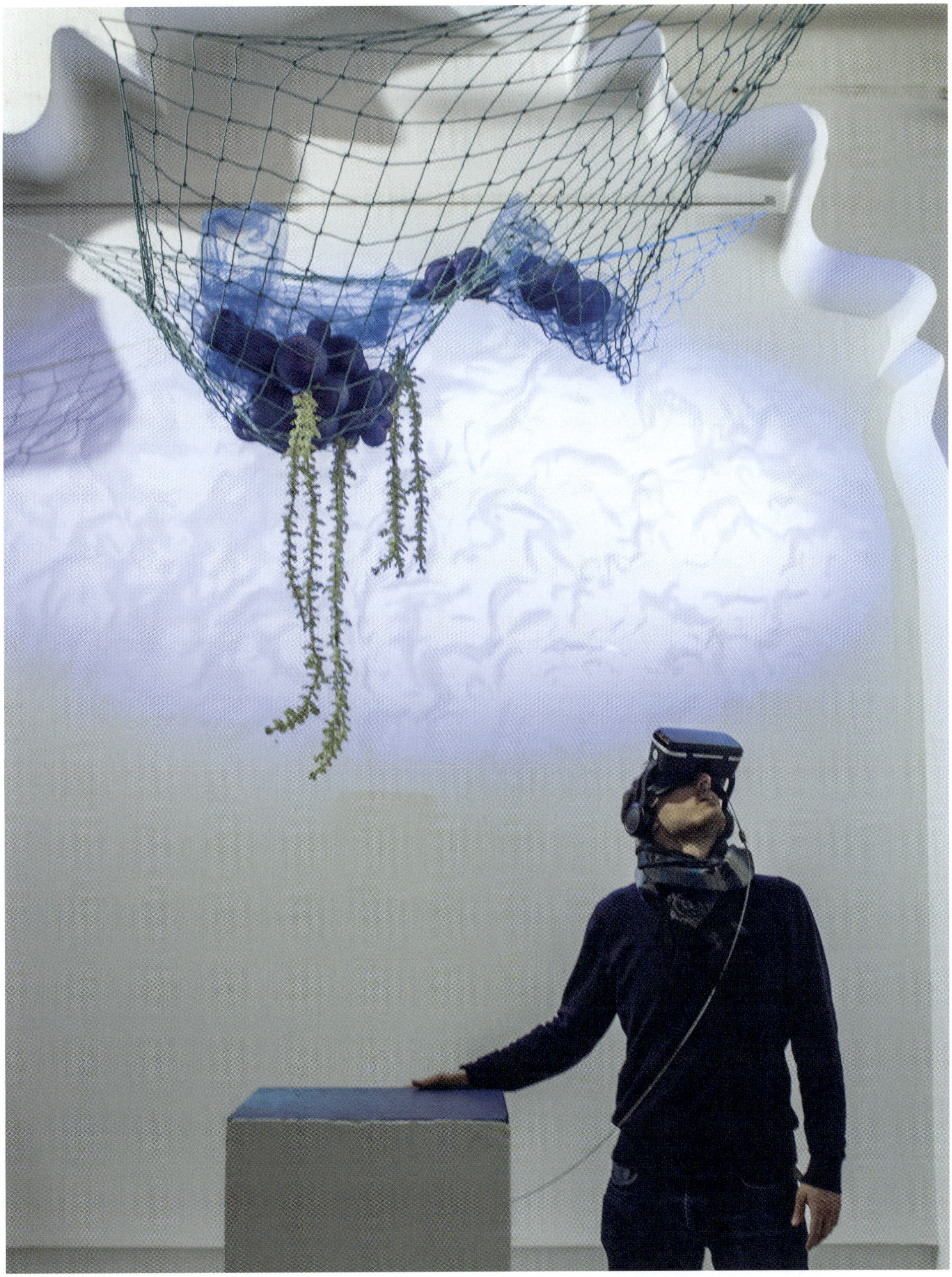

Wedding Journey, 2019, Site-specific immersive installation and 360 VR video

Robertina Šebjanič

Born in Slovenia in 1975

"Art has the power to translate abstract or invisible phenomena into experiences that can be seen, felt, and understood on an emotional level, thereby bridging the gap between scientific data and public awareness."

Robertina Šebjanič addresses both global and local issues related to the ocean, striving to navigate and link these two perspectives. For over a decade, she has dedicated her work to the ocean's complexities, finding it an inexhaustible source of inspiration and inquiry. Torn between a career as a marine biologist and an artist, Šebjanič has successfully merged these paths through her research-based practice. Her artworks are created in close collaboration with scientists, using scientific research as a foundation. By giving artistic interpretations of her findings, she aims to raise public awareness and stimulate curiosity about oceanic life.

Šebjanič's interests lie particularly in microorganisms and the marine aquatic environment, focusing on what is invisible to the human eye. One of her primary goals is to craft narratives that highlight interspecies connectivity and develop strategies for empathy and solidarity, especially with species that remain unseen by humans. This serves as a reminder of the extraordinary biodiversity in the ocean, where new species, including bacteria, microbes, and viruses, continue to be discovered. She emphasizes the unique physics of water, noting that sound vibrations travel five to seven times more effectively underwater than in the air and have prolonged effects. Despite significant advances, much of the deep sea remains unexplored. However, emerging technological tools are enhancing our understanding of these otherworldly spaces. Šebjanič's challenge is to help people grasp the intricacies of what remains largely unknown.

In 2016, Šebjanič created the project *Humalga* (2012-2014) in collaboration with Špela Petrič to investigate the interconnectedness of human and algal communities. This project culminated in the creation of an interactive sound installation that sonified the symbiotic exchange of nutrients and byproducts between these two living systems. This marked the beginning of her intensified focus on aquatic species, further deepened by her discovery of an immortal jellyfish species, *Turritopsis dohrnii*, which can grow from the state of adult back to childhood again in its polyp form. It was the inspiration for Šebjanič's work on the jellyfish species *Aurelia aurita*.

With *Aurelia 1+Hz / proto viva generator* (2014) Šebjanič experimented with the possibility of dialogue between a moon jellyfish and a computer, exploring the boundaries of communication. The concept of non-human languages and interspecies communication allows her to move beyond human-centered perspectives and challenge our understanding of intelligence and consciousness.

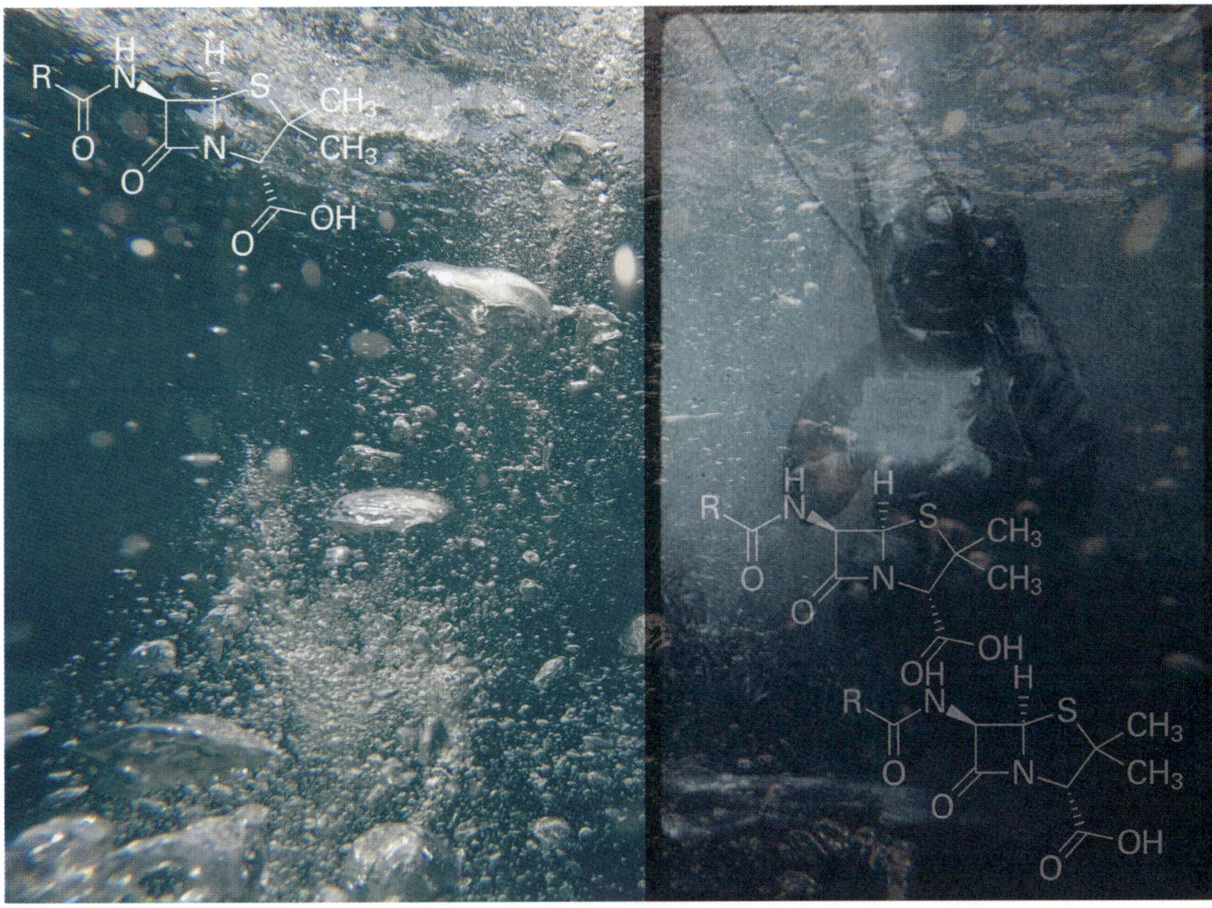

Aqua_forensic, 2019, Site-specific installation, copper pipes and holographic videos

Jellyfish species can communicate with each other through chemical components, but it has not yet been proven that they emit any form of sounds.

In recent years, Šebjanič has developed a framework for working on empathic strategies, inspired by the concept of care. She has conducted several academic workshops and lectures emphasizing ecological issues and interspecies relationships, addressing the common question, "Do you think we can do something about it?" Aware of the non-productive effects of eco-anxiety, she explores how we can navigate our feelings and grief over ecological loss and the perceived powerlessness of individuals.

When not experimenting in a laboratory, Šebjanič is often recording aquatic sounds. For her, sound is a powerful medium for conveying the presence and effects of acoustic water pollution, which seriously affects marine ecosystems. →

She coined the concept of "Aquatocene" to convey this increased yet invisible pollution of the aquatic soundscape. The artist frequently uses hydrophones to capture recordings of underwater soundscapes, careful not to add her noise pollution to the existing environment. She believes that sound can evoke a visceral response, helping people connect emotionally with the issue of water pollution and gain more profound understanding of its significance.

Transforming collected data and research outcomes into artwork involves a multi-layered process. In the projects *Aqua_forensic* (2019) and *Echinodea Future - Adriatic Sensing* (2022), Šebjanič employed scientific methods to collect and analyse water samples from the Adriatic Sea. Collaborating with chemists and environmental scientists, she identified various pollutants, such as pharmaceuticals and other chemical residues. The installation *Aqua_forensic* features copper pipes and holographic videos showing the moment of in vitro experiments of microorganisms dying in solutions 20,000 times weaker than average human doses, highlighting the chemical impact on the inhabitants of the marine world. *Echinodea Future* focused on the morphological changes of sea urchin larvae in coastal area where the levels of oxygen are dropping.

Šebjanič often employs a combination of visual, auditory, and interactive elements to create immersive experiences. Her strategy is to translate complex scientific data into sensory experiences that evoke emotional and intellectual responses from viewers. The large immersive installation *AquA(l) formings* (2021), in collaboration with Sofia Crespo and Feileacan McCormick, allows people to step physically into the artist's research on the noble pen shell, a marine creature she uses as a visual synonym for more-than-human entities. This species has succumbed to disease caused by environmental changes in the Mediterranean. Based on a poetic narrative and AI technology, the artwork reflects on how aquatic life reacts to large-scale changes in the marine environment, such as rising sea levels, increasing water temperatures, and new chemical compositions, aiming to elicit an emotional response from viewers. Šebjanič also incorporates cultural references to help viewers connect to the work and, by extension, to these non-human aquatic species. This is evident in *Atlantic Tales* (2019/20), where she weaves Irish myths into the narrative, creating a resonance that connects ecological issues with cultural heritage.

Through Šebjanič's research and art practice, she seeks to restore a meaningful relationship between all forms of life yet avoiding any form of anthropocentrism. Her multidisciplinary approach offers a deeper understanding and appreciation of aquatic ecosystems.

Shakil Solanki

Born in South Africa in 1997

"My approach to the ocean is informed by a poetic and spiritual sensibility. Most of my work references the musings of Derek Jarman, who considered the ocean as a boundless space for queer imagination and identity, free from the heteronormative restrictions of the land."

Having grown up in Cape Town, with the ocean as a significant part of his environment, Solanki draws from both his surroundings and his cultural heritage. As a third-generation descendent of Indian ancestry, his work integrates elements from Hindu and Persian mythologies, which he recontextualizes to engage with contemporary debates on gender politics, identity, and human vulnerability. One of Solanki's key influences is the work of English filmmaker and artist Derek Jarman, known for his film *Blue* and his writings that celebrate the ocean as a site of liberation. Jarman's retreat from urban spaces to a solitary life in a fisherman's cottage by the sea is emblematic of his desire to explore the ocean's potential as a space of freedom and introspection.

The color blue predominates in Solanki's prints and paintings. Initially, this chromatic emphasis stemmed from the traditional representations of Hindu deities, such as Vishnu, whose color blue refers to the cosmos, or Shiva, the "blue-throated" God. In the mythological narrative *The Churning of the Ocean of Milk*, Shiva consumes a deadly poison that colors his throat with a distinctive blue to save the cosmos from destruction. Solanki revives this myth in his series *A Sea Blooming in the Brown Desert of His Body, Prelude*, wherein the ocean serves as a primordial site of creation. In these works, bodies move in harmony with the ocean's churning rhythms, merging with natural elements in a sensuous, transformative choreography.

For Solanki, the sea represents a liminal space in which dualities –intimacy and trauma, beauty and violence–converge. The color blue, closely associated with the ocean, evokes notions of infinitude and sensuality, reinforcing the complex emotional landscapes that his works seek to explore.

In his series *The Pearl Fishers*, Solanki reimagines the nineteenth century opera of the same name by Georges Bizet. It tells the story of a tragic love triangle and a fractured friendship between two fishermen. The aesthetic of Solanki's paintings draws from the intricate, ornamental styles of classical Persian and Indian miniatures, known for their attention to detail and symbolic richness. However, Solanki situates the narrative within an imagined marine landscape, using the ocean as a symbolic backdrop where the fluidity of bodies can be more fully expressed. The violent conflict between the two friends is transformed into an act of love, their bodies interwoven with the broader movements

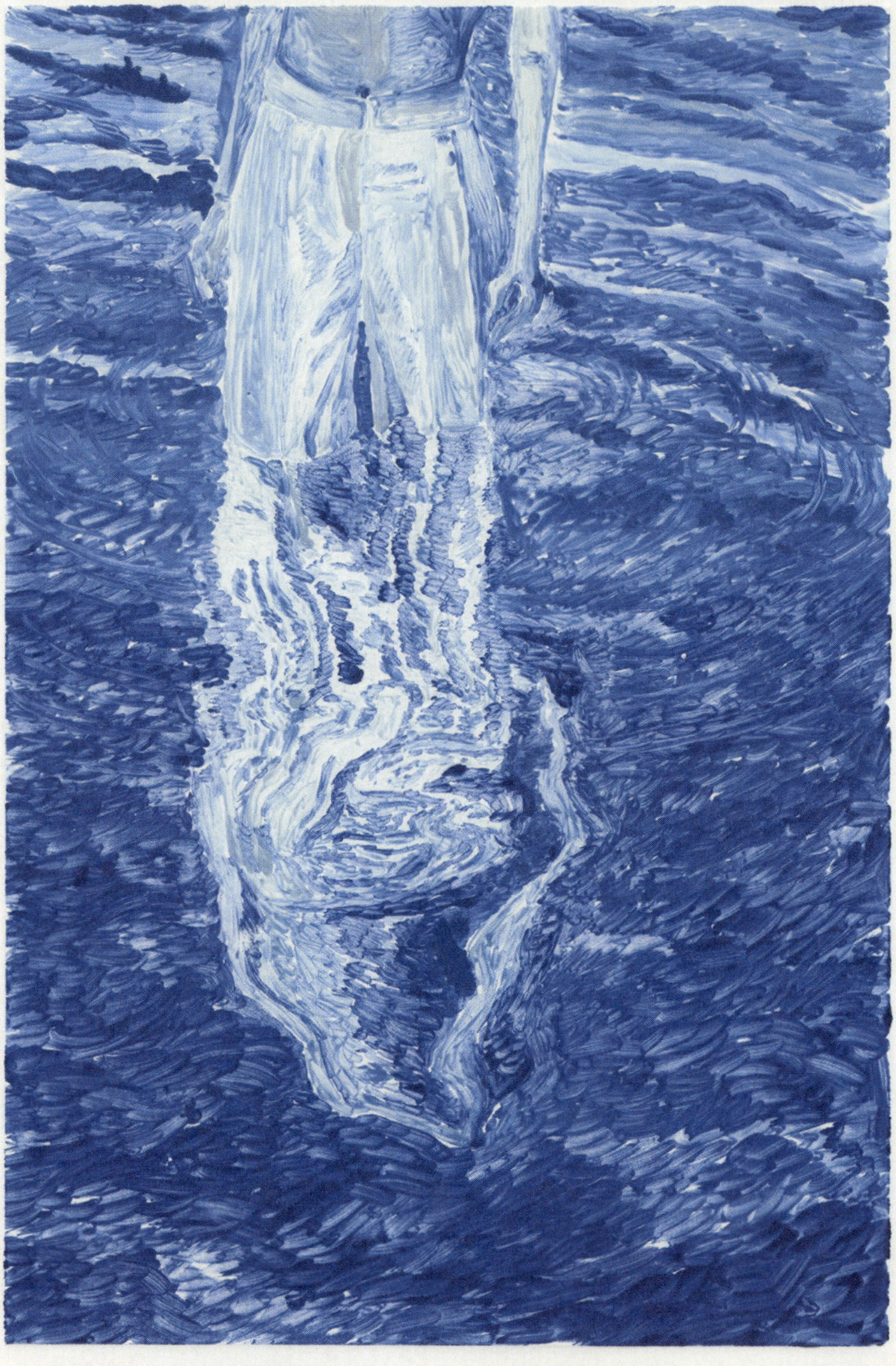

This will slip into the sea, Self portrait in the Lukhanyo trousers, 2021, Oil-based monotype, 36 × 26 cm

of the waves. In this space, distinctions between personal identity and human forms dissolve into the sea, generating an ambiguity that transcends anthropocentric and normative perspectives. From this ambiguity emerges a profound sense of strength, challenging conventional understandings of the body, identity, and relationships.

Pearl Fishers In Azure Seas
Deep waters
Washing the isle of the dead
In coral harbours
Amphora
Spill
Gold
Across the still seabed
We lie there
Fanned by the billowing
Sails of forgotten ships
Tossed by the mournful winds
Of the deep

Derek Jarman, *The Pearl Fishers* (1993), excerpt quoted by the artist.

Lara Tabet

Born in Lebanon in 1983

Lara Tabet grew up in Beirut, the capital of Lebanon, at the sea. As both a medical doctor and an artist, she initially worked in these two spheres separately because trans-disciplinarity is not yet widely accepted in Beirut and there is still little collaboration between artists and scientists. Gradually, however, she began to integrate these two spheres. Initially, she engaged in micrography, which involves taking photographs through a microscope, and her work using tools from synthetic biology and bacteriology subsequently focused on living organisms.

For Tabet, the sea is akin to a talisman. Not only does it serve as a protective entity but it also absorbs everything, including ecological disasters and the consequences of neoliberalism. In August 2020, at the time of the Beirut port explosion, the sea absorbed most of the impact of the blast, shielding the city from further destruction. Fluid and unstable, it represents the political landscape of Lebanon. Simultaneously, with its distinct history and nature, the Mediterranean Sea is like a home to the artist who had to cross it on the journey to her current place of exile.

In 2020, Tabet embarked on an original cartography of the Lebanese seashore, illustrated in the series *Eleven Fragmented Seas* (2020). Walking from south to north along the coast, she collected seawater samples every twenty kilometers. This coastline has been a site of political and ecological contention since the end of the civil war, with increasing privatization of spaces and the proliferation of real estate investment, often conducted illegally and without any regard for the environment. Additionally, the absence of policies of water and waste treatment, mass urbanization, and severe hydrocarbon pollution pose a severe threat to the ecosystem.

The artist applies a precise protocol to her samples to create what she calls a "bacteriography": she cultures the bacteria collected in the samples, first placing them on agar plates and then transferring them onto photographic film, which is also covered with gelatin. By maintaining optimal environmental conditions, the bacteria remain alive and feed on this gelatin. Their activity thus alters the film's chemistry, transforming it into a work of art. The artist observes these processes and halts them at the appropriate moment. She also manipulates the choice of film, exposure to light, and color; however, she emphasizes the importance of allowing the bacteria to perform the work. She notes that the resulting images often resemble the coastline from which they were derived. These images display a profusion of organic lines and curves, shadowy areas resembling lakes, celestial nebulae, and almost maternal forms reminiscent of ultrasound images of pregnant women. The series presents itself almost as a cosmology, encapsulating both life and destruction within a dynamic interplay as the proliferation of toxic substances kills the marine ecosystem. Next to each print, the artist indicates the exact site where the sample was taken and the scientific results of the bacteriological analysis.

Tabet has conceived this work as a dual performance. At the macroscopic and human scale, traversing these private zones constitutes an act of re-appropriation of the territory. At the microscopic scale, it involves decentralizing the human perspective and finding an alternative point of entry into the world that surrounds us.

This series represents the starting point of a more extensive project, *A Personal Cartography of Entangled Toxicities*, which follows the artist's migration and mobility. For instance, in Marseille, she created *Marseille Overflow* after a 2020 storm during which the water was contaminated by *E. coli*. The beaches were all closed in the middle of August (at the peak of summer), and the artist chose to represent the coastline by culturing these bacteria on analog film. The resulting images are strangely beautiful, milky, bordered by thick

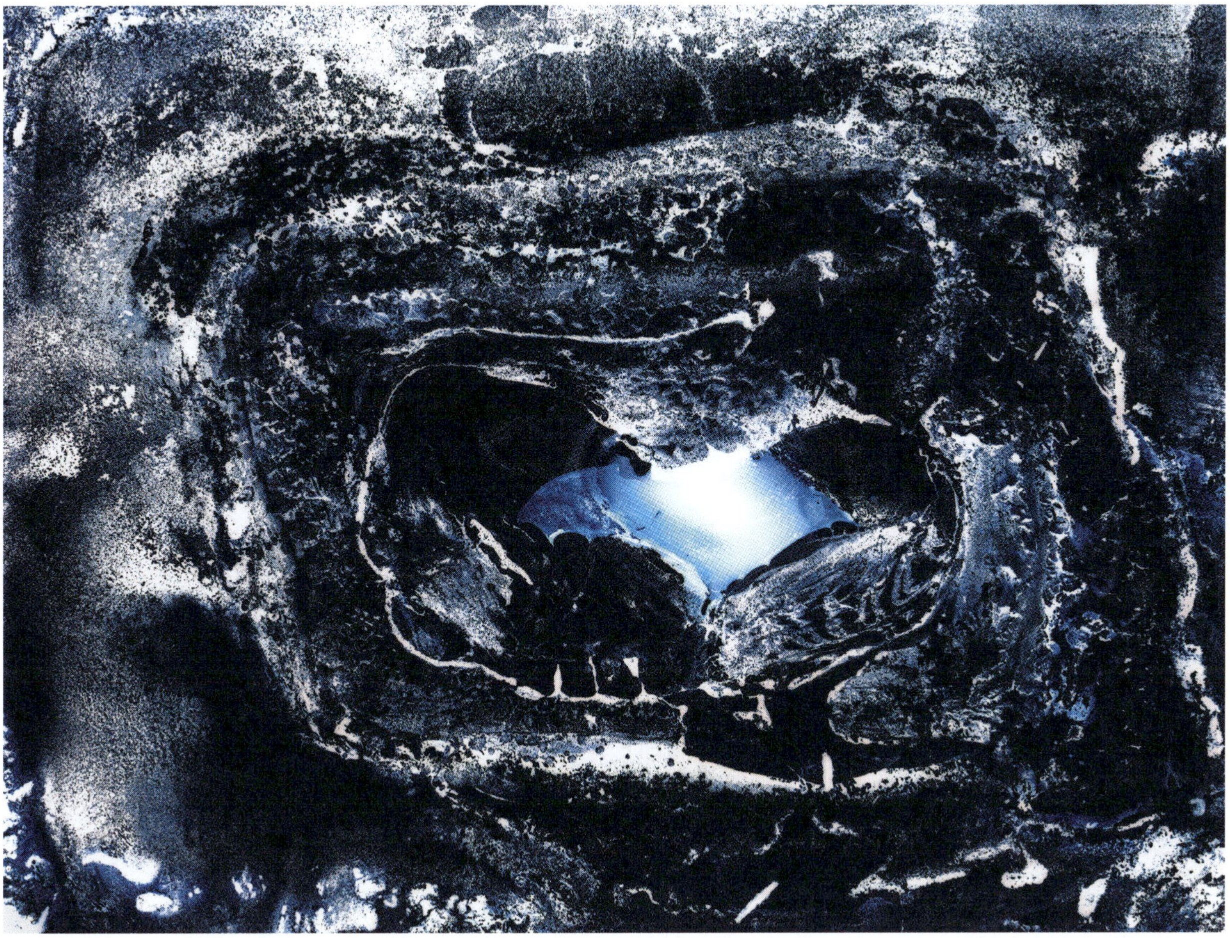

Eleven Fragmented Seas, Sea 01, 2020, Bacteria incubated on medium format color film scanned and printed on color photo paper, 120 × 150 cm

black from which coral-like and aquatic plant forms emerge.

What can bacteria express on photographic film? Ultimately, Tabet questions forms of language for which we do not possess the codes. During her expedition aboard the Tara Foundation's boat, the artist explored another form of manuscript, this time with plankton. *Regnum Marine: A possible Atlas of the invisible species along the way* (2022) also takes cartographic form: from Pointe-à-Pitre to Senegal, the artist collected marine samples treated with a specialized microscope, the FlowCam, which allows biologists to analyze microorganisms and their flow. This instrument generates several types of images. Tabet worked from a particular image representing the negative of the sample, akin to a stencil marking the contour of the plankton. On this occasion, she used a very old photographic process, salt paper printing, which alludes to the salinity of the sea. Thus, the artist juxtaposes cutting-edge technologies with traditional tools, questioning our relationship with technology when it comes to understanding the environment.

Tabet perceives the sea as a milieu rather than a landscape. She chooses to approach it from the inside, from the viewpoint of its living entities, invisible to our eyes. This perspective invites us to change our relationship with the ocean, to attune to a scale that is not ours. For her, revealing the invisible is a way to counteract the effects of the Anthropocene by highlighting its mechanisms. In this regard, her work is inherently political. Biology, in her view, has for too long remained a closed discipline, despite its significant role in people's daily lives and in the impacts of climate change and excessive neoliberalism. Unveiling its inner workings is a form of resistance against biopower.

Alberta Whittle

Born in Barbados in 1980

The water is all

The water is all.

She conjures tremors of life into being.
Each droplet becomes a different way of knowing and feeling.
Our cells, our vessels, our solid flesh will eventually hum their own cadence into liquidity.
But for now, we must try to remember a time when the barriers that divide were once porous.

Lay back and let the water take you.
Your feet two islands.

Put your hand on your belly, softly.
Think of someone you love.
Take a deep breath.
Hold onto that feeling of love (release) (Aaaaaaaaahhhhhhhhhh . . .)

Take another breath, ***hold it***.
Think of the last time you laughed with that person.
Release that breath with a whoooooooossshhh.

Do you sometimes think that you have forgotten how to do this? How to stay alive? How to breathe underwater?

Beyond emerging as flesh, we have forgotten how to survive alongside our oceanic brethren, our sistren, until now we can only move through the world above ground –
guests now.

Every day we take her bounty for granted.
Hungry, thirsty, forever wanting, our parched bodies demand we suckle upon the ocean,
but she is finite.
Taking many names and many forms, I feel you pulsing – Olukun, Mami Wata, Mambo Muntu, La Sirene, River Mumma, Yemaya, Oshun and Yemoo.
She is beginning to dry up.
Her life-giving nectar becomes noxious from withstanding our greed.
Reckless and determined to forget
what we have done,
we become machine.
Mindless,
Greedy and determined to consume, we pick her bones clean thoughtless with desire.

The coral is starving.
Stressed by electric heat that is boiling the oceans, our reefs decay.
Propelling my limbs across this graveyard, I shiver.

I struggle to remember a time when I worshipped water, when my body was not a burden in her arms.
I have become stoppered, slumping in the muck of my desire for more.
My eyes no longer see.
My ears no longer hear.
My touch is a weapon.
Tight and stiff as a corpse, I think about when I was soft and yielding as a puddle she poured me into.

Buoyant in her womb, I once bobbed with her rhythms, never doubting her love – she was all.
This salty mass, endlessly moving, *you can never stand in the same water twice*, keeps me in thrall until become weightless.
I will myself into softness, allowing the currents to launch me between the waves.
My skin tingles, it remembers that a lifetime ago I was once family to kelp.
To remind me, I slip a tendril between my lips and press it against my tongue.
My species has forgotten that we are no more grand than seals, flying fish, ship worms and sea urchins. →

Our fleshy narcissism grips its own reflection in a new plastic liquidity, now slick with oil.

If I had gills, flippers, or spines would I remember what it meant to have fellow feelings for another species?

The kelp tastes bitter.

Hiding in dank pools, the reflection of her gaze is murky.
Oil distorts any clarity.
Thick and gelatinous, the water begins to reclaim me.
Memories of those who have sunk to a watery grave from overcrowded ships, boats, fists, guns and whips are insistent.
No longer brother, mother, sister, father, friend nor lover, we machine only recognize them as chattel.
But the water demands I *see*.

Blinding white coral bleached by trauma.
The feast is over.
Paralysed, I gaze on a graveyard ripe with spirits.

We owe a debt to water that must be repaid.

Originally from Barbados and currently based in Scotland, Alberta Whittle calls on her diasporic heritage to create works that meditate on the journeys, both historical and present, of Black communities across the Caribbean Sea and beyond. Challenging the ongoing erasure of Black bodies, voices, and narratives, Whittle's research-driven practice is informed by an active process of remembering, often resulting in choreographed installations that invoke self-compassion and collective care as integral to building a decolonized future.

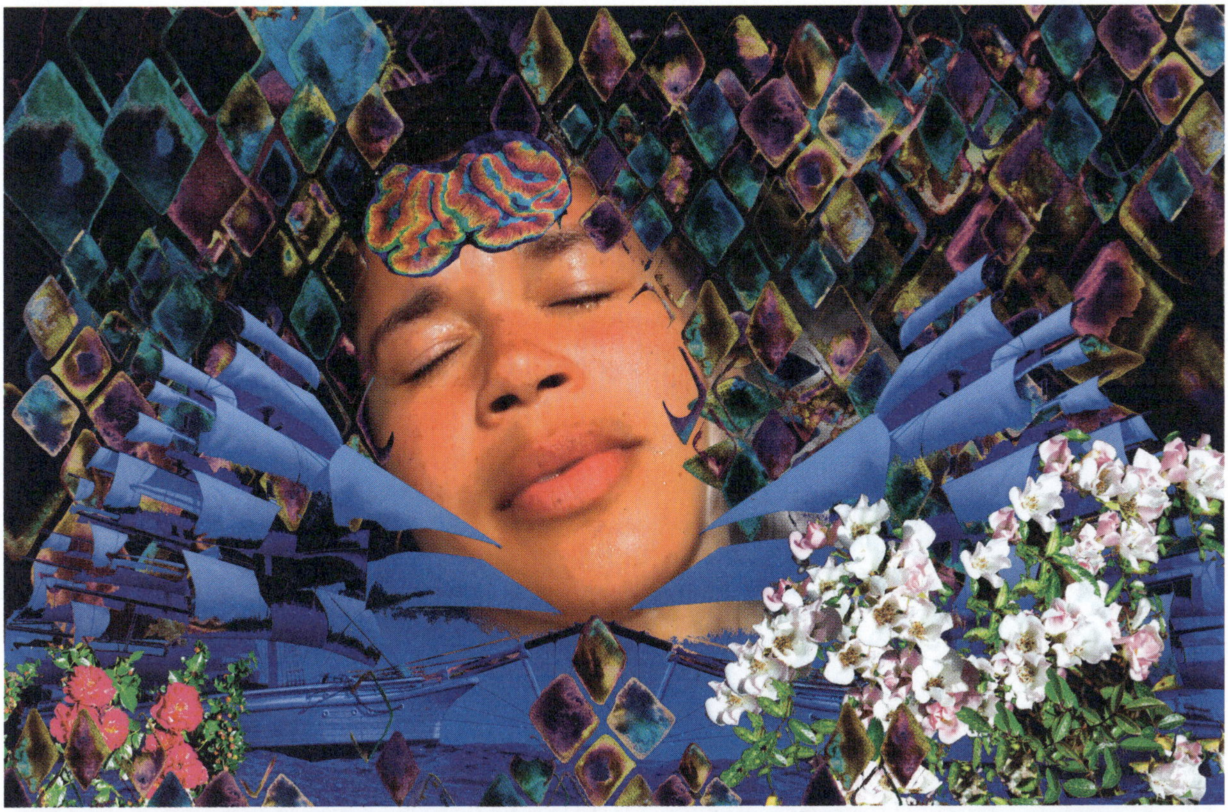

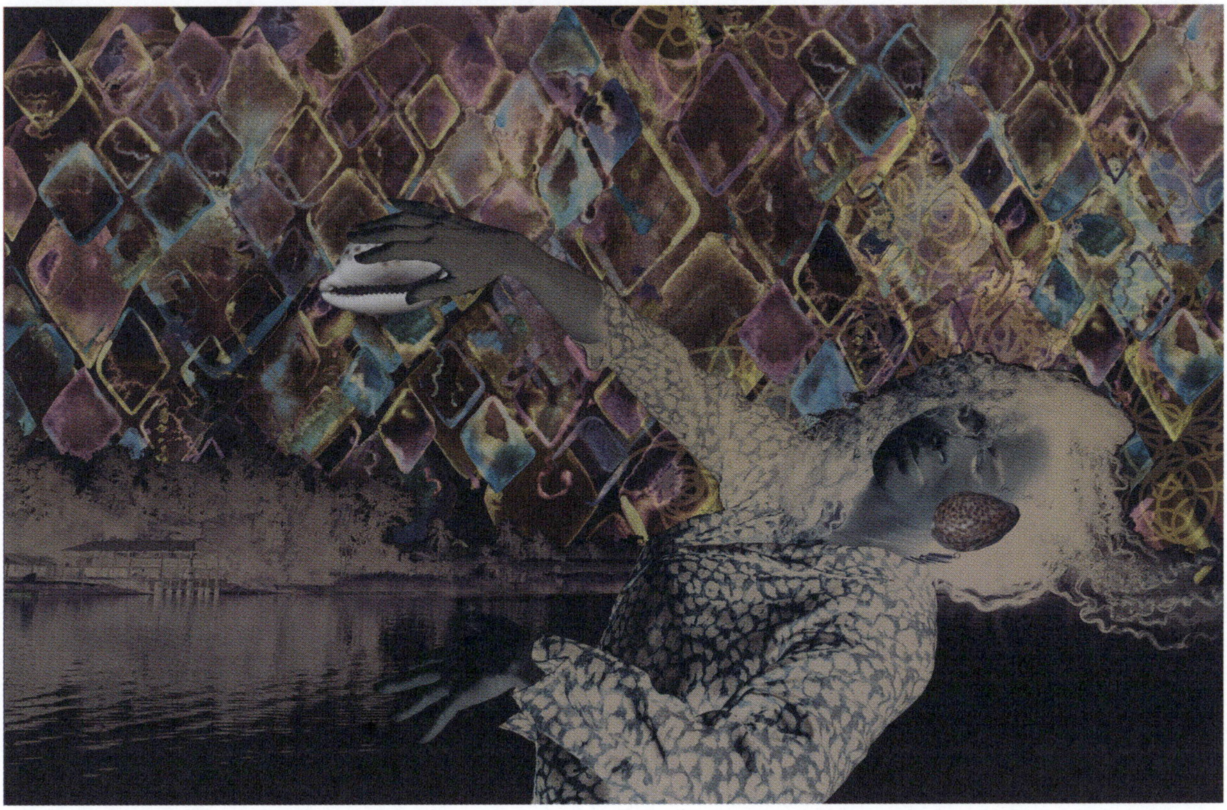

Where The Tide Greets Memory, 2023, *Listening for all that has been forgotten*, 2023, Collages, 80 × 120 cm each

Hélène Artaud
Social anthropologist at the Muséum national d'Histoire naturelle (MNHN), UMR 208, Local Heritage, Environment and Globalisation (PALOC)

The ocean as a method
The place of island artists in the genesis of an oceanic methodology

What would our categories of thought be if the continent and the land had not radically shaped them? What would our research questions be if we had developed them from the ocean? What policies would we have founded if they had been conceived from oceanic perspectives?

The ocean, which has long been invisible, is now generating growing interest, and the novel methodological dimension that this environment introduces is now finding regular echoes in the human sciences. The idea that the qualities of the sea—its three-dimensionality, fluidity, transparency, turbulence and mobility—can provide thought with a repertoire of models more likely to overcome the crises we are experiencing, has been theorized by geographers concerned with "destabilizing the static, bordered and linear frameworks that characterize the geographical structures of territories and time."[1] The novelty of this gesture, however, is relative. As early as the 1970s, post-colonial island literatures were deploying—with the ambition of overturning the dominant imperialist and continentalist point of view—a range of analytical tools directly inspired by the ocean, thus substituting "oceanic ontologies"[2] for the imaginaries of the plantation and agrarian worlds. A few decades later, younger generations of island artists brought about a more concrete oceanic reversal.

The *Tidalectic* (TBA21, Vienna, 2017) and *Mémoires des Hauts-Fonds* (Cité internationale des arts, Paris, 2024) exhibitions are undoubtedly particularly noteworthy examples. Their similarity stems first and foremost from the terminologies used to entitle them. The choice of the title, *Mémoires des Hauts Fonds*, is directly in line with the critical and exploratory gesture that characterizes the approach of island artists and aims to counter the ecological imaginaries imported by the West during its imperialist trajectories[3]. In the Western imagination, the Shoals is a zone of threat and perdition. The painter Géricault immortalized it in a painting that has become archetypal of those spaces in which the foundations of humanity are troubled and shaken. The island artists turn us towards and entirely different memory of these places, changing these shallows into zones of encounter and creative friction. Also borrowed from the oceanic repertoire, the term "tidalectic", which takes its origin in a text by the Barbadian poet Brathwaite, describes this incessant, irregular, generative movement of the waves, inspiring "oceanic worldviews that do not succumb to borders or yearn for certainty in static and established epistemological and ontological positions."[4]

The second element that brings these two exhibitions together, and is a direct result of the oceanic perspective they take, is the amplitude and plurality of the relationships they reveal. The narratives are never reduced solely to the colonized/colonizer confrontation. While this postcolonial dimension inevitably runs through all the works on display—as in the subtle description of thirty years of nuclear testing in French Polynesia in the testimonies of Marie-Hélène Villierme's film *nu/clears stories*—it never takes precedence over others. What characterizes the memories shaped by the ocean seems to be their propensity to multiply encounters, as much as to continually renew their meaning. David Gumbs's liquid works express this metaphorically, encasing and declining the thousand lives of an image through a play of visual vibrations.

The movement of the flow, which practically informs the methodologies of island artists, thus generates unprecedented forms of memory in which linearity gives way to circularity, and memories are updated—like the backwash—thickening with new heuristic contributions. This creative power of the oceans, which assembles, composes, and hybridizes the elements that run through them, prevents the meaning of history from being frozen, as much as the form of its existing elements. It is these resilient humanities, in constant transformation and symbiosis, submerged or just emerging from the ocean, which the sculptures of Jason deCaires Taylor and Thierry Fontaine still provide a glimpse of.

What island artists have sensed, in producing early and more numerous works from the ocean in recent decades, is the need for a different reading of the world. By encouraging the public to discard concepts inherited from the continent, and to replace the idea of borders and essence with that of movement and relationship, they are providing the means to initiate the radical reversal to which future worlds are calling us.

---

1. Steinberg, P., & Peters, K. (2015). Wet ontologies, fluid spaces: Giving depth to volume through oceanic thinking. *Environment and planning D: society and space, 33*(2), 247-264.

2. Mohan, A. (2019). "Amphibious histories": An Interview with Isabel Hofmeyr. *Postcolonial Text, 14* (3 & 4) Winter, T. (2023). A maritime turn and ocean ontologies in critical heritage studies. *International Journal of Heritage Studies*, 1-12.

3. Artaud, H. (2023). *Immersion. Rencontre des mondes atlantique et pacifique*, Empêcheurs de penser en rond, La Découverte, Paris

4. Hessler, S. (2020). Tidalectic Curating. *Journal of Curatorial Studies, 9*(2), 248-270.

Regeneration

Old men ought to be explorers
Here and there does not matter
We must be still and still moving
Into another intensity
For a further union, a deeper communion
Through the dark cold and the empty desolation,
The water cry, the wind cry, the vast waters
Of the petrel and the porpoise. In my end is my beginning.

T. S. Eliot

1 Amina Agueznay
Born in Morocco in 1963

2 Ari Bayuaji
Born in Indonesia in 1975

3 Megan Cope
Born in Australia (Quandamooka) in 1982

4 Jason deCaires Taylor
Born in the United Kingdom in 1974

5 Jérémy Gobé
Born in France in 1986

6 Julia Lohmann
Born in Germany in 1977

7 Moffat Takadiwa
Born in Zimbabwe in 1983

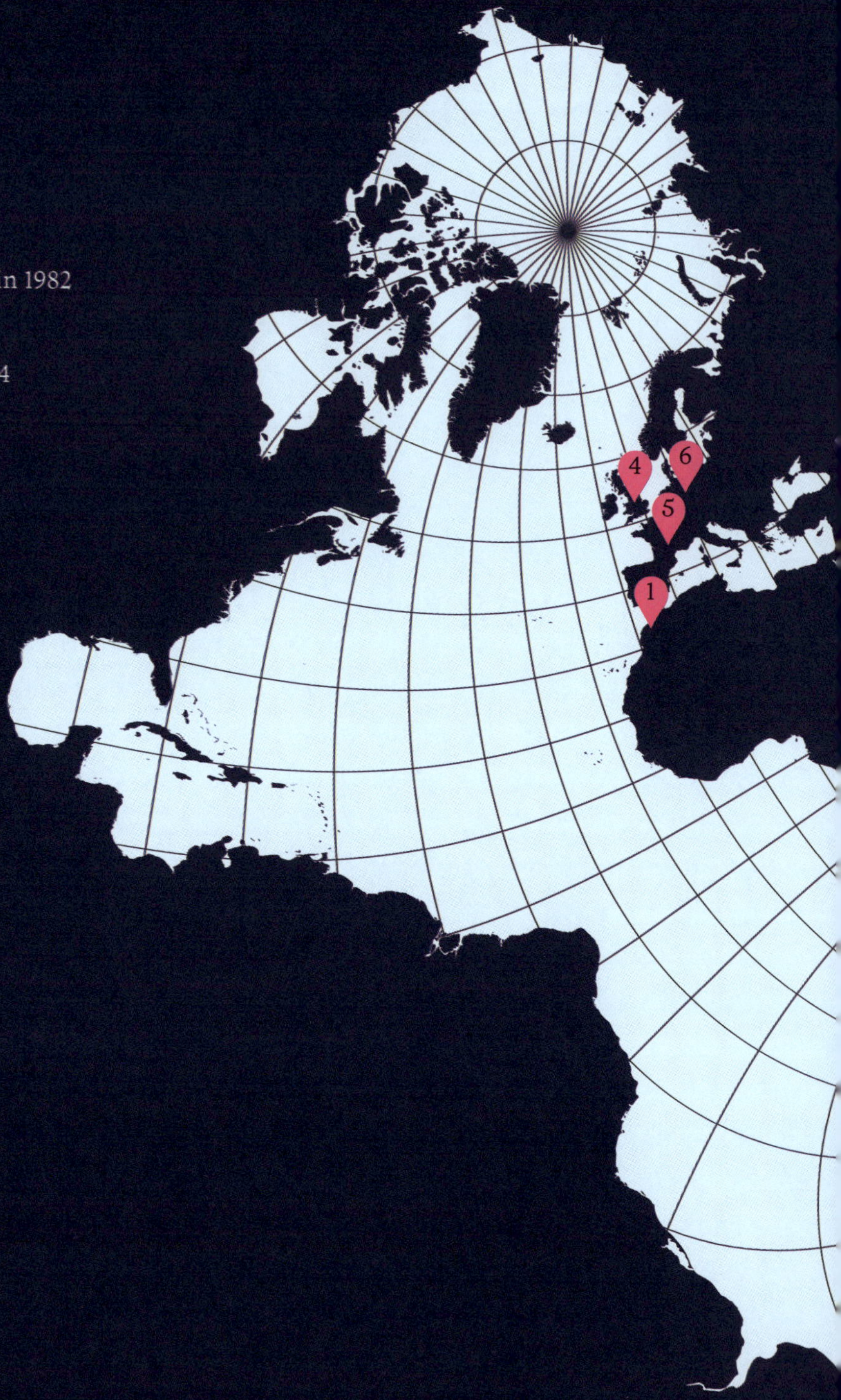

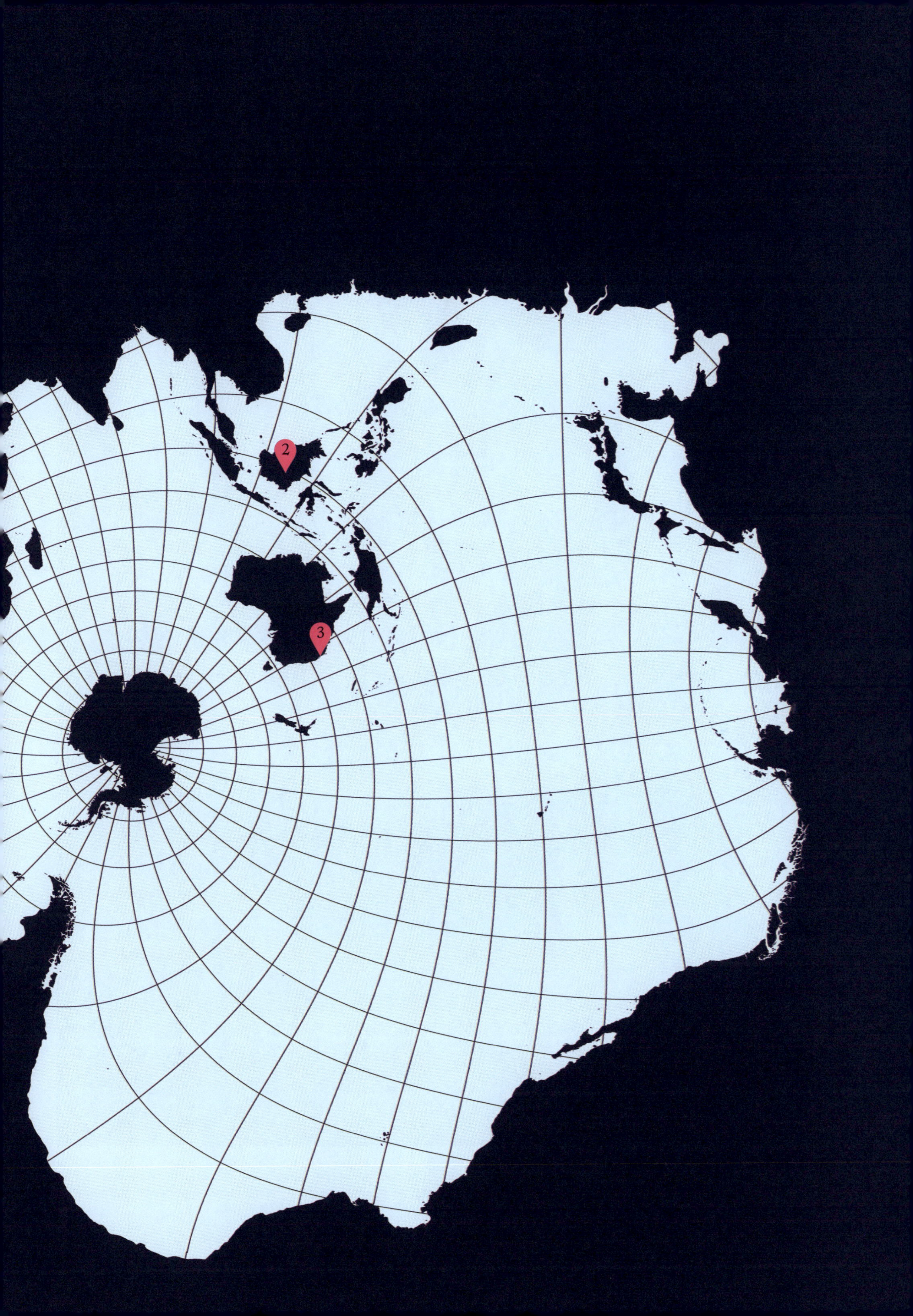
2
3

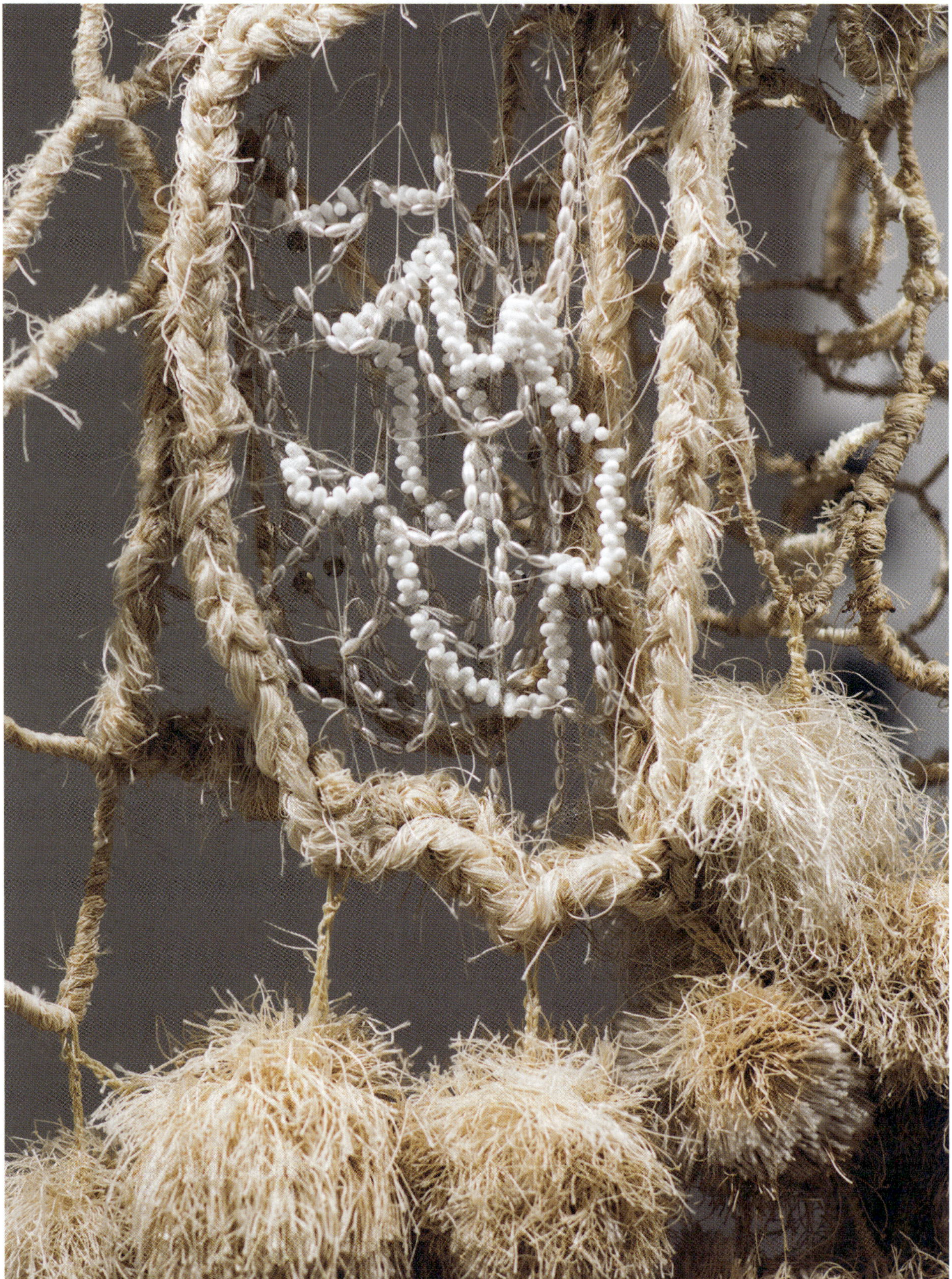

Skin (detail), 2011, Recycled fish net, viscose and nylon thread, cotton thread, sisal twine, stainless steel wire, plastic sequins, glass and plastic beads, paper, stainless steel structure. Crochet, knitting, weaving, macramé, braiding, rope and pompoms, embroidery, randa (traditional Moroccan needlework), beading, sewing, 300 × 680 × 20 cm

Amina Agueznay

Born in Morocco in 1963

Amina Agueznay was born in Casablanca, close to the sea. From a young age, she developed a passion for beachcombing. "I have a deep connection with the ocean," she says, "a bond likely strengthened by my mother, Malika Agueznay, an avant-garde modernist artist whose practice centers on seaweed motifs." Much of Agueznay's work is based on various forms of woven elements–individual modules she reconfigures into original installations, experimenting with different materials to create ambiguities in form and texture. Trained as an architect, she retains from this discipline a keen sensitivity to the specificities of the environments in which she works. Employing recycled fish nets, empty shells, raffia, and traditional rugs, and drawing inspiration from Moroccan heritage weaving and embroidery techniques, she creates unique organic landscapes and installations that serve as powerful depictions of humanity's transformative potential.

Agueznay's artistic practice is largely community-based and strongly influenced by her extensive field research. She often works with with local artisans and weavers to create on-site, often large-scale, artworks. These installations frequently require extensive hours of labor, which provide an ideal opportunity to build relationships, exchange stories, and share skills with the artisans. For her, these participative processes are essential, contributing to the dialogues, transformation, and sharing of the creative energy she seeks to emphasize. As she states: "Today, what we need most is to share energy to address current issues, including climate change."

When she worked on *Skin* (2011), Agueznay invited a group of twenty-five craftswomen who applied their own needlework techniques to weave and transform fish nets collected from a fisherman on Morocco's Atlantic coast. Techniques such as crochet, knitting, weaving, macramé, braiding, and sewing were all integrated to produce a monumental artwork that unfolds like organic skin or a quilt. A symbol of diversity, the piece embodies the shared knowledge and experiences of these women, yet it rejects nostalgia. Instead, the work fosters new collaborations, inventions, and gestures that look toward the future, celebrating human creativity, adaptability, and sustainable recycling processes.

Casablanca Green (2015) was an interactive large-scale installation composed of thousands of mussel shells, arranged on the floor by shape, size, and color, resembling a mosaic. Agueznay collaborated with local artisans who pierced each shell and threaded them together using green wool–the color of seaweed–and employing crochet techniques. They created repeating patterns that form a giant tapestry. The idea of using mussel shells originated from

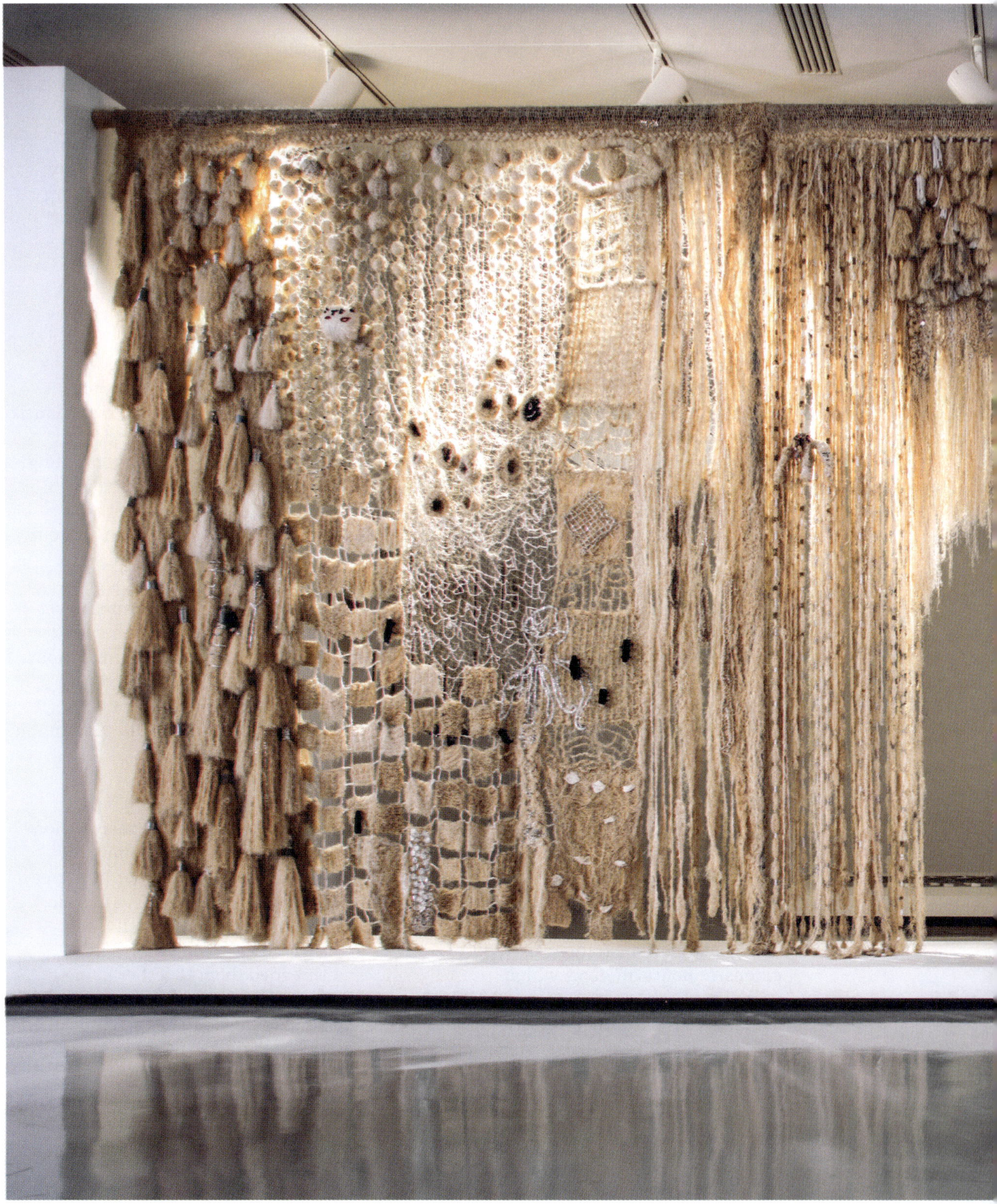

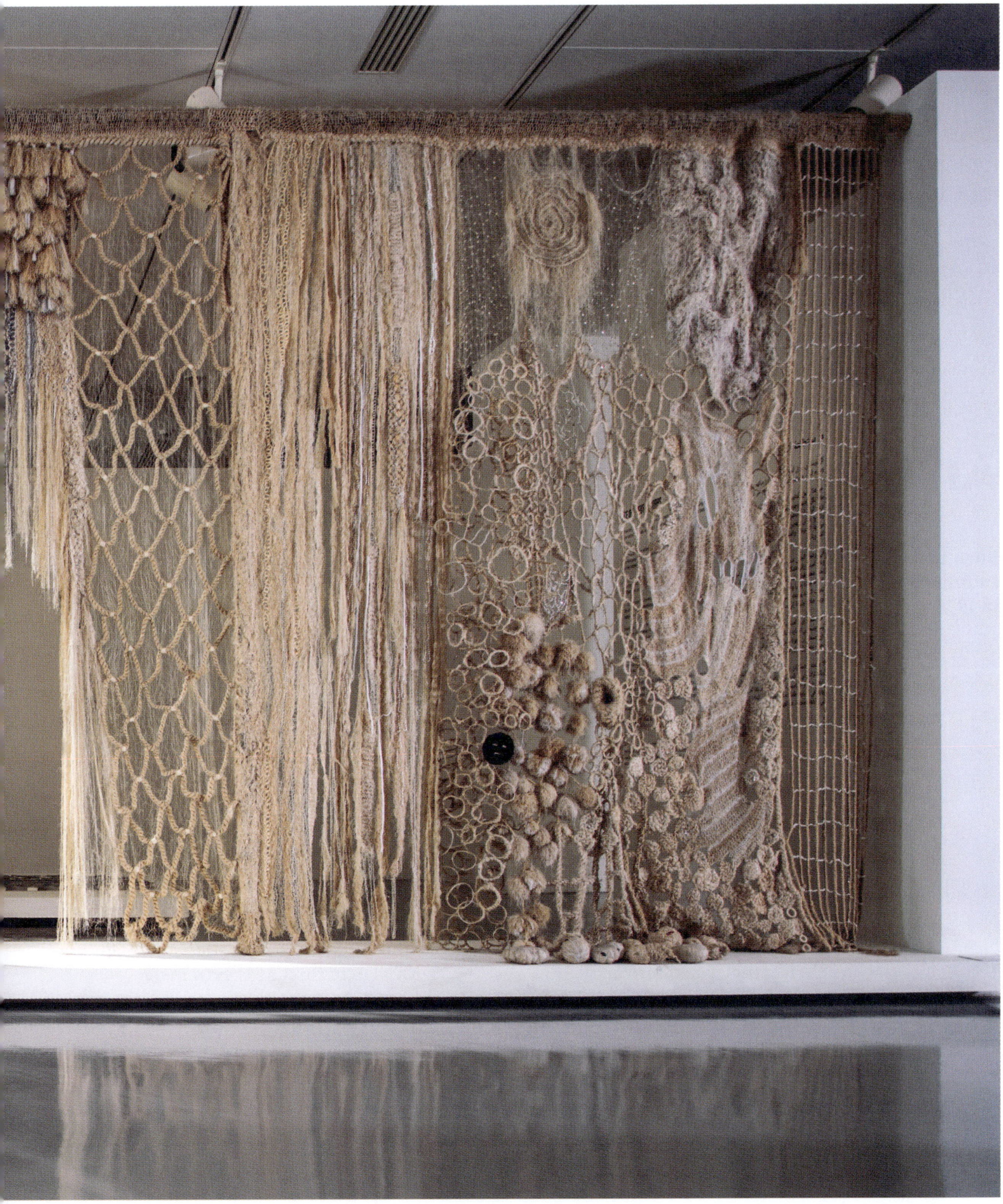

Skin, 2011, Recycled fish net, viscose and nylon thread, cotton thread, sisal twine, stainless steel wire, plastic sequins, glass and plastic beads, paper, stainless steel structure. Crochet, knitting, weaving, macramé, braiding, rope and pompoms, embroidery, randa (traditional Moroccan needlework), beading, sewing, 300 × 680 × 20 cm

a previous trip along the Atlantic coast, where Agueznay observed villagers collecting mussels for sustenance. After extracting the meat, the shells were discarded, leaving the coastline littered with accumulated waste.

In the final phase of the installation, Agueznay invited the public to walk across the artwork, crushing the tapestry beneath their feet. The amplified sound of cracking shells created an evocative moment, as participants heard the crunching until, suddenly, there was silence. For the artist, the work is a response to the severe pollution in Casablanca and along its shores. More than anything, the piece serves as a cathartic gesture, one that seeks to release and destroy what she refers to as "dark energy" through a collective transformative process. What remains after the destruction are the small green constructions—threads of hope—that endure and survive this act of deliverance.

Ari Bayuaji

Born in Indonesia in 1975

Ari Bayuaji was born in a small town in East Java, far from the ocean, but throughout his childhood, his parents regularly took him on holiday to Bali, where he spent much of his time by the sea. Over the years, he observed significant changes to the beaches, which were becoming increasingly polluted with plastic waste. In 2013, he began documenting this phenomenon, particularly in the mangrove forests of Sanur, Bali. *Weaving the Ocean Project*—an ongoing, community-based initiative—focuses on the collection of abandoned plastic ropes, strings, and fishnets found in the mangroves and the sea and which are then transformed into artworks. The principles of recycling and upcycling lie at the core of the artist's practice. Drawing inspiration from diverse textile cultures, Bayuaji employs traditional weaving techniques to give a second life to materials typically regarded as waste.

The installation *Still Life (The Treasure)* (2023), for instance, includes a rubber tire, plastic threads, corals, coins, stone beads, wood, and woven plastic. From these discarded items, Bayuaji creates a contemplative space, with the tire transformed into a symbolic sphere reminiscent of Zen Buddhist artworks. In addition, the artist incorporates organic materials from the beach, such as corals and seaweed, into his work. His signature pieces often involve the weaving of cotton threads with plastic. Weaving with plastic threads is extremely difficult, he says. This is why he needs to include cotton threads in his tapestry, as they give a smoother texture to the fabric. In works such as *The Sparkling Waves under the*

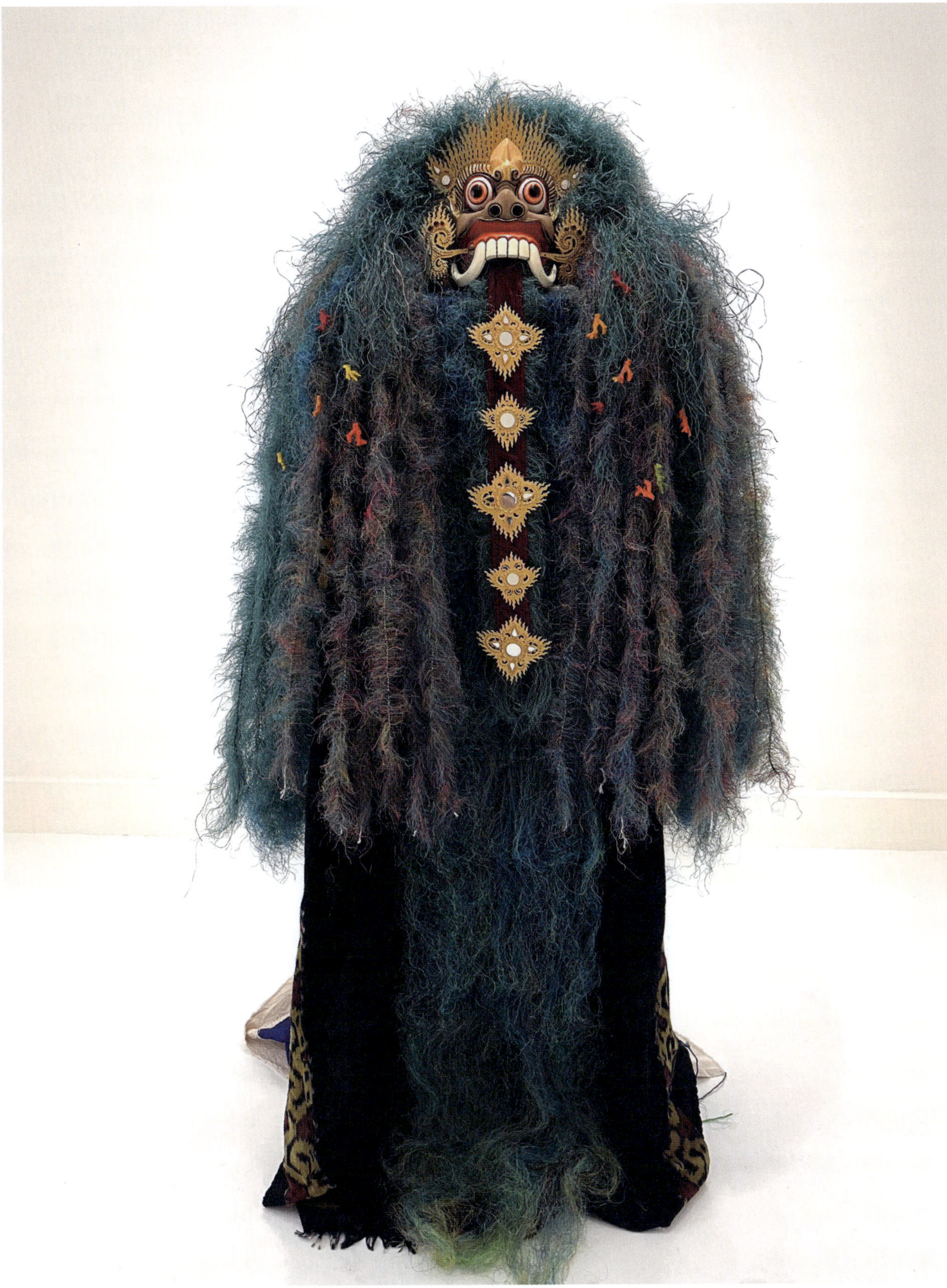

Rangda, 2022, Mixed media including plastic thread, 180 × 45 × 50 cm

Full Moon (2023), which evokes the shimmering surface of the sea, it is nevertheless almost impossible to imagine that plastic threads have been interwoven into the textile.

There are tougher challenges involved in the *Weaving the Ocean Project*, because Bayuaji, and the community of women involved in the project, needed to work with plastic ropes. These ropes, which are collected from the mangroves, undergo a labor-intensive process of being washed, dried, and unraveled by hand to obtain the finest possible threads. These threads are then either woven using traditional Balinese techniques on a loom or employed in macramé or fringe work.

The inception of this project began during the Covid pandemic, when Bayuaji was in lockdown in Bali. He began to clean the mangrove forest with the help of the community who needed money and food to survive, and used his own money to pay them. Over time, he recognized the potential of the plastic ropes as a medium for his art. The project has continued to grow, with an increasing number of people participating. "They are very happy to be part of it, as they can make a better living while acquiring new skills that will be of great benefit to them," the artist says.

Much of Bayuaji's artwork draws inspiration from the seascape, culture, and history of Bali. He has a particular interest in masks and frequently incorporates references to them in his work, using them as a means to question humanity's relationship with the sea and, more broadly, with nature. For instance, *I Went Deep into the Ocean, It Was Wonderful* (2022) resembles a menacing ritual mask, created from found plastic threads, barnacles, and rubber. This sculpture gazes back at the viewer with a sense of irony. Another piece, *Rangda* (2022), refers to the role of law enforcement during the Covid pandemic, when police officers urged people to wear masks to protect themselves and others.

The figure of Rangda is rooted in Balinese mythology and is known as the queen of the Leyak, a demonic figure that is the incarnation of Calon Arang, a legendary witch who caused devastation in ancient Java in the late tenth century. According to legend, Calon Arang was a widow who had mastered the art of black magic and was often blamed for crop destruction and the spread of disease. To this day, Rangda masks are used by the Balinese in sacred Hindu ceremonies, particularly in dance performances. The figure serves as a reminder of past tragedies, reinforcing the importance of nurturing both nature and culture to ensure a peaceful and prosperous life.

Megan Cope

Born in Australia (Quandamooka) in 1982

Megan Cope is an Indigenous artist of the Quandamooka nation, who live in the Moreton Bay region of Queensland, northeast Australia. Her multidisciplinary artistic practice and cartographic works focus on themes of identity, climate change, land tenure, and migration. Known as the People of the Sand and Sea, the Quandamooka have inhabited the Moreton Bay archipelago for over 25,000 years and their culture is deeply intertwined with the marine environment.

Cope's site-specific installation, *Kinyingarra Guwinyanba* (2022), which translates to "place of oyster rocks" in the local Aboriginal language, is inspired by the traditional oyster culture and reef formations of the region. Prior to British colonization, these expansive oyster reefs, composed of discarded shells, were large enough to constitute underwater cities along the Quandamooka coastline, serving as navigational markers. A few years ago, Cope began researching this cultural heritage, investigating the consequences of the colonial destruction of the reefs, which was driven by the development of the lime burning industry. She also learned about the critical ecological role of oyster reefs in maintaining the balance of the marine ecosystem. Her large-scale artwork aims to reconnect with this indigenous knowledge and revive the oyster culture, which she sees as a restorative process. After extensive collaboration with former oyster farmers, marine biologists, and members of the Quandamooka community, Cope constructed a living

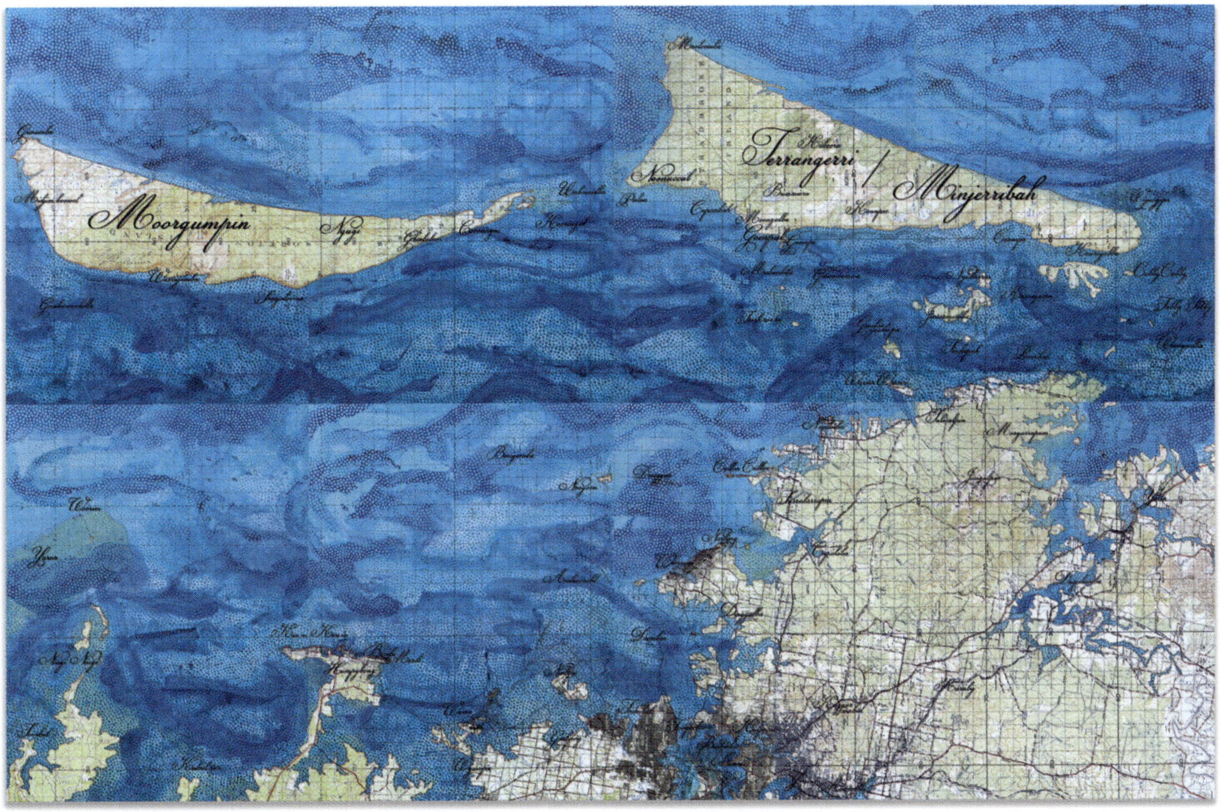

After the Flood IV, 2024, Acrylic and ink, military maps, cotton rag mounted on board, 200 × 300 cm

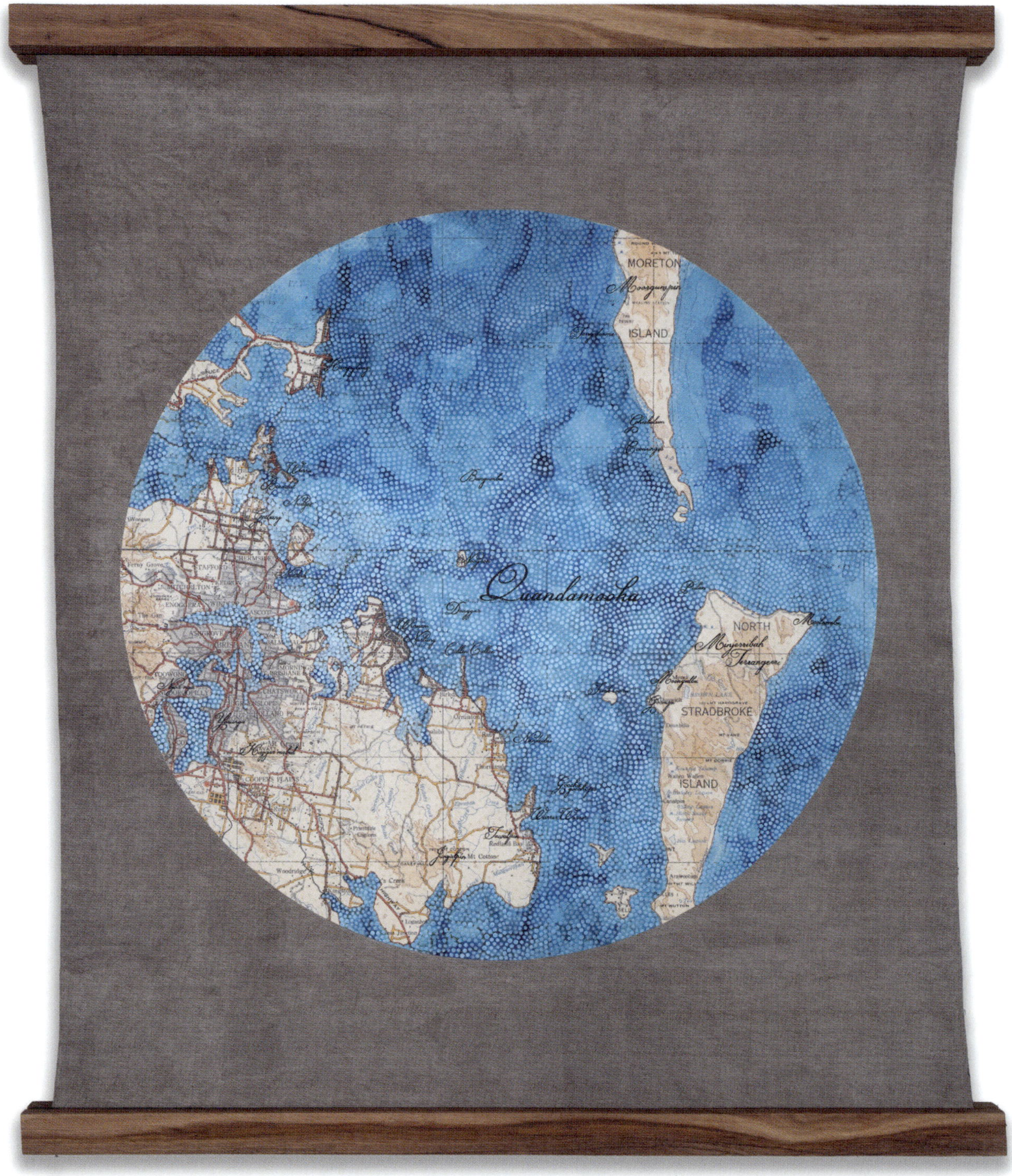

After the Flood I, 2024, Acrylic and ink on military map, mounted on linen with Blackbutt Timber, 180 × 151 cm

sculpture located in the intertidal zone of North Stradbroke Island, within the Moreton Bay region. As time progresses, she hopes to facilitate the restoration of the oyster reef. This installation has already inspired additional artworks and initiatives, stimulating constructive and thought-provoking conversations about regenerative aquaculture, oyster cultivation, cultural heritage, and climate change.

Cope has recently increasingly responded to the impacts of climate change on the east coast of Australia, which has been strongly affected by rising sea levels and flooding. In 2022, she experienced a devastating personal loss when her studio and much of her artwork and archival material were destroyed in the Lismore floods, one of the largest flood events in modern Australian history with floodwaters reaching a height of 14.4 meters.

After the Flood (2024) is a series of four cartographic works that draw upon historic military maps to interweave historical, political, and geographical narratives against the backdrop of projected future coastlines after a rise of five meters of the sea level. Issues such as land reclamation and land rights are revisited in light of these anticipated landscape transformations. By juxtaposing collective memories of the past, current multifaceted challenges, and future projections, the artist questions our shared sense of belonging, inviting us to renew our relationship with the land and sea.

Jason deCaires Taylor

Born in the United Kingdom in 1974

"Ninety percent of reefs worldwide are in danger, the chemistry of the sea is increasingly changing and over 44,000 marine species face extinction. But facts and figures are not an easy way to deal with these issues, and I believe that emotional responses might be stronger. Artists and storytellers have a very important role to play in changing how we feel about the sea and our natural world. Through my work, I try to tell stories that can connect with people emotionally and help them change how they relate to the sea." (Jason deCaires Taylor's statement on his website: underwatersculpture.com)

Jason deCaires Taylor grew up in Malaysia, where his close proximity to the sea provided him with the opportunity to dive regularly and to explore its depths. For him, the sea represented the unknown, a space to discover as an explorer. He studied sculpture at the London Institute of Arts, where he focused on outdoor and public art. He has always wanted to connect his artistic practice with the ocean, and the opportunity arose

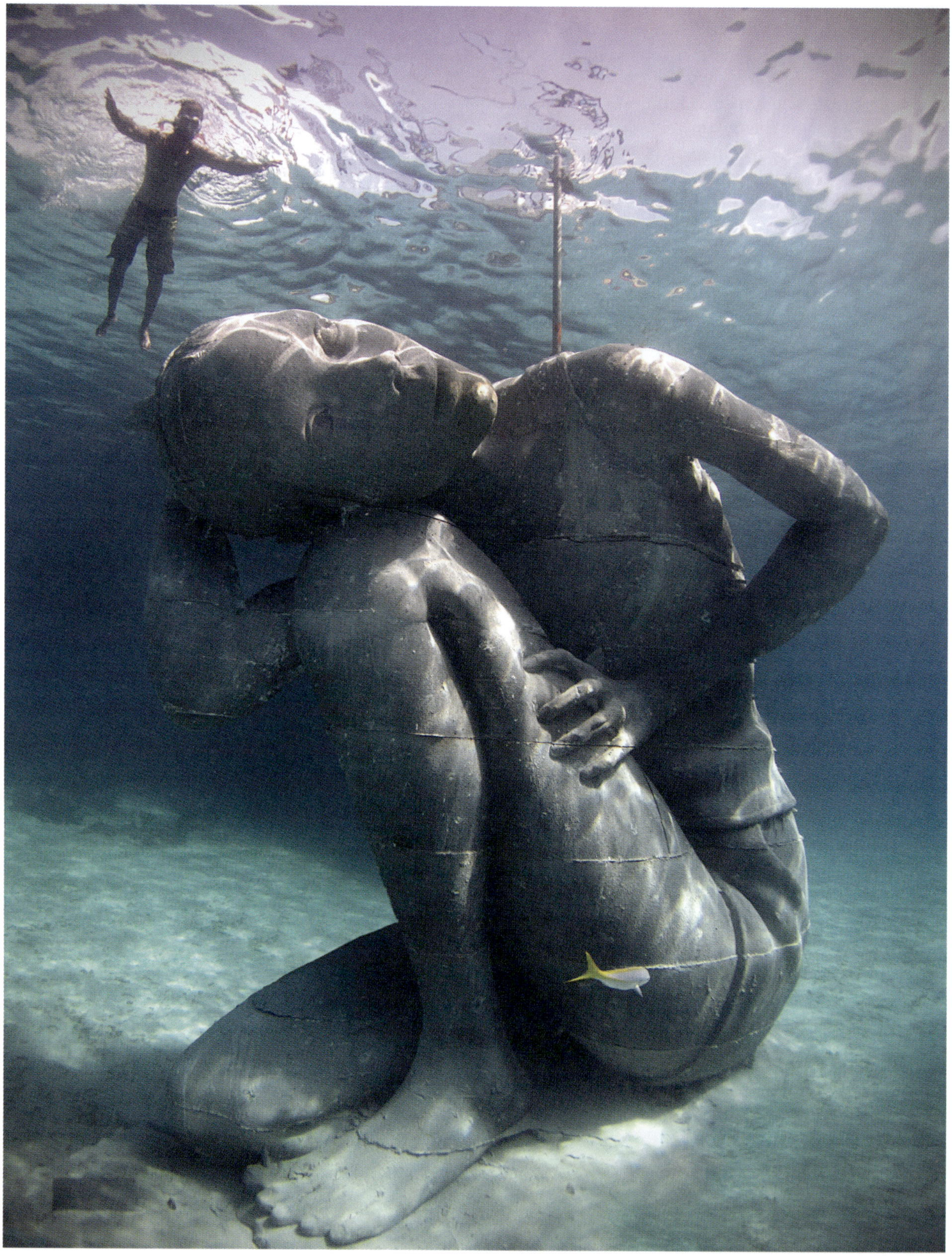

Ocean Atlas, Bahamas, 2014, Stainless steel, pH neutral cement, basalt and aggregates, 5 m

in 2006 when he was commissioned to create a project in the West Indies, off the west coast of Grenada, in response to the damage caused by hurricane Yvan. For this project, now listed as one of National Geographic's Twenty-Five Wonders of the World, he created seventy-five underwater sculptures. They were strategically positioned at depths of five to eight meters below sea level, inviting divers to explore and swim around them. Over time, these artworks have not only aided in the restoration of the marine habitat ravaged by the cyclone but also serve as a protective sanctuary for the site.

Through this project, deCaires Taylor came to realize that working underwater necessitates a multidisciplinary approach that combines art with science, history, and socio-economic perspectives. To ensure sustainability, the sculptures are crafted using pH-neutral materials, and deCaires Taylor has conducted numerous experiments to discover the most ecologically friendly components and techniques. Over the years, he has also collaborated closely with scientists who installed devices within some of his sculptures: one, *The Listener*, immerged in Cancun, Mexico, is fitted with a hydrophone that records the sounds from the local marine environment. A lonely standing figure made with hundreds of ears gathered by local students during a workshop, this sculpture is used by marine biologists to monitor the sound activities of the underwater national park in which it is located.

In 2009, he contributed to what became the first underwater museum, the Museo Subacuatico de Arte. Progressively, his practice has become centered on three core pillars: conservation, culture, and tourism. In 2019, he opened The Coral Greenhouse, an underwater museum located in the John Brewer Reef in Australia. The museum consists of a large triangular architectural structure, designed to withstand the pressure and water movements of the marine environment. Its purpose is also to serve as a sanctuary for smaller marine species, providing them with a secure space where they can seek refuge from larger predators. Within this striking structure, a series of figurative sculptures are crafted to house and nurture growing corals and a diverse array of marine creatures. These sculptures portray children engaged in various activities, such as sitting at their school desks or tending to coral gardens. They symbolize the potential and promise of the younger generation, encouraging viewers to envision a future where humanity and nature harmoniously coexist.

For each project, deCaires Taylor spends a long time on site, meeting with the local communities and scientists, inquiring about their situation and relationship with the sea, conducting research, and collecting stories. His sculptures often portray the local people whose faces and bodies he casts. For *Ocean Atlas* (2014), he casts the body of a Bahamian girl, transforming her into a contemporary Colossus, in reference to the Greek myth of Atlas, one of the Titans. Instead of bearing

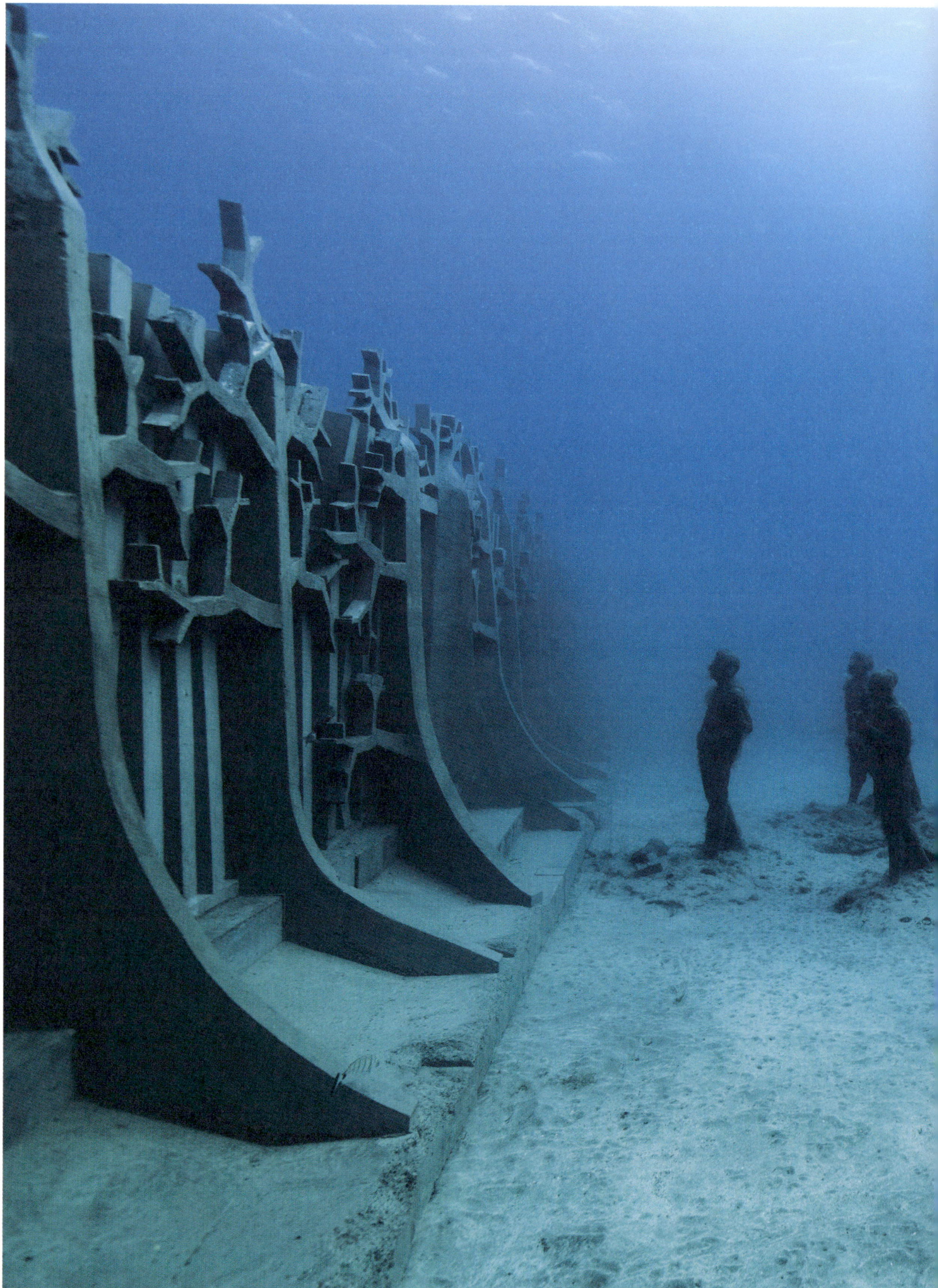

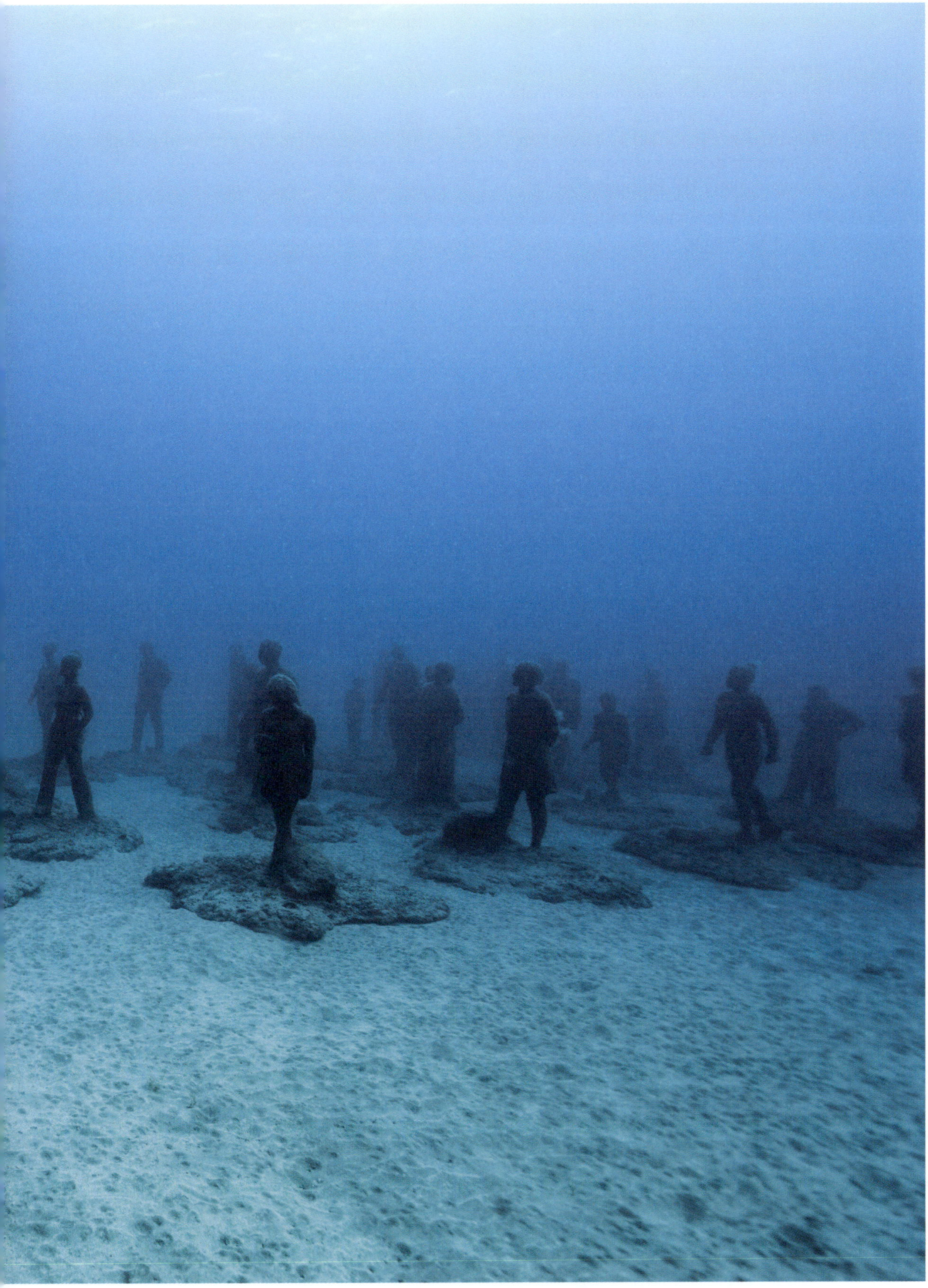

Crossing the Rubicon, Museo Atlántico, Lanzarote, Spain, 2016,
200 life-size human figures in an oceanic gyre, Stainless steel, pH neutral cement, basalt and aggregates

the weight of the Earth, she carries the burden of the ocean above her, symbolizing the pressing issues faced by the ocean in the era of the climate crisis. This monumental underwater sculpture was installed at a depth of five meters, drawing attention to water pollution along the Bahamian coast, particularly the longstanding leak from a nearby refinery.

In addition to paying homage to local schoolteachers, fishermen, women, and students, deCaires Taylor also celebrates local researchers actively involved in marine conservation. *Ocean Sentinels* (2023) features marine experts like Professor Charlie Veron, known as the godfather of corals. In the artwork, his body undergoes a transformation, with encrusted holes on his skin and his legs melded within tentacular rocks. These sculptures embody the artist's vision of a fusion of human and marine life.

Some artworks are more visible and only partially submerged, such as *The Rising Tide Haugesund* (2021), a series of horses and riders located along the coast of Norway. This tidal artwork features a series of horses and riders, drawing inspiration from the imagery of the four horsemen of the Apocalypse. Positioned at a level where the water rises to the waist of the riders, these sculptures face the sea in a manner that evokes ambiguity. They appear to be both waiting for something and serving as guardians of the calm bay.

Upon closer examination, it becomes apparent that the heads of the horses have been replaced by oil well pumps, transforming these majestic creatures into symbols of extraction and exploitation. Through this juxtaposition, the artist prompts us to reflect on our shared responsibility and passive attitudes towards environmental destruction. deCaires Taylor poignantly reminds us that approximately twelve percent of the oil in the ocean originates from oil spills.

deCaires Taylor consistently emphasizes the significance of introducing a broader audience to marine environments and effectively communicating the pressing issues that the ocean faces today. In his view, the solution lies not in developing more technologies, but rather in transforming our modes of thinking and our relationship with nature. He advocates for a shift in perception, urging us to view the ocean not as a mere resource to exploit but as a precious realm deserving of our respect and protection. He aspires to reshape our values and belief systems, reestablishing a harmonious connection with the natural world, while reminding us that we, too, are an integral part of nature.

Jérémy Gobé

Born in France in 1986

Jérémy Gobé's art practice revolves around connecting people and knowledge from different disciplines, with a focus on reviving ancient know-how in innovative ways to preserve talents, equipment, and traditional factories. One notable example of this is his work in Épinal home to an historic print-making factory famous for its illustrations, where he initiated a program that brought together contemporary artists with the old image-making factory. Through this collaboration, they rediscovered the art of printmaking using the factory's old machines.

Gobé strongly advocates for the continuation of past practices, emphasizing the importance of avoiding abrupt disruptions and instead favoring the transformation and adaptation of techniques and know-how. Simultaneously, he draws inspiration from nature.

His fascination with the ocean began in 2010 when he stumbled upon corals at a charity shop. Intrigued by their structure, which he likened to lace, Gobé found a connection between the diminishing living coral material, essential to the planet's natural cycle, and the gradual disappearance of the industrial production of cloth and traditional knowledge in France.

During his research, Gobé discovered that the pattern of the traditional bobbin lace known as *point d'esprit* from Puy-en-Velay, created over 400 years ago, closely resembled the skeleton of one of the corals he was studying. This led him to the idea of using lace to help coral regeneration. After conducting research, Gobé found that Puy cotton lace met the necessary criteria for a substrate capable of capturing coral larvae: roughness, flexibility, and transparency. In addition, the lace offered the advantages of being biosourced, biodegradable, and biomimetic.

In 2017, Gobé established a research protocol to study the compatibility between lace and anemones, which yielded promising results. Then, in May 2019, he created his first sculptures and structures using the pattern of the labyrinthine brain coral of Neptune, made from ecological concrete. His experiments demonstrated that these structures provide a conducive environment for the attachment, long-term fixation, and reproduction of cnidarias, such as corals and anemones, without releasing any toxic substances.

Gobé's art practice encompasses two interconnected aspects: his research and socially-engaged works, primarily focused on ecological purposes, and his standalone artworks that support his research. He is a self-taught artist who conducts extensive online research and experimentation in his studio to expand his knowledge and skills.

Recently, Gobé has collaborated with the aquarium of Paris, leading to a very constructive and mutually beneficial

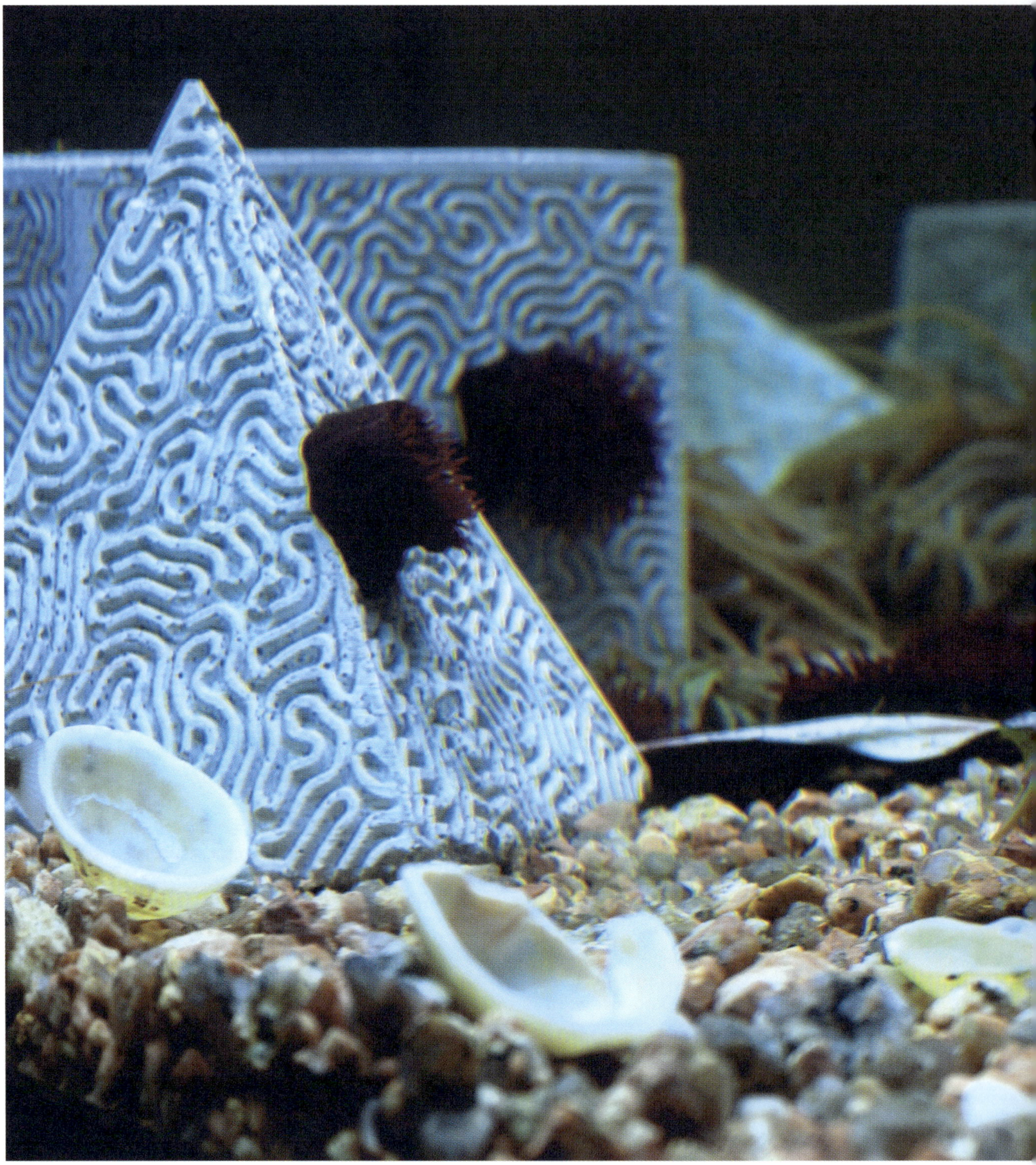

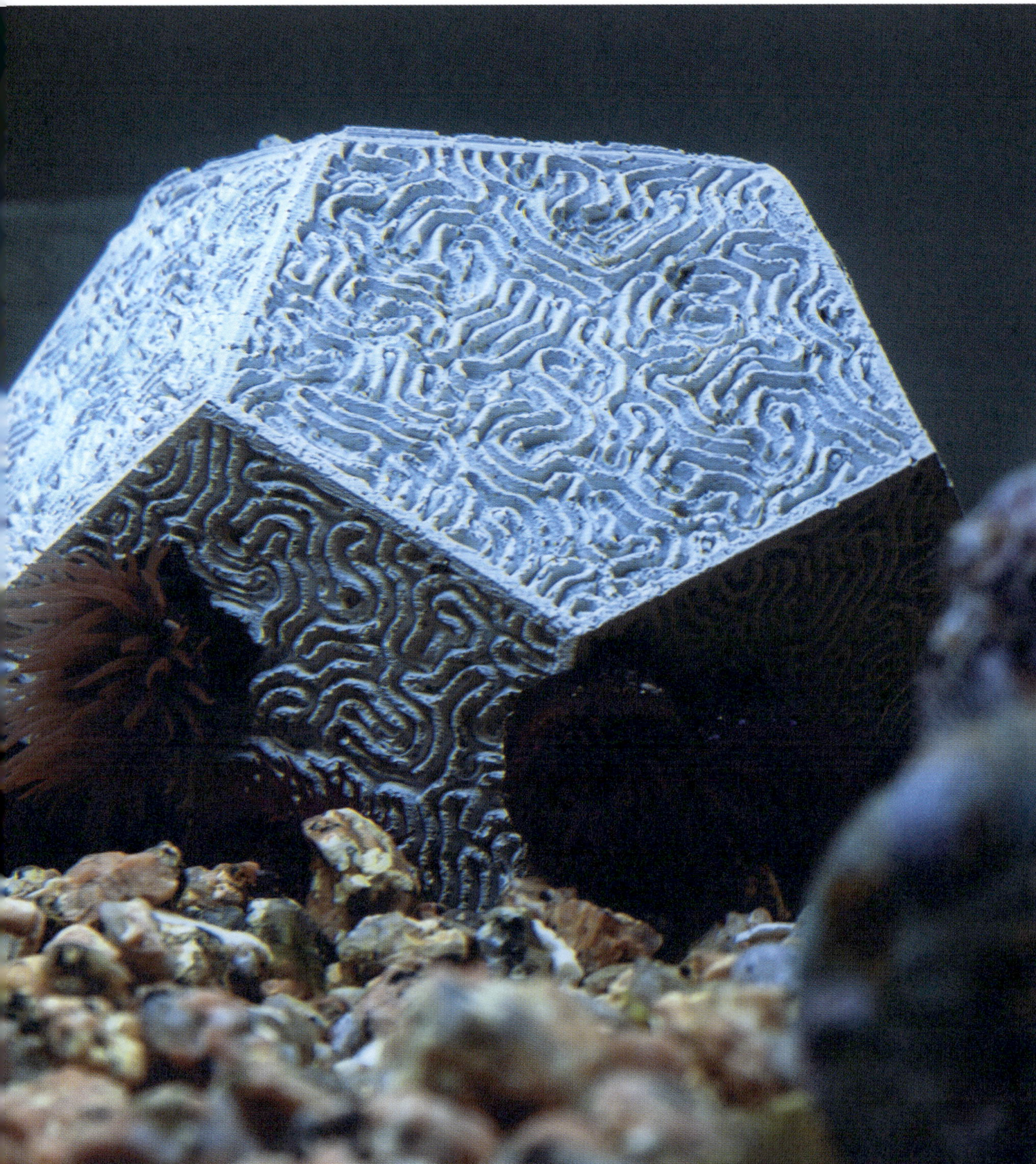

Corail Artefact, Solides de Platon, 2019, CCA1 eco-friendly concrete developed by the artist and eco-friendly pigments

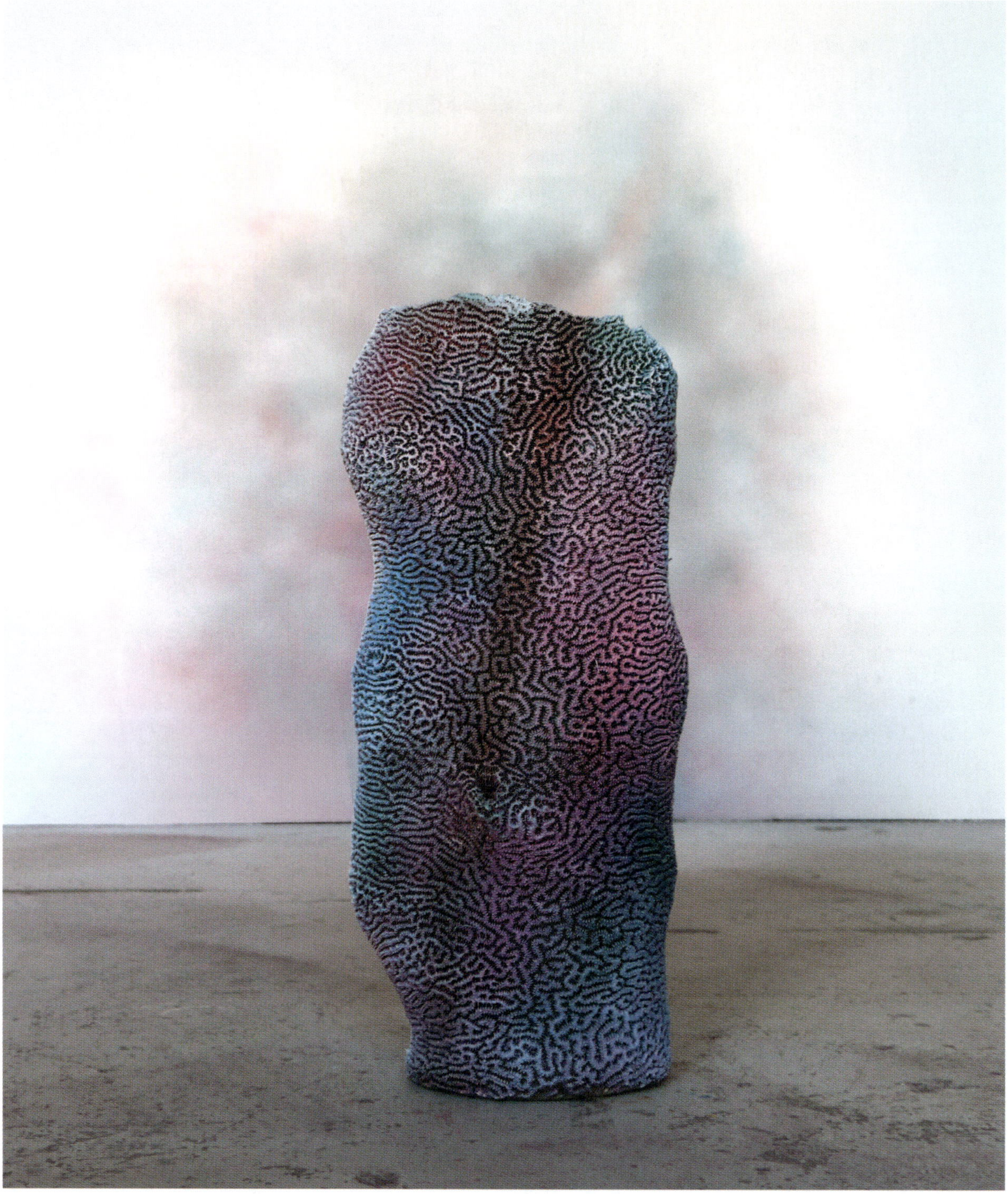

Corail Restauration, Variation 13, 2016, Plaster, ecological paint, coral skeleton, 30 × 70 × 20 cm

long-term relationship between the artist and the scientific team of the aquarium. Besides his permanent sculpture *Corail Artefact, Maille cerveau de Neptune* (2023), the artist suggested that the aquarium team test his research on coral reef restoration by immerging some of his pieces, notably from the *Plato* series.

"Jérémy Gobé has drastically transformed our way of working and our techniques. He has disrupted our habits and opened our perspective." (Conversation with Etienne Bourgouin, marine biologist and jellyfish curator in charge of aquariology at Aquarium de Paris, May 2024).

Coral is a collection of small polyps that absorb calcium from the water to build an exoskeleton. The Aquarium houses very old corals that can give birth to new colonies. One way to rapidly multiply them is through cloning, similar to the grafting of plants. A piece of living coral is broken off and then implanted on another structure in a favorable environment. This reproductive technique is used to populate display tanks and to facilitate exchanges with other aquariums.

Before collaborating with Gobé, the Paris aquarium team worked with inexpensive, basic but sometimes environmentally unfriendly materials. For example, a cyanide-based adhesive was used to attach the coral branch to its new support during the cloning process, which is toxic locally. The grafted part would necrotize but quickly be covered by new coral growth. In an aquarium this operation has a success rate of nearly one hundred percent under favorable conditions (in the natural environment, it is typically around ten percent).

Fast-setting concrete was favored as an artificial base for reefs and coral habitats. These materials are commonly used worldwide, both in aquariums and for constructing artificial reefs. It's not uncommon for a seaside hotel that has undergone extensive construction, resulting in the destruction of part of the coastal ecosystem, to submerge concrete structures to rapidly regenerate coral colonies for tourists to admire. Cost-effective materials are preferred. It's worth noting that constructing artificial reefs requires several tens of tons of concrete, making cost a crucial factor.

With his research on new materials, Gobé has challenged these practices and sparked discussions at the aquarium of Paris. He has provided solutions that are currently being implemented. He has patented an eco-friendly cement for artificial reefs. The structures he has created were submerged in the tanks for over six months for testing, and they have proven to be perfectly compatible.

While some types of eco-friendly cements existed before, Gobé's practice and commitment have encouraged the teams to modify their working methods. Today, they are collaborating with him to facilitate his approach and make this material more accessible for wider use. The main obstacle currently lies in financial constraints. Scaling up production is necessary to reduce costs. For Bourgouin, this approach also allows them to engage in discussions with the public about eco-friendly materials and the restoration of the ecosystem. →

Furthermore, Gobé has invented a completely innovative and eco-friendly cloning process that eliminates the use of adhesive. He has designed perfectly fitted structures in which the coral can integrate mechanically. He refers to this as a "plug": it is either a hole in which the coral branch is inserted or semi-flexible clamps that close around the coral to secure it and enable its growth. The artist uses 3D printing to create these pieces. According to Bourgouin, this system is revolutionary.

Julia Lohmann

Born in Germany in 1977

Since 2007, Julia Lohmann has been utilizing seaweed as the primary material for her large-scale sculptures and artistic practice. Seaweed, a fast-growing macroalga, forms dense underwater forests and derives essential nutrients for its growth from seawater, the atmosphere, and sunlight. This aquatic organism plays an essential role as a renewable resource in the marine environment. It not only absorbs significant amounts of carbon dioxide and produces oxygen, but also acts as a natural cleanser of the oceans, reducing ocean acidification while providing a vital habitat and food source for the various species residing within them. Additionally, seaweed helps combat coastal erosion by mitigating wave force and serving as a natural barrier.

"This material grows not at the expense of others, but to their benefit," says the artist, who remembers collecting flotsam with her father as a child and making sculptures from them. "I didn't grow up in close proximity to the sea, but we would often visit during the holidays with our caravan. Throughout my life, I have always cherished the sea for its profound ability to provide perspective. The vast expanse of the ocean extends the boundaries of our world beyond our individual selves. As I reflect upon my childhood, I realize that the sea was a source of healing for me. However, as I've gained more knowledge and understanding, I've come to appreciate that this relationship needs to become mutual, and my work be aimed at ocean literacy and regeneration."

Lohmann fell in love with algae in 2007 during an artist residency in Japan. Her intent was to explore the relationship between human beings and marine life, and to study how we use the sea to sustain ourselves. Prior to this trip, she had been working with leather. However, upon discovering seaweed, she immediately recognized certain similarities between the two materials. This realization inspired her to imagine a leather-like substance that could be used without the need for chemicals and could be sustainably replenished—a material that grows up

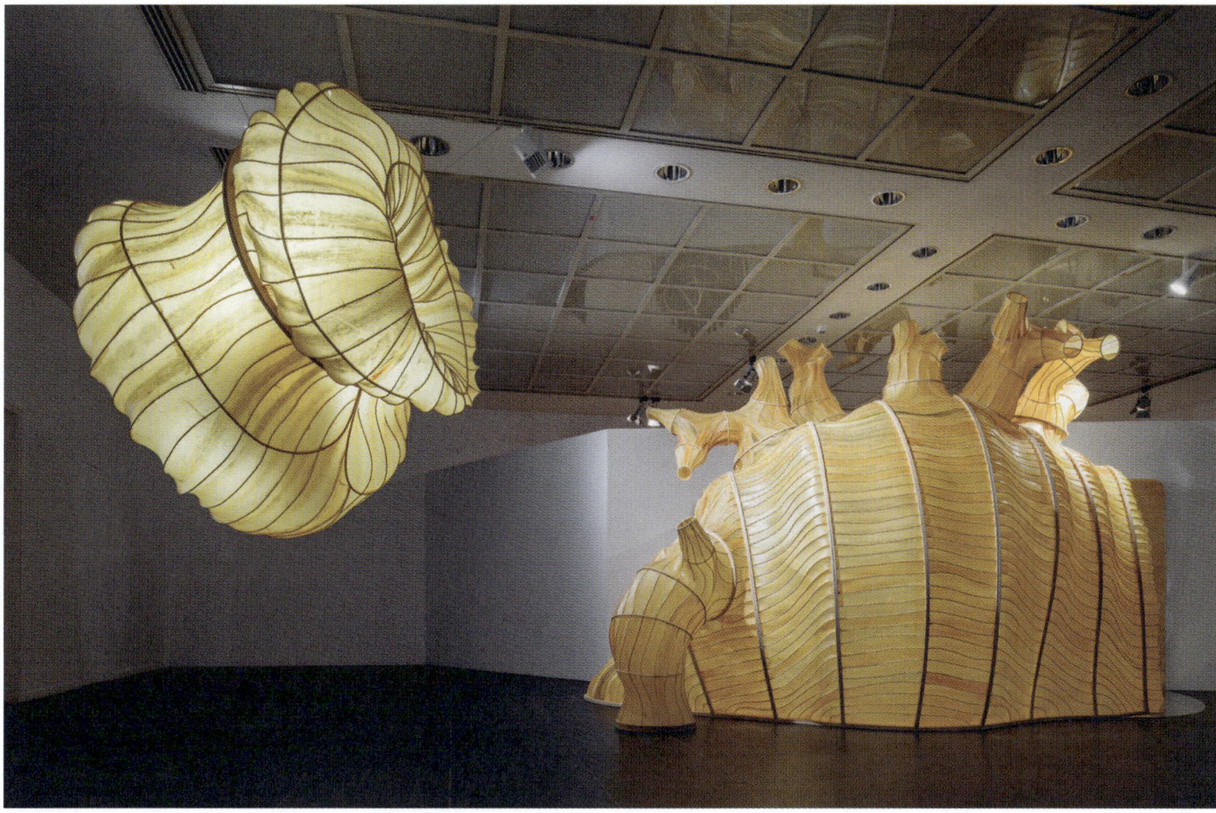

Corpus Maris II, 2023 and *Hidaka Ohmu*, 2020, Algae glued to rattan and laminated wood, 150 × 150 × 130 cm and 535 × 325 × 224 cm

to six meters long and thirty centimeters wide in just one year. She brought back ten kilograms of seaweed from Japan and began experiments to turn it into a biomaterial. Realizing that these experiments required more in-depth research, Lohmann enrolled in a PhD to study them and explore the potential application of such a material. Durin her studies, she founded the Department of Seaweed at the Victoria & Albert Museum in London in order to showcase seaweed and elevate its status as a valuable material, alongside traditional textiles and silverware. In 2013, she exhibited her first seaweed sculpture at the V&A, the *Oki Naganode*.

There is often an educational side to the artist's artworks. The giant sculpture *Hidaka Ohmu*, (2020), for instance, is used as a space for conversations around ocean literacy and protection for Lohmann's growing Department of Seaweed. Made with algae glued to rattan and laminated wood, this walk-through installation makes visitors feel as if they are inside an ecosystem while they learn about the properties of algae. Its title refers to Hidaka, a region on the Japanese island of Hokkaido where kelp originates, and *ohmu*, the name of a creature from *Nausicaä of the Valley of the Wind* (1984) by filmmaker Hayao Miyazaki. →

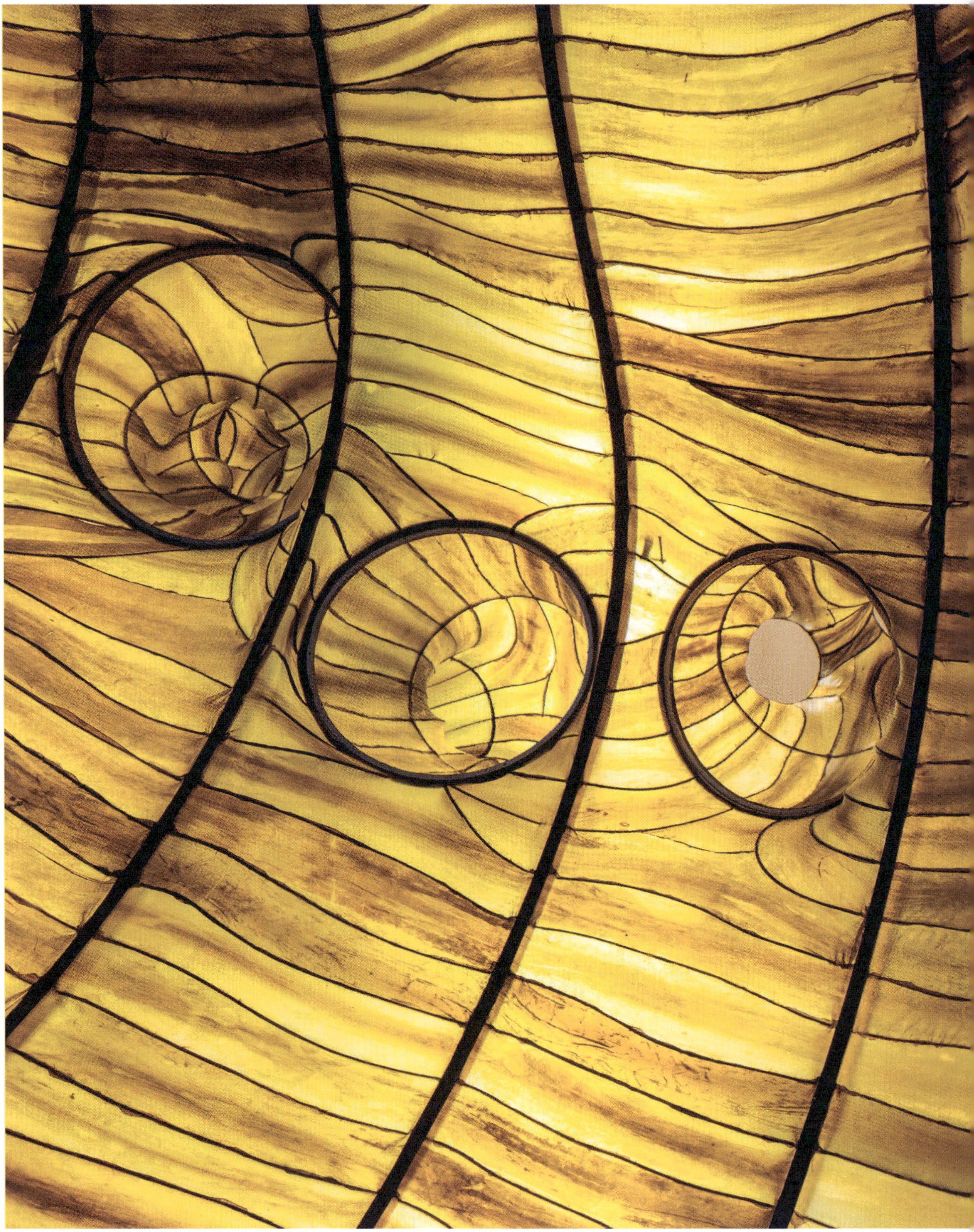

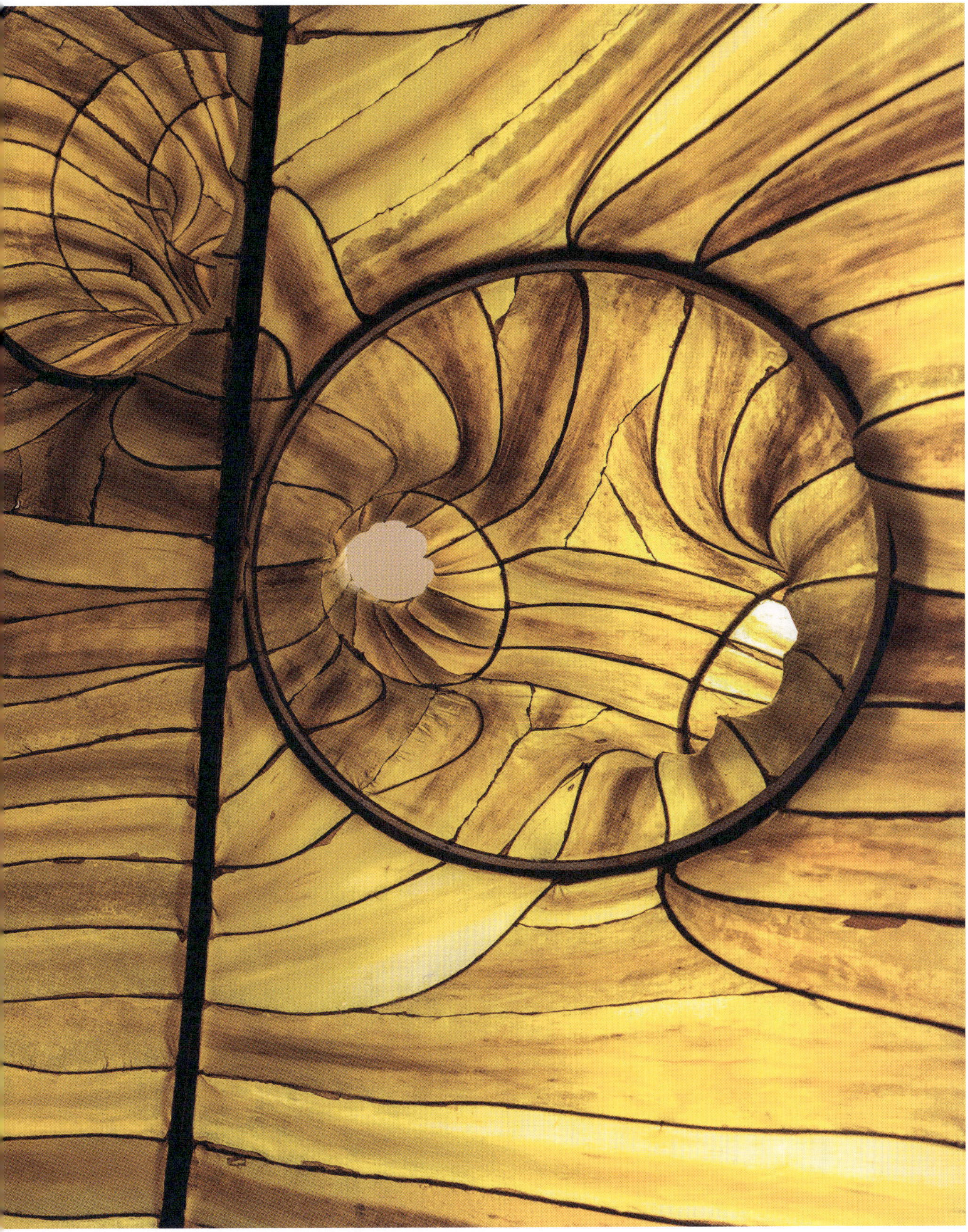

Hidaka Ohmu (detail), 2020, Algae glued to rattan and laminated wood, 535 × 325 × 224 cm

In the film, this insect protects the forest and restores the ecosystems that humans have disturbed.

Lohmann's sculptures are inspired by organic life and marine species, but yet they are transformed and almost mythologized. *Corpus Maris I* (2023) takes its shape from polyps and jellyfish but seems to grow out of the wall and to expand in the space. Over time, her artworks change, and the artist plays with these natural metamorphoses that occur due to the marine material's interaction with humidity, temperature variations, and light. Originally green because of its chlorophyll content, *Corpus Maris I*'s leaves turned to yellow and almost golden hues as the chlorophyll degraded. The algae also tends to dry and contract, like artworks that constantly change as though they were alive.

"What captivates me is how the sea facilitates a profound connection with my body while simultaneously encouraging me to look far beyond myself. It's an incredible balance. Furthermore, I've discovered that even a single piece of wet seaweed can evoke the very essence of the sea, infusing any room with its presence. This alga carries within it the spirit of the ocean itself." When looking at these giant installations, viewers can indeed feel as if they are being plunged underwater. The artist creates conditions for fantastic encounters.

Today, Lohmann collects seaweed from various sustainable algae farms, mostly in Sweden and in Japan. In her opinion, collecting seaweed contributes to a mutually beneficial outcome: as brown algae grow, they tend to absorb excess nutrients from the water, which can be detrimental to the environment. By farming local seaweed species and harvesting them for use as biomaterials, we can remove some of these nutrients and create a win-win situation.

"It is fascinating how this material can offer not just a resource to us, but also another aspect of how to be in this world." The artist's motto is "Knowing-Acting-Caring." For her, design and artistic practice bring these three aims together. She wishes to develop a collective spirit to create constructive, connected, and inclusive activism. Lohmann collaborates with experts and researchers who study the environmental benefits and impacts of seaweed. Yet, as an artist, she feels she remains freer than any scientist. "As creative practitioners, we have an incredible freedom in practices, in modalities and approaches, and we can use different methods, compile them, and mix them almost like a DJ. When we are working with others, we don't have to abide by any specific rules."

It is extremely challenging to turn invisible effects into something a larger audience can understand. How can the increased acidification of the ocean, for instance, be shown? Lohmann likes to take the example of Professor Sam Dupont, who has been working on this topic for many decades at the University of Gothenburg. His idea was to invite people to eat shrimps "from the future" so that they would realize that food will be altered as a result of the impact of climate change. He bred them in water conditions simulating those of 2030 and 2050 if humans continue to destroy marine habitats as we do now. Similarly, Lohmann's artworks are often immersive, sensual, and tactile, favoring tangible and emotional experiences.

"Science provides us with factual information, yet it requires support from other fields to create the changes we need. True transformation arises when science intertwines with individuals capable of implementing it, revealing to us the reasons we should care, the ways in which we should care, and why it holds significance in our lives. It becomes evident that the combination of science and politics alone cannot universally mobilize everyone. People yearn to understand their role, to comprehend the significance of altering their behaviors, and to recognize their impact within the broader context. It is through creative practices that we can forge connections on this profound level."

Moffat Takadiwa

Born in Zimbabwe in 1983

Known for his use of discarded materials, particularly consumer waste, Moffat Takadiwa creates large-scale installations that explore themes of postcolonial identity, environmental degradation, and the global circulation of goods. His work speaks to both the local and the global, connecting Zimbabwe's social, cultural, and historical context to broader discussions about sustainability, identity, and global capitalism.

Takadiwa's large installation *Zip it* (2024), for instance, exhibited in the Pavilion of Zimbabwe at the Sixtieth Venice Biennale, is composed of remnants from a clothing factory, including zippers, tags, and buttons. This artwork resembles the profile of a cargo ship and symbolizes how the ocean bears witness to African narratives, illustrating the journey of the continent's resources. In the complex network of global trade, the sea acts as both a conduit and a witness to the movement of goods. Raw materials, often extracted from Africa, embark on long transoceanic journeys to industrialized nations, where they are processed into consumer products. These goods, replete with added value, are then shipped back to African markets, sometimes as waste or obsolete commodities nearing the end of their life cycle. The ocean thus becomes a silent archive, carrying the remnants of these exchanges—of resources, labor, and capital—across borders. In this sense, the sea is not merely a geographical entity but a metaphorical space where histories of exploitation, colonialism, and modern-day capitalism converge. The relentless flow of goods between Africa and the rest of the world underscores a form of economic dependence that continues to shape the continent's postcolonial reality. →

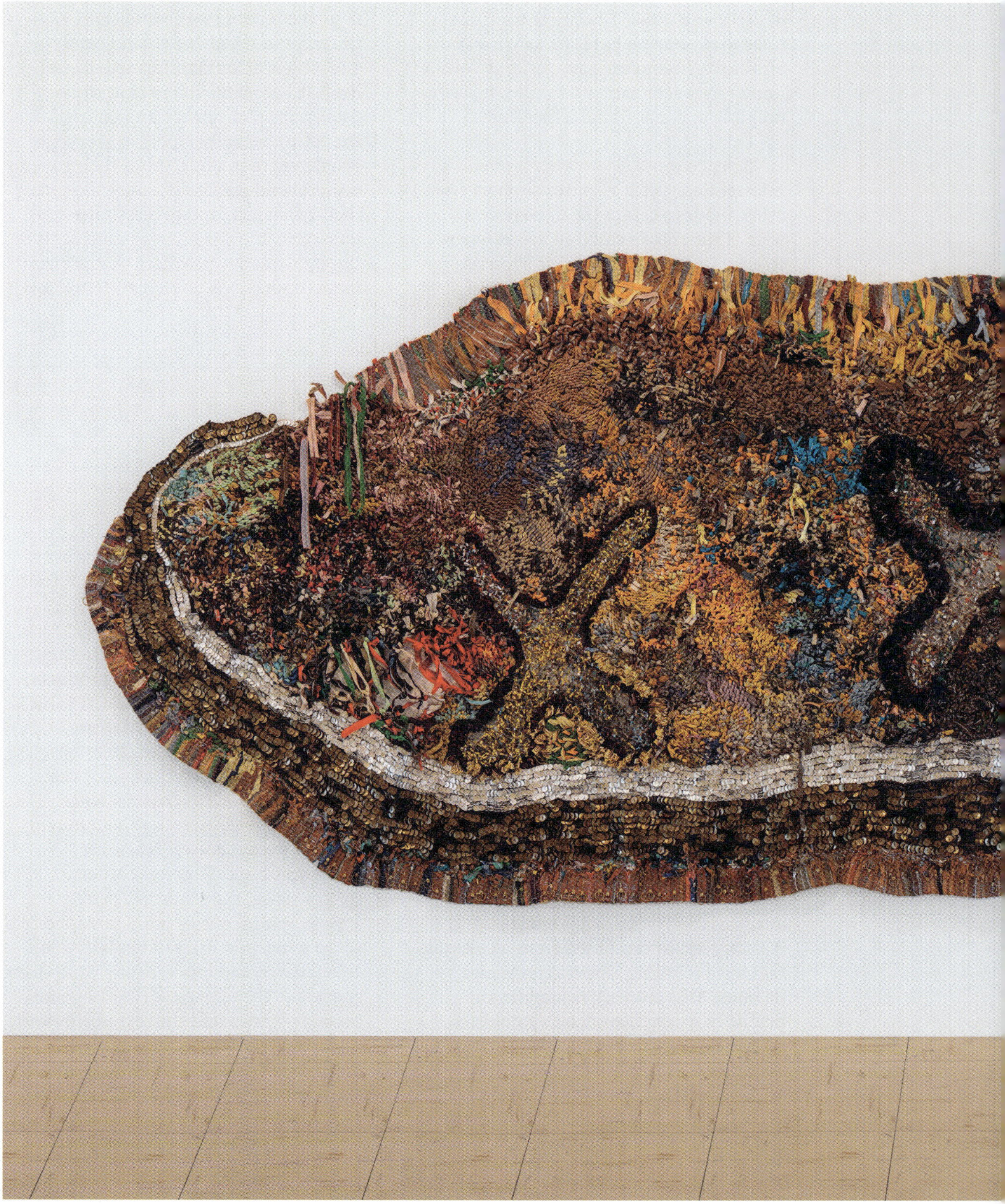

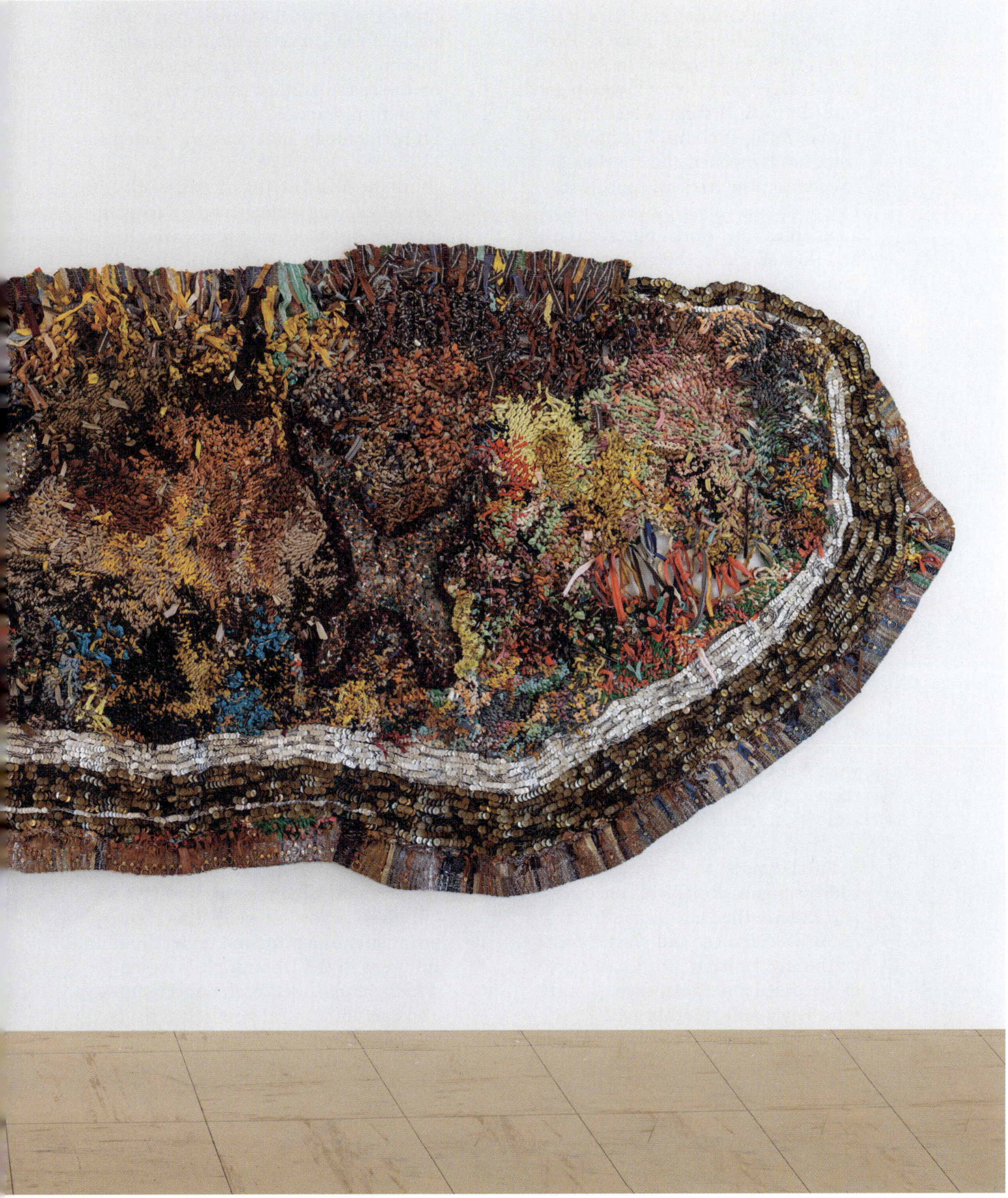

Zip it, 2024, Fabric, zippers, buttons and staples

Takadiwa's exhibition *Tales of the Big River* in Gennevilliers, close to Paris, and significantly located in France's largest river port through which goods enter Europe, further delves into this theme. As he explains, "In light of what has happened and continues to occur on the African continent, I view the sea as both a vessel and an enabler." His work captures the ebb and flow of water, the rhythms of global industrial activity, and the intertwined lives of objects and materials. Through these movements, interactions, and transformations, the artist portrays not only economic and environmental power relations but also our collective memory, shaped by centuries of global trade and resource extraction.

This collective memory is reactivated in Takadiwa's other installations, which often engage more directly with the dark histories of colonialism and the transatlantic slave trade. Works such as *The Water Vessels* suggest a potential for reconciliation and a more hopeful future. Through these pieces, the artist imagines the possibility of reconfiguring cultural and historical consciousness, proposing that even within the legacies of exploitation, there may be room for transformation and regeneration.

Furthermore, these installations address urgent ecological concerns, particularly the long-term effects of toxic substances and plastic waste on the environment. The artist's use of discarded materials, such as bottle tops, keyboards, toothbrushes, and toothbrush tubes, serves to highlight the temporality of industrial products and the impending environmental crises they pose. By focusing on the life cycle of these objects—from production to disposal—he prompts reflection on the sustainability of current industrial practices and their impact on both human and ecological systems.

At the heart of Takadiwa's work lies a deep engagement with ancestral African knowledge systems, which form the foundation of both his artistic style and thematic concerns. These systems offer alternative ways of understanding the relationship between humans and nature, particularly in terms of coexistence and sustainability. A defining feature of ancestral African architecture, for instance, is its capacity for coexistence and its ephemeral nature, utilizing materials that can be seamlessly reintegrated into the natural environment. Furthermore, these architectural designs often exhibit curvilinear forms, minimizing friction with surrounding natural phenomena.

An illustrative example of this cultural ethos can be found in the vernacular greetings among many African communities, such as the Korekore language, a Bantu language spoken by the Shona people of Zimbabwe. A typical exchange begins with "Mamukasei" (Good morning), to which the response is "Tamuka, kana mamukawo" (We are only well if you are well). This greeting encapsulates a profound understanding inherent in the Ubuntu philosophy, which emphasizes interconnectedness and communal well-being. It reflects the belief that individual wellness is intrinsically linked to the wellness of others, fostering kindness

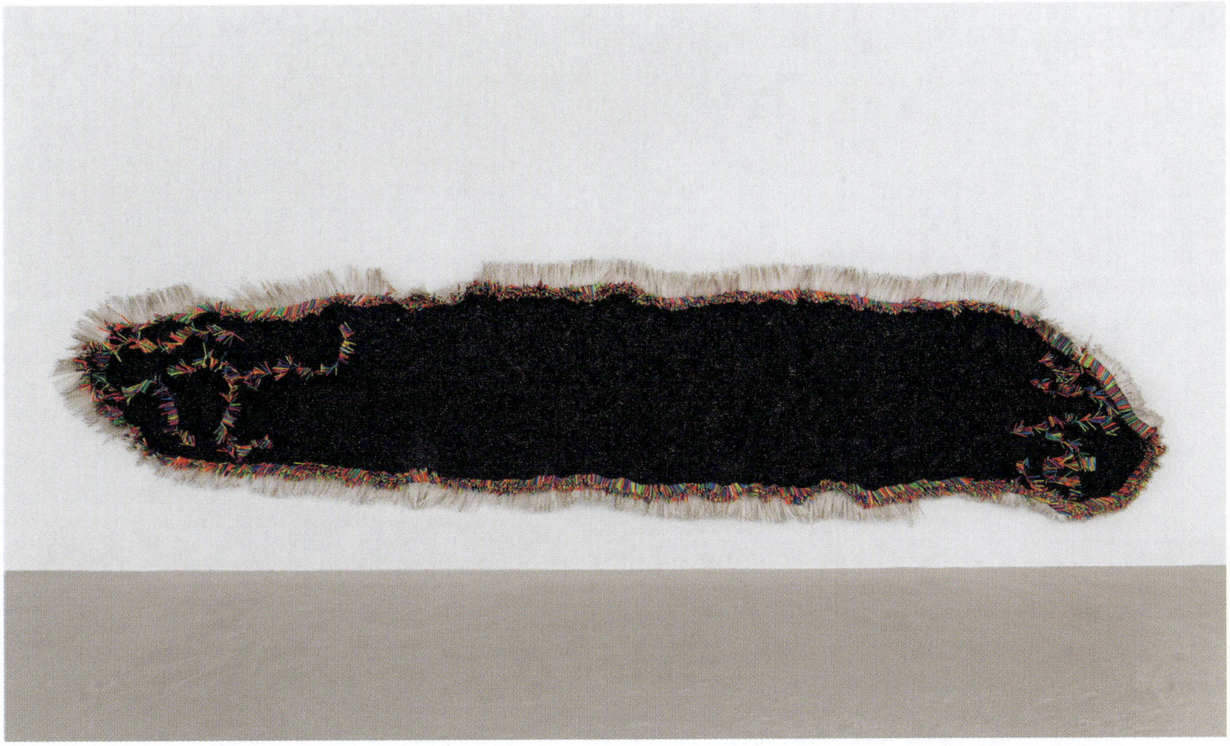

The Water Vessels, 2024, Computer keys, pen barrels and bucket handles in plastic, 180 × 775 × 35 cm

and mutual care. This cultural perspective encourages the sharing of resources and objects, effectively addressing issues related to growth and accumulation often associated with modernity. Moreover, such practices promote environmental stewardship, as collective care for each other inherently extends to the care for our shared environment, thereby contributing to solutions for urban waste management challenges.

Jennifer Morris
CEO of The Nature Conservancy (Arlington, United States of America)

The Fate of the Ocean

More than fifty years ago, in 1972, the crew of the Apollo 17 spacecraft took the first photograph of Earth as a whole from space. This image, known as the Blue Marble, is one of the most iconic and reproduced photos in history and something I reflect on when thinking about our ocean.

The most dominant element visible is the blue expanse that connects us all—circulating oxygen, nutrients, water and weather across the globe like a vital organ. From the shores of the Mediterranean near Nice—where the UN Ocean Conference convenes in 2025—to the tidal wetlands of the Potomac River where I live in Washington, DC, the global ocean touches all of our lives.

When we talk about addressing the climate and biodiversity crises, land protection goals often take center stage. But terrestrial and marine ecosystems are not separate—they work together in a delicate balance. Half of the oxygen we breathe—however far inland we live—comes from the ocean. The ocean is a source of food and income for more than three billion people, the vast majority in developing countries. Marine and coastal habitats like reefs and mangroves provide a first line of defense against sea level rise, climate-charged storms, and increased flooding and erosion. Forty percent of the global population—over three billion people—depend on fish for protein, yet more than one-third of global fisheries is overfished, and another third cannot sustain further pressure.

Thankfully, there are solutions that benefit both marine ecosystems and people's livelihoods. With The Nature Conservancy's goal to conserve four billion hectares—or more than ten percent of the world's ocean area—to contribute to the global commitment of thirty percent protection by 2030, we are laser focused on implementing durable ocean conservation at scales that matter.

We are working for a healthy and resilient ocean that supports all the people of our planet. We have over fifty years of experience working across the whole ocean, locally and at scale, and we go to where the problems exist, using science, and expertise, and leveraging partnerships to implement lasting solutions.

Close to two-thirds of the ocean and almost half of the planet's surface are international waters—areas beyond the control of any one nation. International policy is critical for progress. That's why The Nature Conservancy supports the ratification and implementation of the High Seas Treaty—a critical piece of achieving 30x30 for the ocean and ensuring that the high seas are protected (protecting at least thirty percent of the ocean by 2030).

Since 2016, our Nature Bonds program has helped developing countries use debt solutions to create long-term sustainable financing for conservation, including establishing Marine Protected Areas and supporting improved ocean protection and management across 238 million hectares of ocean—that's an area larger than the Tasman Sea and almost the size of the Caribbean.

Innovative solutions like insuring coral reefs against damage caused by tropical storms and hurricanes provide critical funds for reef restoration and management efforts. And investing in natural climate solutions, like securing the future of fifteen million hectares of mangroves globally by 2030 through the Mangrove Breakthrough, is vital to halt biodiversity loss, safeguard the health of the ocean, and benefit coastal communities.

To address illegal, unreported, and unregulated fishing at sea, we launched the Tuna Transparency Pledge, a global initiative uniting retailers, suppliers, and governments to advance one hundred percent on-the-water monitoring across all industrial tuna vessels by 2027. A lack of transparency at sea allows unsustainably and illegally caught fish to enter our supply chain undetected. Accurate and reliable data is essential to ensure that the seafood on our plates has been harvested in compliance with fishery laws and social standards. The Tuna Transparency Pledge represents a bold, achievable step in transforming the health and sustainability of our ocean.

We also need to think and act locally for the millions of people around the world who live near and with the ocean. Lasting marine conservation—whether protection, restoration, or more effective management—means working alongside local communities every step of the way.

In many places, Indigenous communities are leading this effort, such as in the Great Bear Sea, Canada, where seventeen First Nations, the Government of Canada, and philanthropic investors through the Enduring Earth partnership closed a Project for Finance Permanence to fund conservation and stewardship of marine territories. This Indigenous-led effort will deliver an estimated $750 million during almost twenty years to conserve one of the richest and most productive cold-water marine ecosystems on Earth, while bolstering sustainable marine-based economies and creating thousands of jobs. Indigenous and traditional knowledge has stewarded the ocean for generations and these communities are critical partners in marine conservation.

While the ocean faces serious and increasing threats, it is also one of our biggest allies against climate change and a lifeline for sustaining the planet. I am reminded of the awe of the ocean when I visited the pristine Palmyra Atoll, a low-lying coral island in the Pacific protected by The Nature Conservancy and the U.S. National Fish and Wildlife Service. As an avid snorkeler and diver, seeing the thriving coral reefs, abundant sharks, and rich biodiversity was like seeing what the ocean used to and should look like—a testament to the name Dr. Sylvia Earle gave this place: a hope spot.

There is great hope in the ocean and we must value and sustain it. From the depths of the high seas to coastal communities, the ocean connects us all, and it's time to recognize that the ocean's fate is intrinsically linked to our own.

Metis Projects

Sometimes a mermaid indicates in a dream a ritual to bring her into our world.
Settle in while I tell you my story.
You have the right not to believe me, and I won't hold it against you.
I think I would have certainly reacted like you.
I can't tell you the day when all of this began;
I just know that I dream of this woman,
The lady in the middle of a stretch of water,
Her presence both reassuring and unsettling.
Have you ever felt a presence inhabit you to the point of no longer being able to distinguish the real from the dream?
Have you ever fallen in love with a vision?
Have you ever wanted to live in your dream to be close to a silhouette?
Do you feel, as I do, the call of the water?

Lhyliann Therance Ralff

1. *Coastal Whisper*
Bangladesh

2. *Blue Exile Art Project*
Turkey

3. *Dualidad Marina*
Costa Rica

4. *Filets du Futur*
Martinique, France

5. *La Mer et l'enfant*
Democratic Republic of the Congo

6. *Transmigrations*
Morocco

7. *Vivre ensemble*
Togo

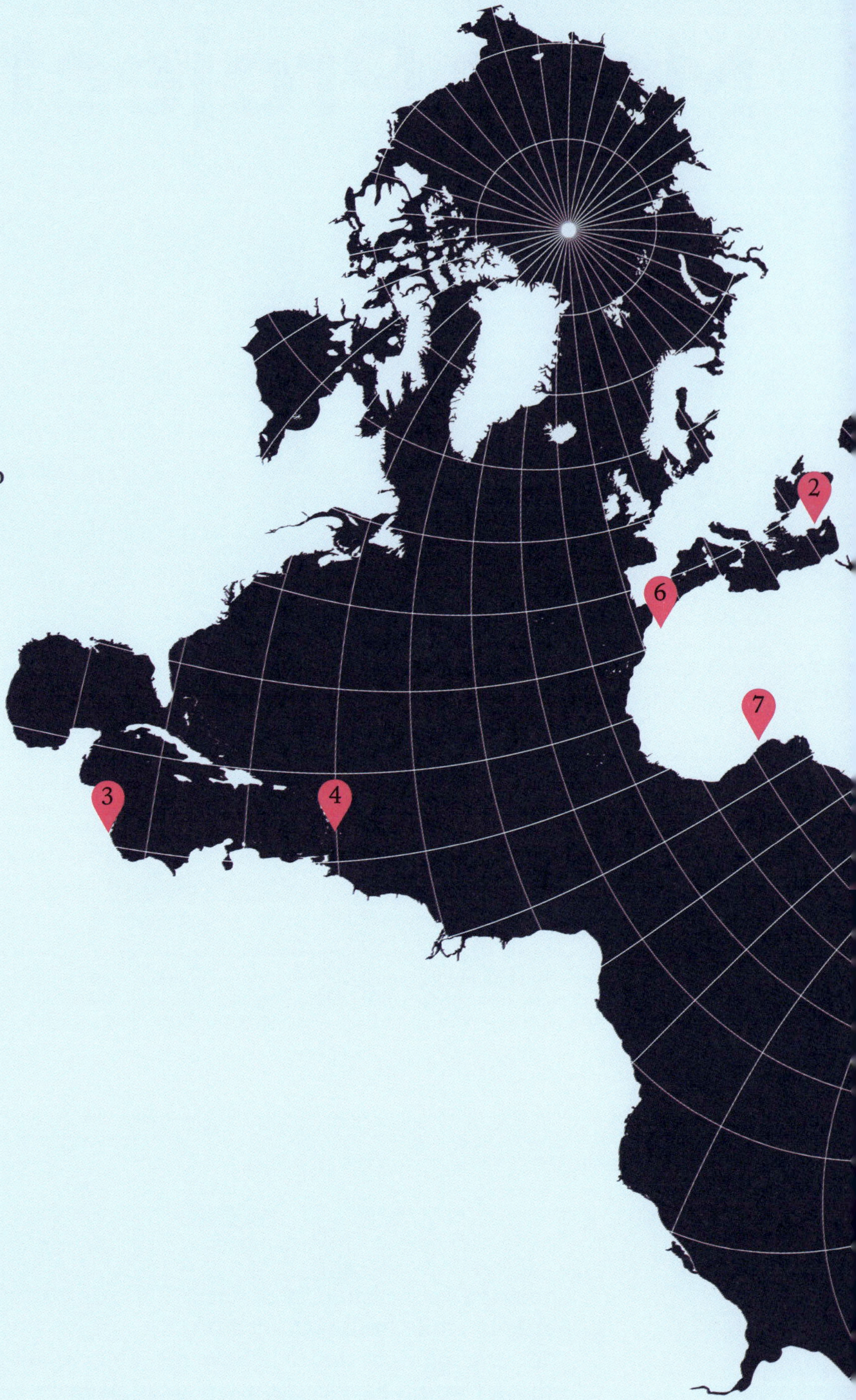

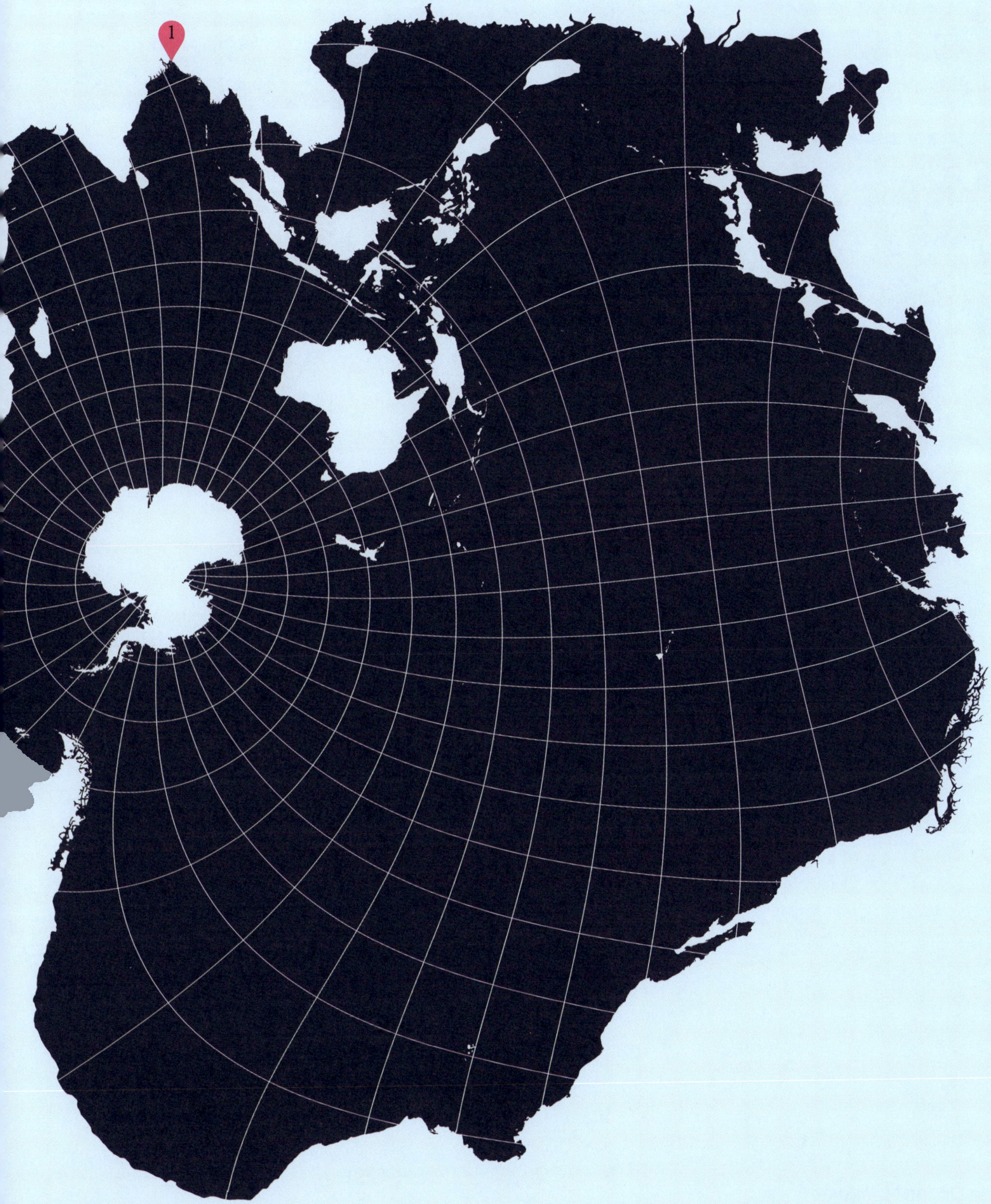
1

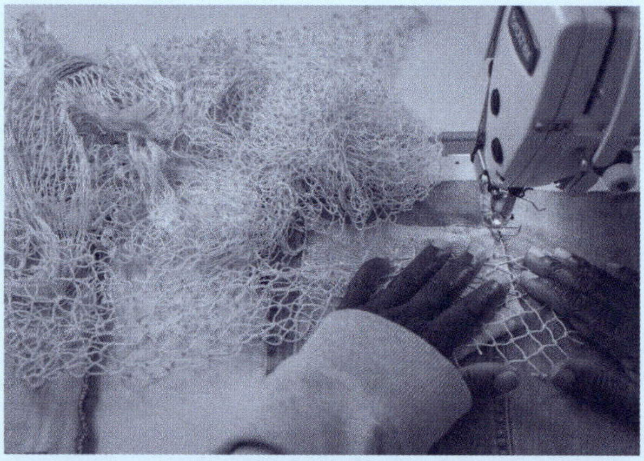

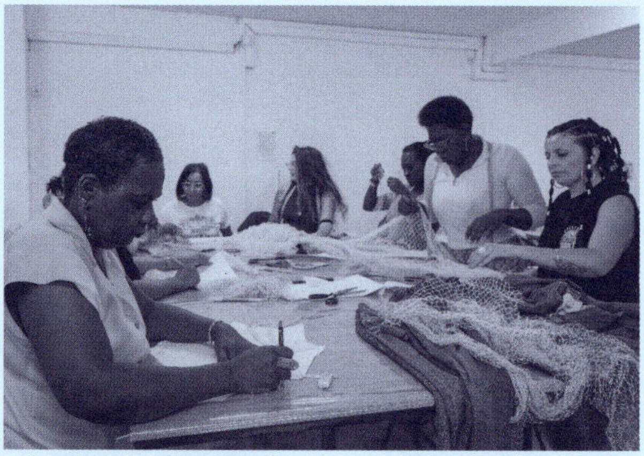

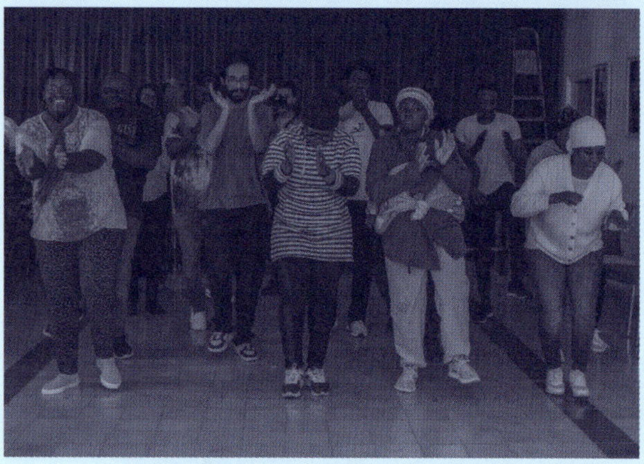

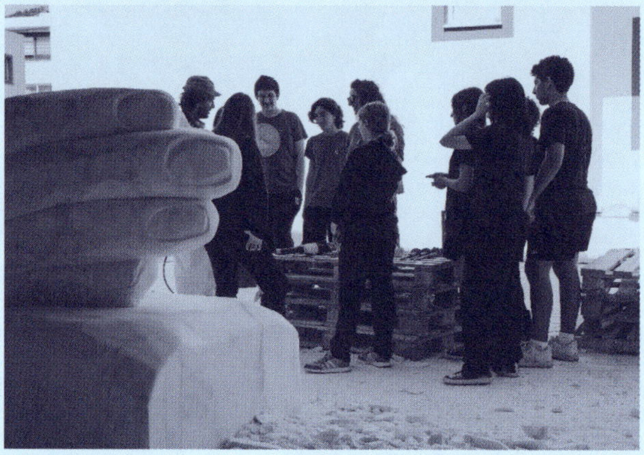

Filets du Futur, Martinique, France
Transmigrations, Morocco

Filets du Futur, Martinique, France
Blue Exile Art Project, Turkey

Created in 2021 under the impetus of the French Development Agency, the Metis Arts and Development Fund anchors its activity in countries facing major social and environmental challenges. Metis turns to the arts and artists as a response. It funds projects that lead to the creation of works of art that are committed to and inspired by concrete social realities. These projects are implemented with and for the people in developing regions. They provide space for emotions and creativity. They encourage individual and collective change. Metis is about "sharing emotions, creating transformations."

More specifically, a work of art produced as part of a Metis project is the result of a distinct approach, conceived by the local actors involved with the support of the Metis fund. Each project brings together three categories of stakeholders:

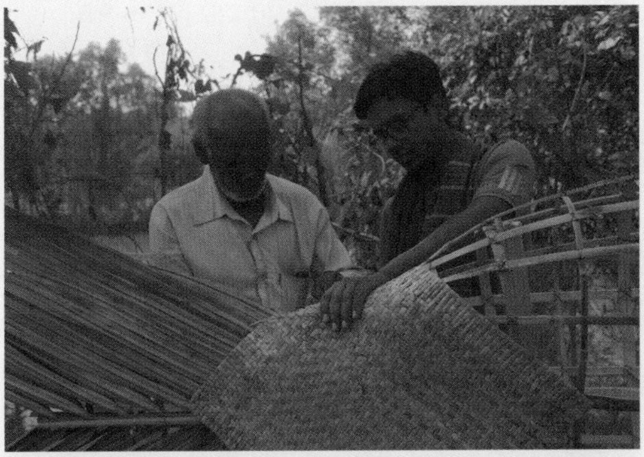

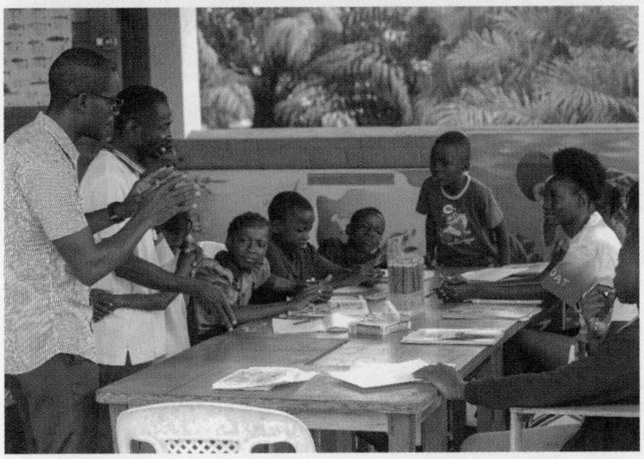

Birth of a Wish Fish, Bangladesh
La Mer et l'enfant, Democratic Republic of the Congo

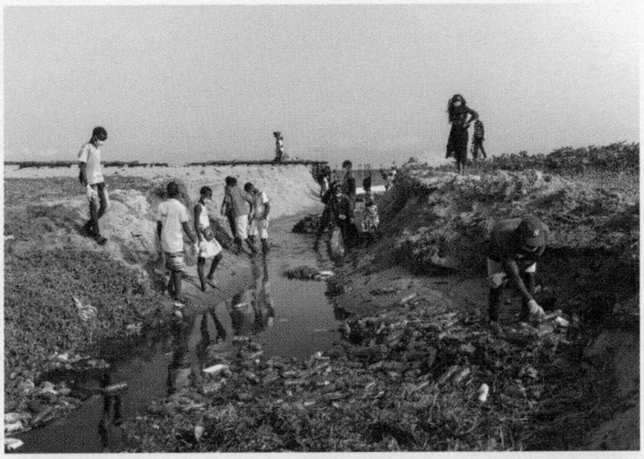

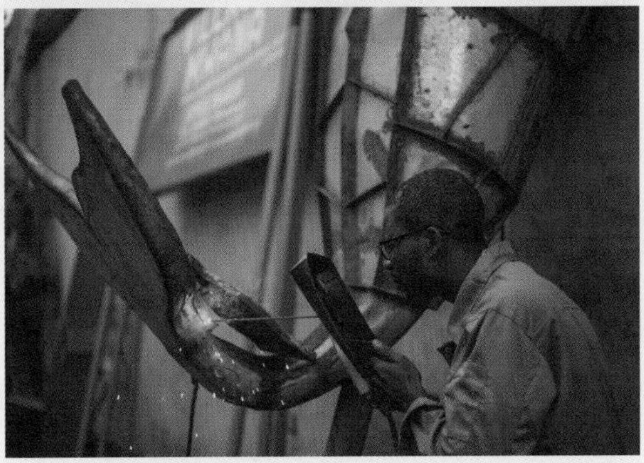

Vivre ensemble, Togo
La Mer et l'enfant, Democratic Republic of the Congo

an artist, a local community, and a development agency–an NGO, a local authority, a company, a state decentralized service, a hospital or even a school network. Together, they identify and address a development issue that affects them and, even more so, the community. The project always begins with the encounter of the three groups of participants, through a joint artistic practice. This decisive stage paves the way for the creation phase. What emerges–a work of art often surprising and unexpected–is then presented during a moment of restitution and circulation.

This chapter showcases some of the projects conducted as part of the Metis Diapason program in 2024 and 2025, dedicated to Oceans. The works created, the voices of the artists who have pledged their commitment to these local endeavors, enrich the selection presented in this book and offer a new perspective, always plural.

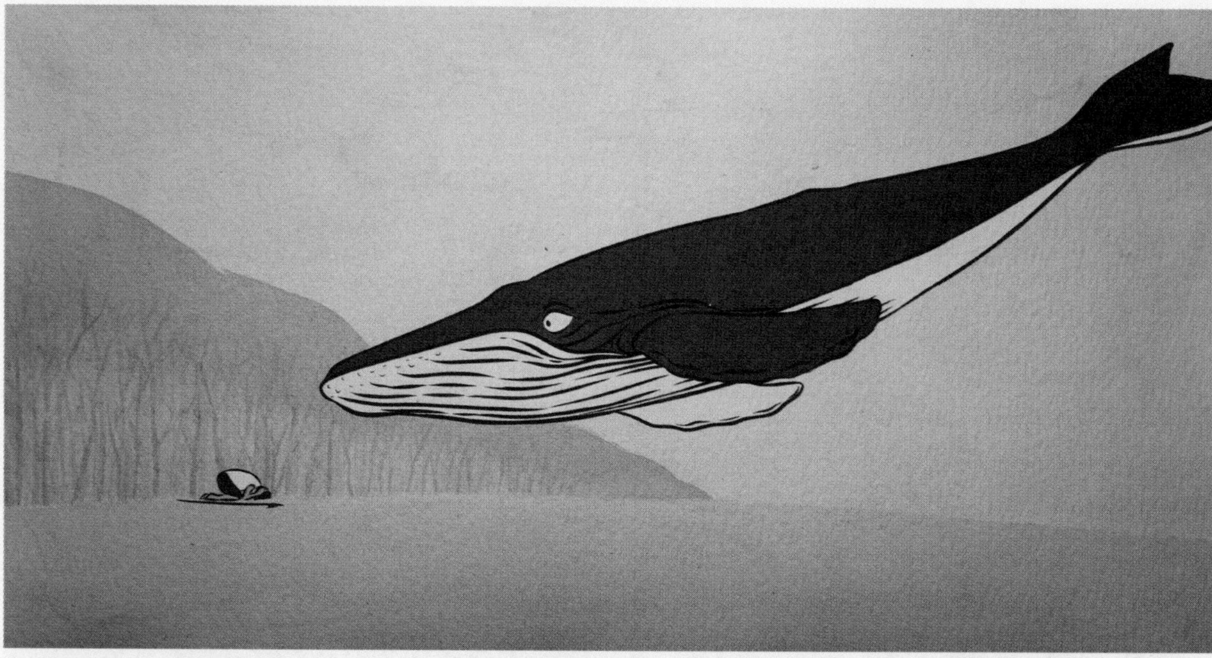

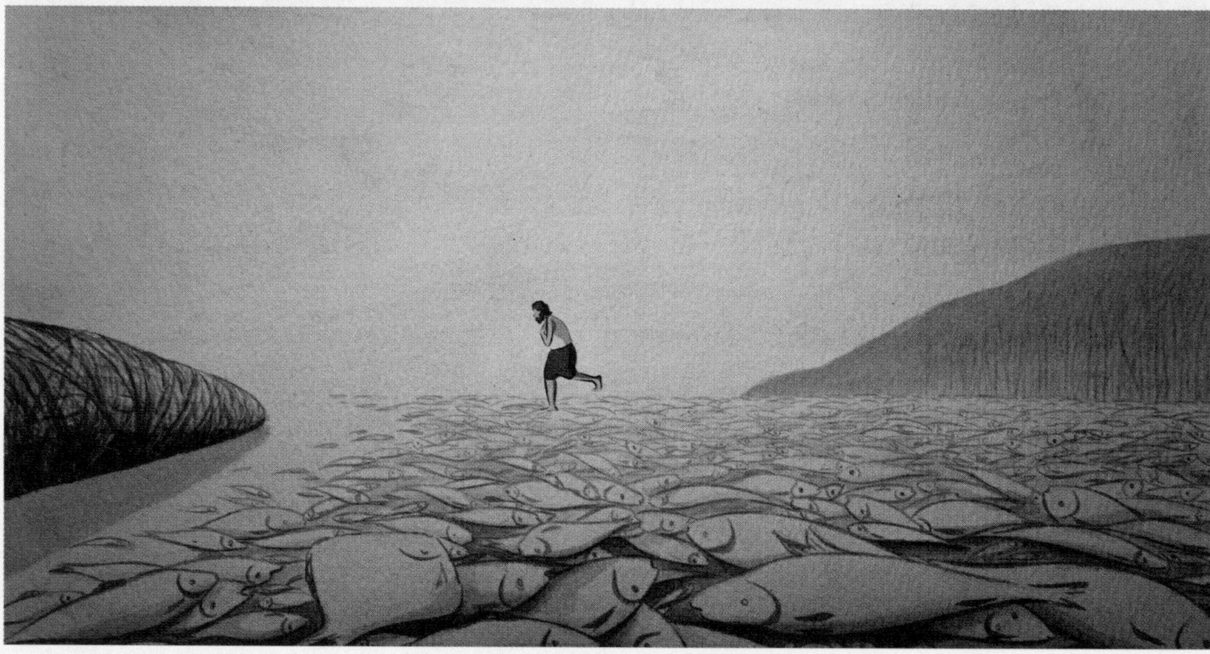

Song of Whales, 2024, Video, 12 min

Coastal Whisper

Bangladesh

Coastal Whisper **is a long-term research project led by Afroza Hossain Sara along the coastline of Bangladesh. It originated with** ***Song of Whales*****, an artistic and collaborative initiative, and culminated in** ***Birth of a Wish Fish*****, a collaboration involving nine artists, Afroza and residents from West Dhangmari nestled within the vast mangroves of the Sundarbans. The project integrates ritualistic performances and aims to valorize local coastal traditions and ecosystems.**

Song of Whales (2024) is a 2D animated short film that centers on the mournful song of a mother whale. The narrative follows a fisherman who bears witness to the cataclysmic uprising of the ocean and its inhabitants. As the fish, crows, crabs, sky, water, and even the spirits of deceased creatures join in the mother whale's grieving over the loss of her calf due to human-inflicted harm, the fisherman initially attempts to move on. However, he ultimately decides to immerse himself in a dream of a future where nature will once again provide him with an abundance of fish, as it had in the past. The film was inspired by the washed-up carcasses of whales observed along the shores of Cox's Bazar, a city on the southeast coast of Bangladesh. As the project developed, it incorporated real-world events, such as the mass stranding of millions of small fish, into a spiritualized narrative intended to evoke an emotional response from the audience.

In continuity with this film, Afroza is now engaged within the local community in the coastal areas around Sundarbans to exchange and activate debates about marine issues with people from other parts of the country, including thought leaders, policy makers, and the youth. From these conversations, they will create a collective artwork, *Coastal Whisper*, with the support of the Metis Fund.

For Afroza, it is urgent to change our attitude towards the ocean. She says that despite the abundance of research, articles, essays, and various other informational materials published over the years, a notable lack of collective knowledge regarding the ocean and "her children" persists to this day. In this context, the introduction of art as a medium to convey knowledge can have a transformative effect, as it allows people to experience and internalize the information in a more visceral and impactful manner. Through the empirical knowledge gained from this experiential engagement, individuals are more likely to open their hearts and minds to other forms of knowledge sources.

"While reading, watching, or other forms of knowledge gathering helps people know the information, it doesn't have much impact. We have got to understand that the majority of people already have a lot on their plate to deal

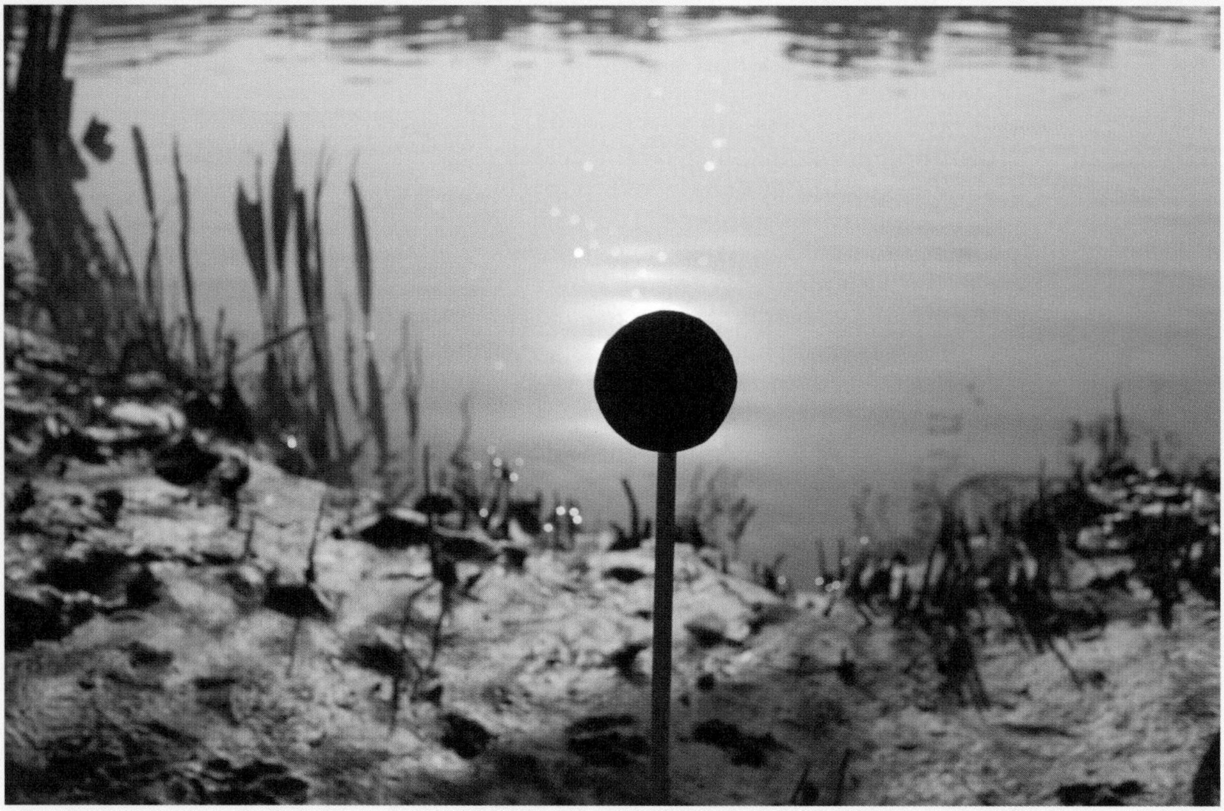

Nojor Phota on the bank of Dhangmari canal, 2024, Blank ink, paper, bamboo stick

with at home, work, and in society. Thus, when art comes in and helps people experience the knowledge, it is magical."

In Bangladesh, coastal people routinely witness the washing ashore of millions of dead fish, while fishermen at sea return with dwindling catches, and those operating in rivers struggle to harvest migratory fish. In response to this scenario, aquaculture and pond-based fish farming have become prevalent. However, these farms are repeatedly devastated by the onslaught of cyclones. Despite these challenges, the coastal communities maintain a deep understanding and reverence for the laws of nature, expressing their gratitude through the cultural practices they have developed. For these local populations, the ocean is both a provider and a destroyer. The Sundarbans, a mangrove-dominated delta formed by the confluence of the Ganges, Brahmaputra, and Meghna Rivers in the Bay of Bengal, serves as a mediator, mitigating the ocean's wrath while also furnishing food and other vital resources. Fishermen and forest harvesters have dedicated household shrines for Bonbibi, the revered protector of the Sundarbans, with both Hindu and Muslim populations believing in her watchful guardianship.

For Afroza, it is imperative to engage direcly with and learn from the coastal communities themselves. The preexisting mindset of raising awareness among these communities and inspiring behavioral change must be reevaluated. These coastal populations have already demonstrated a profound understanding and stewardship of their natural environment. The critical first step is to learn from their modes of thinking, their connections, and their interdependent relationships with nature, which are rooted in gratitude. Subsequently, it is necessary to amplify the knowledge and voices of these communities, with the aim of inspiring global policymakers to enact nature-centric policies that can effectively mitigate humanity's active and passive participation in harming marine life and disrupting the ocean's balance. Additionally, it is urgent to reconnect the hearts of people from other areas with nature, and to facilitate the opening of their minds to the growth of a collective, naturalist intelligence.

Blue Exile Art Project

Turkey

The *Blue Exile Art Project* is an environment- and eco-oriented art project, initiated in 2022 by the NGO Project Posidonia in partnership with the Bodrum Maritime Museum and the Blue Aegean Sea Rotary Club. Overseen by Mert Gökalp, a marine biologist and underwater documentary specialist based in Istanbul, the initiative involves the creation of five large-scale boat sculptures made by five local artists.

Caroline Ha Thuc *How would you describe the daily issues linked to the marine environment that you are facing near Bodrum?*

Mert Gökalp The Bodrum Peninsula transitioned from a sponge diving hub to coastal tourism after the industry collapsed in the 1980s due to a disease affecting the local sponges. Early eco-friendly tourism (small hotels, local charters) later gave way to mass tourism, with resorts and cruise ships proliferating. Government-backed, unchecked development ignored environmental assessments and local opposition, degrading natural spaces. In the 1990s, fish farms emerged, polluting northern waters until they were relocated offshore. Unfortunately, instead of creating natural reserves in these coastal areas, the vacated coastal zones were then overtaken by new resorts. Despite early sustainability efforts, profit-driven expansion prioritized infrastructure over ecology.

CHT *How do you measure this impact on the local biodiversity?*

MG I grew up in Istanbul but spent most of my time here, in Bodrum, where I feel I belong. My first free dive was when I was four, collecting seashells from the bottom and, since then, I have developed a passion for the sea. I know the coastline and the underwater geography of this place very well, perhaps almost as well as the sponge divers. I can say that I am a pioneer in witnessing the destruction of the underwater world around here. How can I live a normal life while watching this?

You need to understand that water here has always been paramount. Now there is also a shortage of drinking water, and coastal dredging is causing more destruction. The resorts are dredging the coastline and destroying seagrasses while their landscaping and the marinas and beaches they are creating are destroying the natural beaches.

Overall, we are witnessing the disappearence of a whole ecosystem.

CHT *Your PhD focused on sponges, but now you are studying a very local, essential type of seagrass called* Posidonia oceanica*. What role does it play in this ecosystem?*

MG *Posidonia oceanica*, which is endemic to the region, is unique to the Mediterranean Sea. This seagrass is paramount for many reasons: it provides an important habitat and hunting and breeding ground for a large variety of marine species. It protects the coastline from erosion, absorbs excess CO_2 thus mitigating of climate change, produces oxygen, and sustains fisheries and tourism with the ecosystem services it provides.

What people need to understand is that if they cut it, if this seagrass begins to disappear, then the quality of the water will deteriorate, because the water is pristine thanks to it. However, tourists want perfect beaches, and they have an ideal image of white sandy beaches, like in the Maldives. The hotels are ridding these areas of Posidonia, or covering them with platforms for sunbeds. Without light, photosynthesis cannot take place, resulting in the death of these plants.

The project I am working on now is to map the areas where *Posidonia* grows in order to protect these zones. I believe this is a way to officially acknowledge their existence, and to better communicate about these areas, which should be better protected. These maps can inserted into Mediterranean Sea maps and given to boat captains to prevent them from anchoring their boats there. Hopefully, one day, we will be able to create a marine park.

CHT *You are also involved in an art project that aims at protecting these areas.*

MG The Blue Exile Art Project involves the creation of large-scale

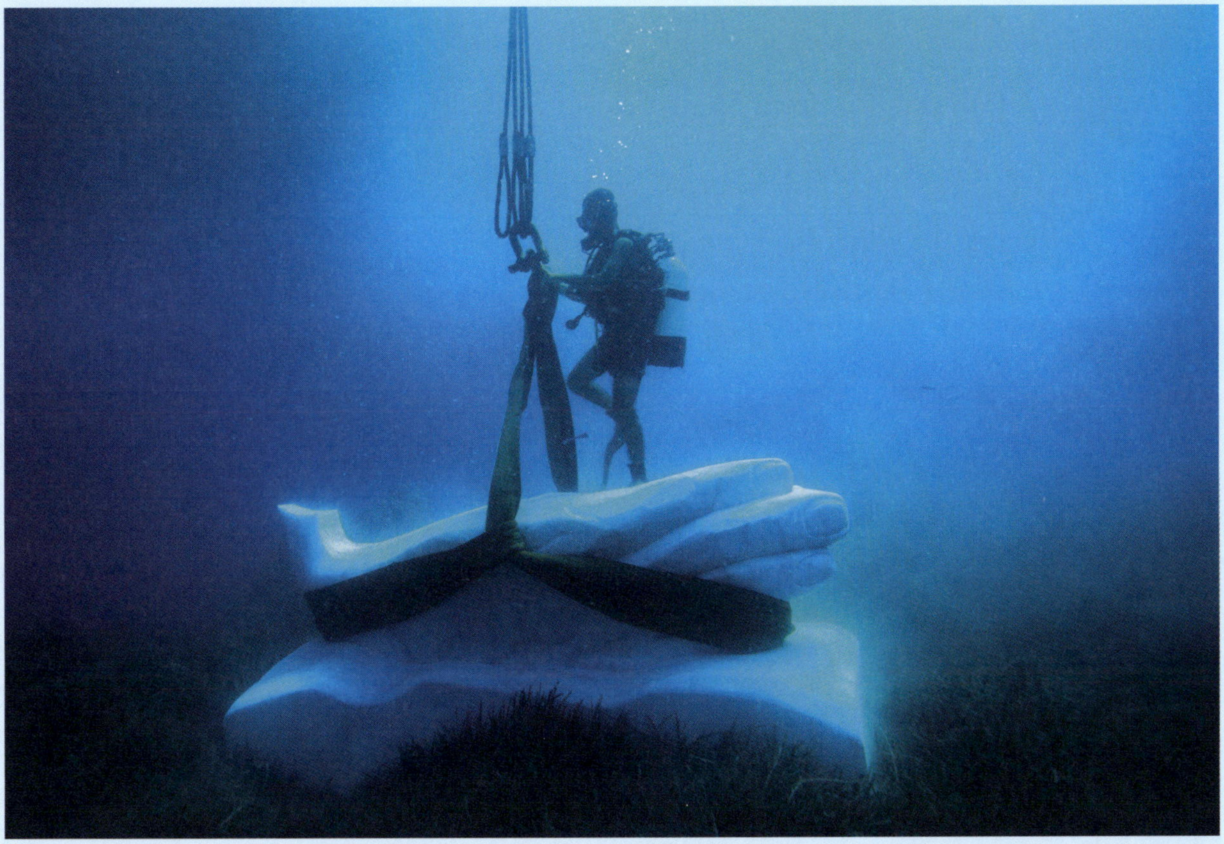

Blue Exile Art Project, 2024

boat sculptures made by local artists. Once submerged, those sculptures will be permanently visible under the sea, as an artistic approach to helping to raise awareness of the issue of seagrass meadow conservation and of the local biodiversity at large. In addition and thanks to the support of the Metis initiative and the French Consulate in Ankara, five artists have created various boat styles representing Bodrum's seafaring history.

CHT *How do you feel about introducing alien elements underwater, even if it is for a good cause?*

MG Of course, as a scientist, I cannot tell you that this is totally eco-friendly. No one can say that any material placed underwater will be ecofriendly as, fundamentally, it does not belong to the sea. But we can find low-impact and marine flora- and fauna-friendly materials. Moreover, you need to accept compromises if you want to move forward. We cannot ban tourism from the coast, right? We want the tourist industry to respect the coastal habitats and value seagrasses. We want the sector to be satisfied, and we need to find a middle ground. We are now debating the best material

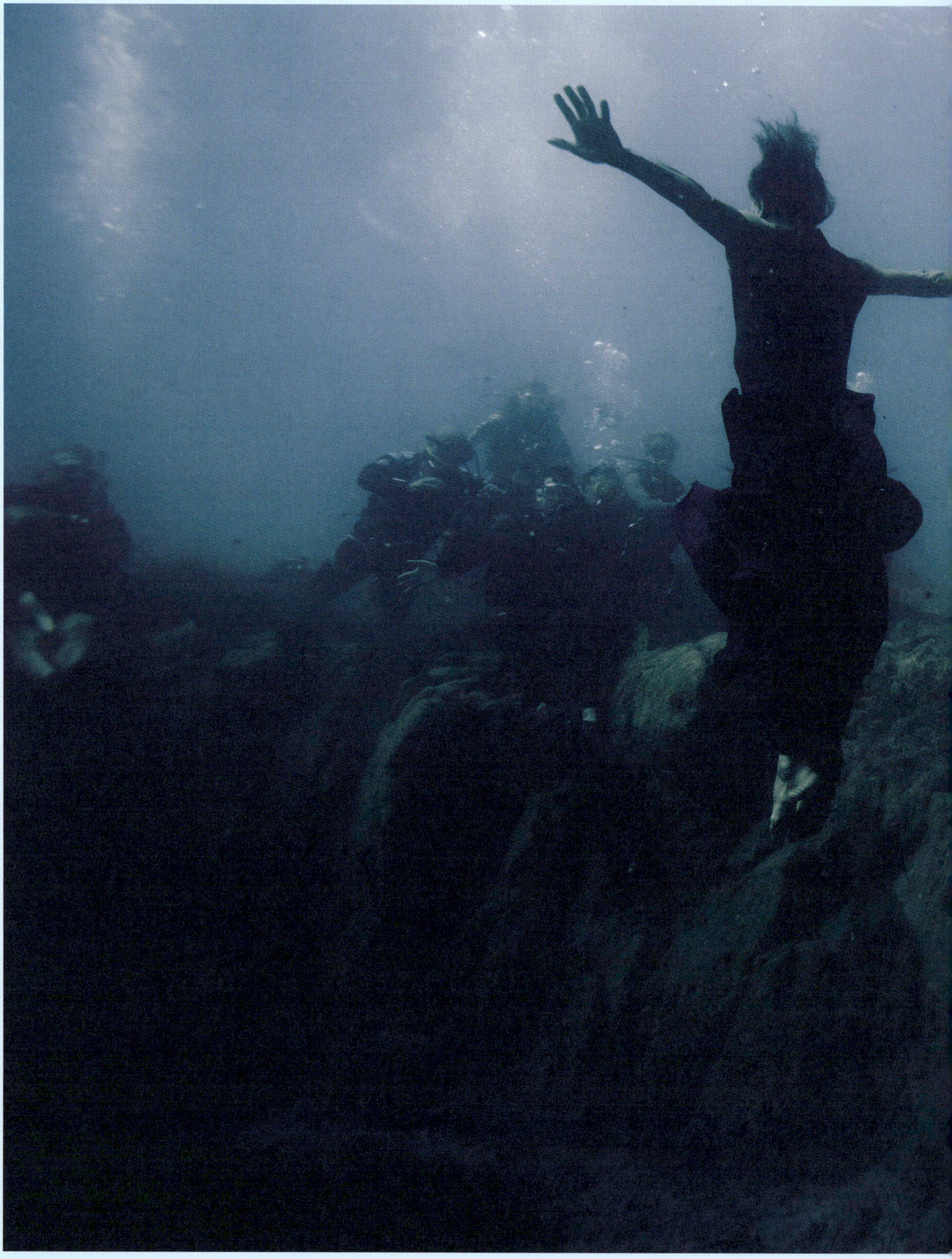

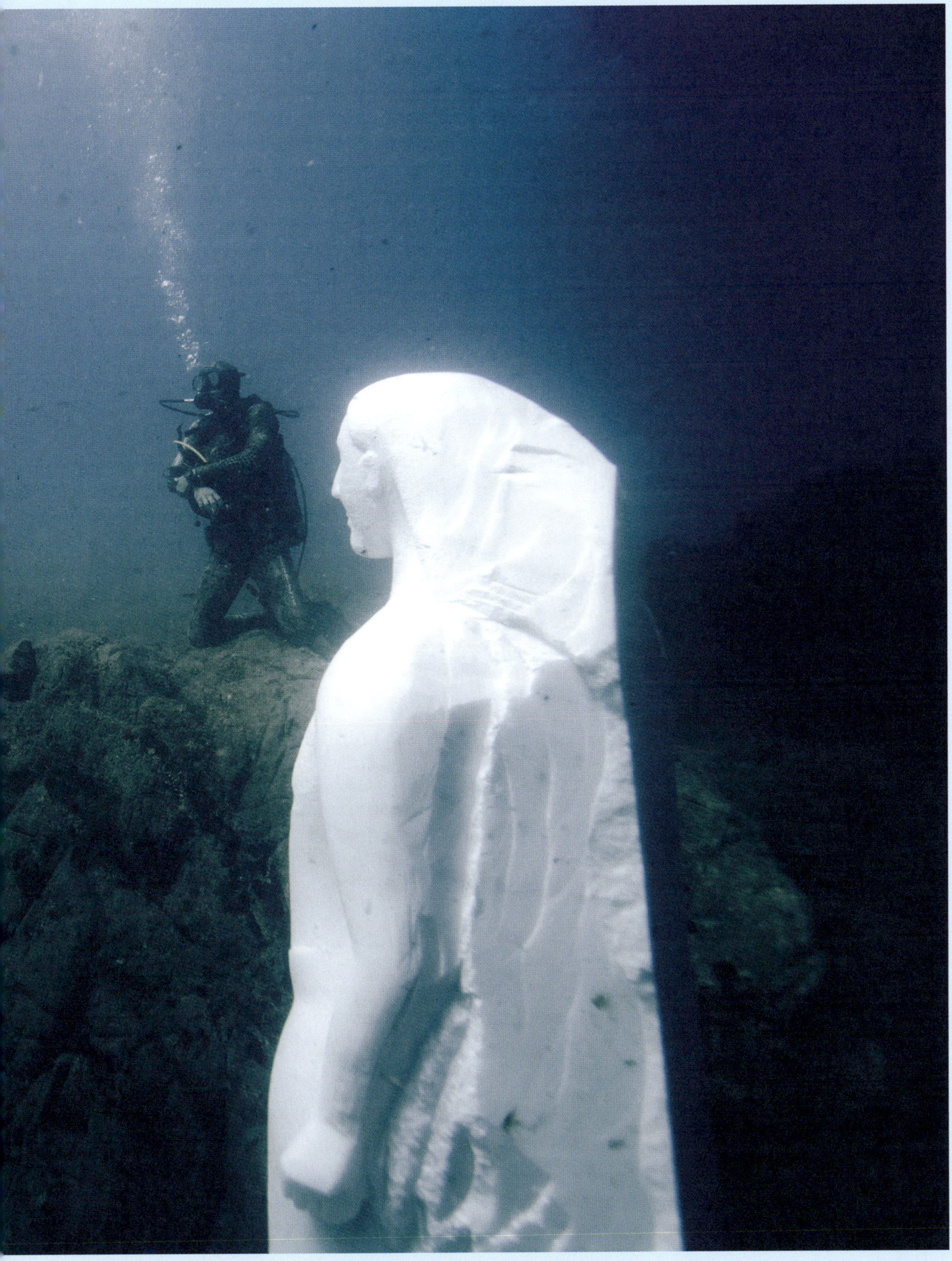

Blue Exile Art Project, 2024

to be used for these sculptures, which will have cavities for animals to build their habitat.

CHT *Why do you think that art can offer relevant opportunities to address these complicated ecological issues?*

MG I define myself as an activist, a documentary filmmaker, and a scientist. As a scientist, even though we know how to identify issues and often understand how to solve them, our voices are seldom heard. Academic papers are read by a limited audience, and government bodies frequently disregard our findings. The only solution is to be on-site, engaging directly with people to change their perceptions.

To draw attention to these critical issues, I employ all available means. As long as people fear the sea, harbor misconceptions about seagrasses, or view the ocean merely as a resource, progress will be limited. I see myself as a bridge between marine life and the public.

Dualidad Marina

Costa Rica

Implemented by a team of scientists and by new media artist Diego Esquivel, the *Dualidad Marina* Project aims at intervening in the public space in San Jose, Costa Rica's capital, in order to raise awareness about current issues in coastal and marine ecosystems. It consists chiefly of a video mapping projected on the façade of the National Museum and includes discussions, workshops, and knowledge-sharing activities among various local communities.

Located in Central America, Costa Rica is bordered by the Pacific Ocean to the west and the Caribbean Sea to the east. Its name, Costa Rica, meaning "rich coast," reflects the abundance of its coastal regions. However, the local culture remains predominantly continental. A former Spanish colony, Costa Rica gained independence in the mid-nineteenth century. At that time, the official national narrative highlighted the European origins of its inhabitants, marginalizing indigenous peoples and Afro-Caribbean communities who arrived during and after colonization. "Our identity has been constructed inwardly, with a denial of the sea and despite our profound dependence on maritime transport," says Fabián Coto Chaves, Executive Director of the Huellas del Cambio Foundation which is overseeing the project.

It is only recently that Costa Rica has begun to pay attention to the need to protect coastal and marine ecosystems and to emphasize their essential role in preserving life and combating climate change. Today, however, these ecosystems, particularly mangroves, are at risk of collapsing due to the transformation of productive systems. This would have severe ecological and economic consequences. For example, mangroves act as natural barriers against extreme weather events and are essential for the reproduction of many species.

Working with scientists and artists, Coto Chaves's engagement with marine issues is threefold. First, he seeks to raise awareness about the immense potential of Costa Rica's marine environment and coastal regions. "It is important for the country to decentralize and recognize this part of the territory, along with its communities," he emphasizes. Second, he aims to highlight the issue of chemical pollution, which is severely impacting the sea and coastal areas. Monoculture farming, such as pineapple production, and the heavy use of pesticides are

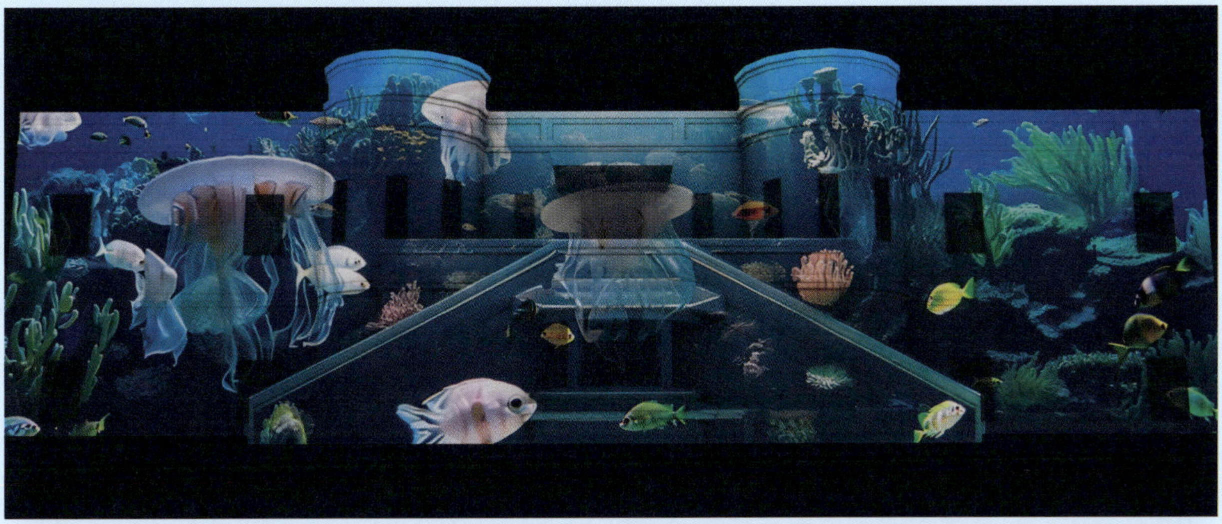

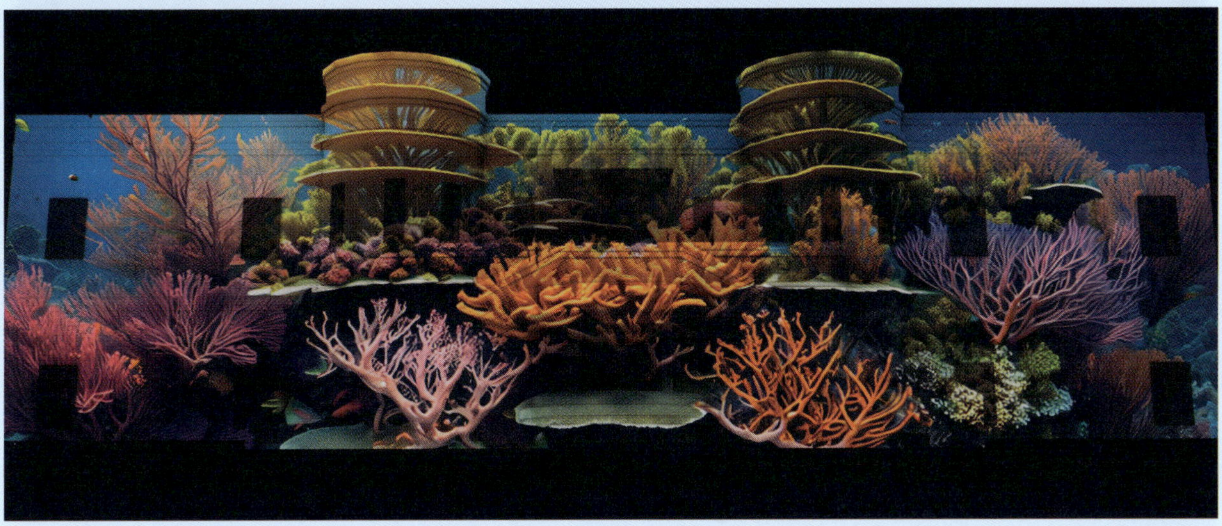

Diego Esquivel, *Dualidad Marina*, 2025, Video projection

contaminating the water. This pollution seeps into the soil, flows into rivers, and eventually reaches the oceans. Finally, Coto Chaves wishes to address the national issue of the sewage system, because only half of the country's urban waste is being properly treated. Currently, untreated waste often ends up in the sea.

For the Metis project *Dualidad Marina* Chaves has collaborated closely with scientists from the Center for Research in Marine Sciences and Limnology (CIMAR) and artist Diego Esquivel. Their community-based initiative focuses on two major issues: supporting the fishing communities along the Pacific coast and addressing the deforestation of protected areas along the Caribbean coast. Recently the clearing of a forested area that, according to court rulings, was supposed to be reintegrated into the Gandoca-Manzanillo refuge, was authorized. Coastal communities on both sides of the country are thus under threat from the development of mass tourism and environmental degradation. Chaves hopes that the project will spark debate about these pressing issues and, more broadly, encourage people to recognize the value of Costa Rica's shores and its rich natural environment.

Filets du Futur

Martinique, France

For the project *Filets du Futur: Sustainable Fashion and Second Chances* (2024–2025), Martinican artist Jaddict collaborated with Spanish fashion designer Alba Royo to create a clothing line from discarded fishing nets, sourced from local fishermen and sewn by women undergoing professional reintegration programs.

"For me, the sea is not merely a backdrop; it is a living, rich entity that nourishes my imagination and shapes my creations."

Jaddict is a self-taught artist from Martinique, trained in graphic design and 3D modeling. As a child, she observed her grandmother practicing sewing and patchwork, which today serves as a source of inspiration for her work, which blends design, collage, and textile creations. Central to her artistic practice is upcycling, with the ocean serving as a constant wellspring of inspiration.

"My interest in fishermen's nets stems from their strong presence in local culture. Essential tools, they are often discarded when they become too worn. This represents a significant source of pollution, with many fish

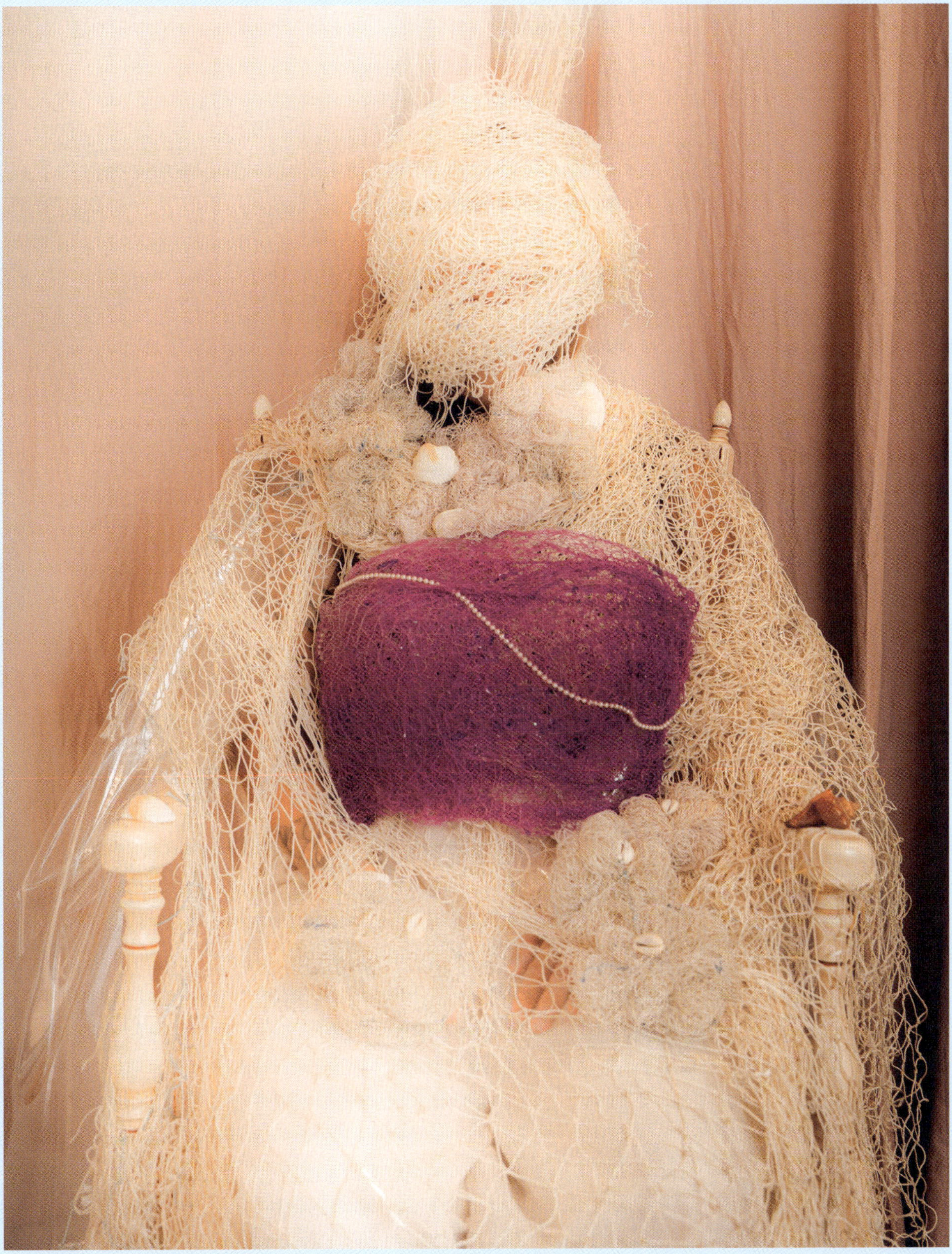

Filets du Futur, 2025

and turtles dying from entanglement in these abandoned nets. As an artist, I see in them an opportunity for transformation. They carry a history, a memory, and offer countless creative possibilities."

For the Metis Project, the garments are crafted using other repurposed materials that lend structural rigidity to the designs, such as recycled denim and sails also donated by fishermen. Even inflatable buoys are incorporated as accessories. For the fashion show, Jaddict expanded her creative vision by designing a collection of masks made entirely from fishing nets.

This collection is the culmination of an extensive collaborative effort with a group of women in situations of extreme vulnerability, who are receiving sewing training at the Made in Women workshop, managed by the non-profit organisation D'Antilles et D'ailleurs. This local group seeks to create a synergy between sustainable fashion, social reintegration, and environmental preservation in Martinique.

For Jaddict, listening to these women—most of whom come from neighboring islands and face significant isolation—is a central aspect of her work. The artist drew inspiration for her collection from interviews with them. For some, the initiative also provides an opportunity to learn sewing techniques while sharing their stories and aspirations, fostering dialogue on environmental issues. Social, economic, and ecological challenges are often deeply interconnected and, as she emphasizes, must be addressed holistically.

The fishing net, as a conceptual starting point, becomes a medium for broader discussions on themes such as travel, connection, solidarity between communities, and pollution—an urgent issue in Martinique, where numerous coastlines and rivers are severely contaminated. The accumulation of waste poses threats to local health, depletes resources such as fisheries, and undermines the tourism industry. Frequent strikes at waste management facilities, combined with a lack of environmental education, have led to the transformation of certain natural sites into garbage dumps, with the abandonment of refrigerators, armchairs, and washing machines in public spaces, causing lasting pollution to ecosystems.

According to Jaddict, the younger generation is becoming increasingly aware of these challenges, but significant efforts are still required to shift collective mindsets. This is where art can play a transformative role: embodying discourse, evoking emotions, and engaging citizens. For Jaddict, it is equally important to bring art to the people, as seeking out art is not always an intuitive process. She advocates for a proactive, community-oriented approach, involving schools and local organizations to enhance the visibility of artistic works and to create meaningful points of interaction with the population.

La Mer et l'enfant

Democratic Republic of the Congo

***La Mer et l'enfant*, a title that plays on the French homonyms for "mother" and "sea," is an art project initiated by painter and sculptor Rémy Gaylord Mongo-Etsion in collaboration with children from Pointe Noire, Republic of the Congo. Together, they have constructed a large-scale sculpture utilizing recycled materials that reflects the children's perceptions of the sea.**

Caroline Ha Thuc *You were born in Brazzaville but currently reside in Pointe-Noire, a major port city in the Congo. What are the primary environmental challenges faced by this coastal urban area?*

Rémy-Gaylord Mongo-Etsion In Pointe-Noire, the coastal environment faces significant challenges, including plastic pollution, mangrove degradation, and overfishing. These issues are further aggravated by a lack of environmental awareness and education. While residents are aware of the ocean's importance in their daily lives, they do not always grasp its broader global implications. However, initiatives such as those led by Renatura Congo are helping to shift mindsets through grassroots actions and educational programs. As an artist, my role is to strengthen the connection between the community and their environment, utilizing art as a universal language.

CHT La Mer et l'enfant *(2024), a giant metal whale adorned with reclaimed materials, was created as part of the Metis Diapason project. What were your objectives for this work?*

RGME This project was conceived as a bridge between art, environmental awareness, and education. In partnership with Renatura Congo, we conducted workshops in primary schools to gather children's perspectives on the ocean. These interactions allowed me to create an artwork that reflects both the innocence of youth and the urgency of ocean conservation.

My objectives were twofold: to foster collective awareness and to mobilize both local and international stakeholders to invest in sustainable initiatives. In this context, art becomes a catalyst for change. The whale sculpture is displayed in a prominent public space, the Rond-point des Chevaux in Pointe-Noire, a very busy area. My hope is that this initiative will inspire similar efforts and provoke reflection in each viewer.

CHT *How did you collaborate with the children?*

RGME The children were co-creators of this artwork. During the workshops, I asked questions designed to stimulate their imagination, such as, "What does the ocean mean to you?" or "If the ocean could speak, what would it say?" Their responses were often surprising and poetic. In parallel, we discussed the importance of preserving marine

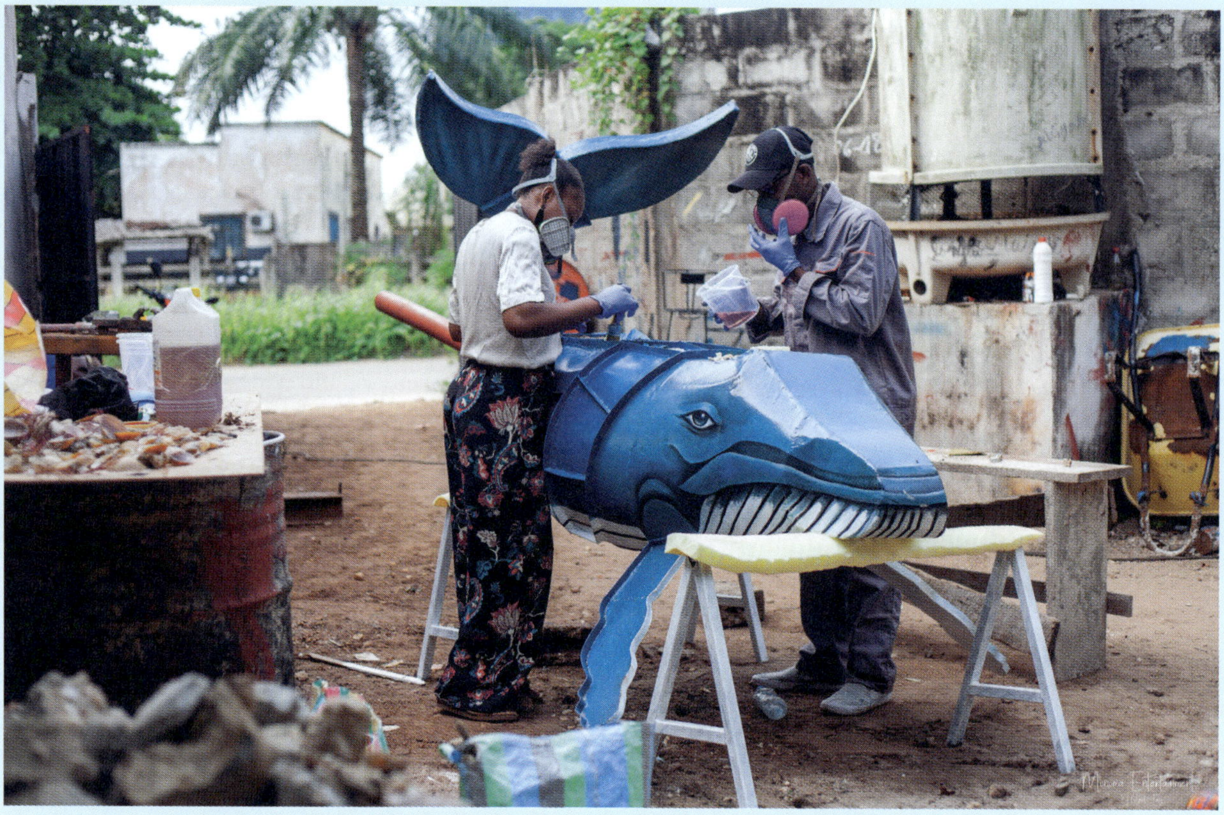

La Mer et l'enfant, 2025

turtles, coral reefs, and ocean ecosystems to raise awareness about local biodiversity. As an artist, I sought to integrate and transform their ideas, drawings, and suggestions into the final artwork, culminating in the creation of this giant whale. For instance, their simple yet powerful depictions of fish, waves, and marine creatures were incorporated into my sculpture and paintings.

CHT *How has this collaboration influenced your artistic practice?*

RGME This project not only allowed the children to express their creativity but also profoundly enriched my own artistic practice. Their spontaneous and hopeful perception of the ocean challenged me to reconsider my approach to art. I sought to preserve this freshness and authenticity in every piece I created.

Additionally, I aimed to demonstrate that art can serve as a vehicle for intergenerational dialogue. The resulting artwork is thus a synthesis of my experience as an artist and the children's boundless imagination.

CHT *What challenges do you encounter when working in schools?*

RGME The primary challenge lies in capturing the children's attention while conveying complex environmental messages. Children in Pointe-Noire are often more preoccupied with immediate realities, such as their daily lives, than with global environmental issues. Thus, it is necessary to employ creative pedagogical tools, such as drawing and interactive workshops, to engage them effectively. Moreover, the lack of resources in some schools makes organizing these activities particularly challenging. Despite these obstacles, these difficulties strengthen my resolve to provide children with opportunities for reflection and learning about their environment.

Transmigrations

Morocco

Based on the stories and interviews of migrants, *Transmigrations* is a multi-voiced project that examines the complex dynamics of migratory flows and the related environmental, social, and technological challenges they address. The initiative has been undertaken by Youssef El Idrissi, a multidisciplinary artist, cultural activist and researcher based in Casablanca, Morocco, whose practice bridges ecology, art, technology, and socio-political questions.

In recent years, El Idrissi has been exploring and developing critical reflections and works centered on water as an element, symbol, natural resource, and instrument of power. His interest lies particularly in the metaphors and concepts associated with water that provide tools for rethinking power dynamics and, more broadly, for decolonizing imaginaries and modes of thought. For instance, he engages with concepts such as hydro-colonialism and tidalectics. The latter term, proposed by Barbadian poet and scholar Kamau Brathwaite, invites us to approach questions of identity, time, and ontology in a sensitive and cyclical manner, akin to the ebb and flow of tides, with movements that are fluid and rhythmic.

El Idrissi previously worked in Tangier, Morocco, where he learned about the recent urban, industrial, and demographic transformations of the area. Tangier is now home to one of the world's largest port complexes, Tanger Med. Situated at the gateway to Europe via the Strait of Gibraltar, the city also attracts a substantial population of migrants. These developments profoundly affect local ecosystems and biodiversity. In addition, the issue of water–its traditional management and distribution–has become increasingly pressing.

Due to its unique geographic location, Morocco faces critical water-related challenges. The country frequently experiences periods of drought, prompting local populations over the centuries to develop collective water management practices that minimize usage and waste. For example, communal wells have historically been utilized for both irrigation and drinking water. However, colonial interventions disrupted many of these systems, as colonial powers appropriated water resources for their own agendas. In contemporary times, instances of neo-colonialism in water management persist, with capitalist enterprises exploiting water resources to the detriment of local communities.

Since the late 1960s, the Moroccan government has implemented an ambitious policy of dam construction aimed at providing water for agriculture, the largest consumer of water in the country, while ensuring a reliable water supply for the population.

As of 2024, approximately 140 dams exist throughout the country; however, extensive droughts pose significant challenges to their operational efficiency. This dramatic situation has forced the country to redirect its strategies towards the oceanic horizon to mitigate uncertainties related to water access, and consequently those related to agriculture, food security, and even sociopolitical balances. Massive seawater desalination plants have been developed to provide water to the major cities, despite their negative ecological impacts on the environment and high-energy consumption.

These issues are central to *Transmigrations*. The project creates a space for dialogue and reflection on the environmental, social, and technological issues surrounding human migration—whether forced or voluntary—and the strategies of adaptation in a context where resource scarcity is becoming a global reality. In this framework, the sea continues to symbolize a horizon toward Europe while also emerging as a new resource—a body of water to be filtered—whose exploitation poses significant environmental challenges.

In 2023, El Idrissi created an in-situ installation on a beach in Casablanca, featuring a silver ladder painted with astrological and occult symbols. This work was conceived as an "offering" to all the souls who perished in both the external and internal seas while seeking a better life in more welcoming territories. For the artist, the "internal sea" refers to a psychological sea, in which migrants also drown due to depression or social pressures.

"The ethos of my work revolves around complex questions—whether political, mystical, or environmental—and I believe that one of the roles of art is to illuminate what people (or authorities), whether consciously or unconsciously, attempt to omit, distort, neglect, or suppress. It is also about urging authorities and industries to change habits that severely impact the environment and the mental health of both human and non-human ecosystems. For me, art represents a movement toward collective liberation and heightened awareness."

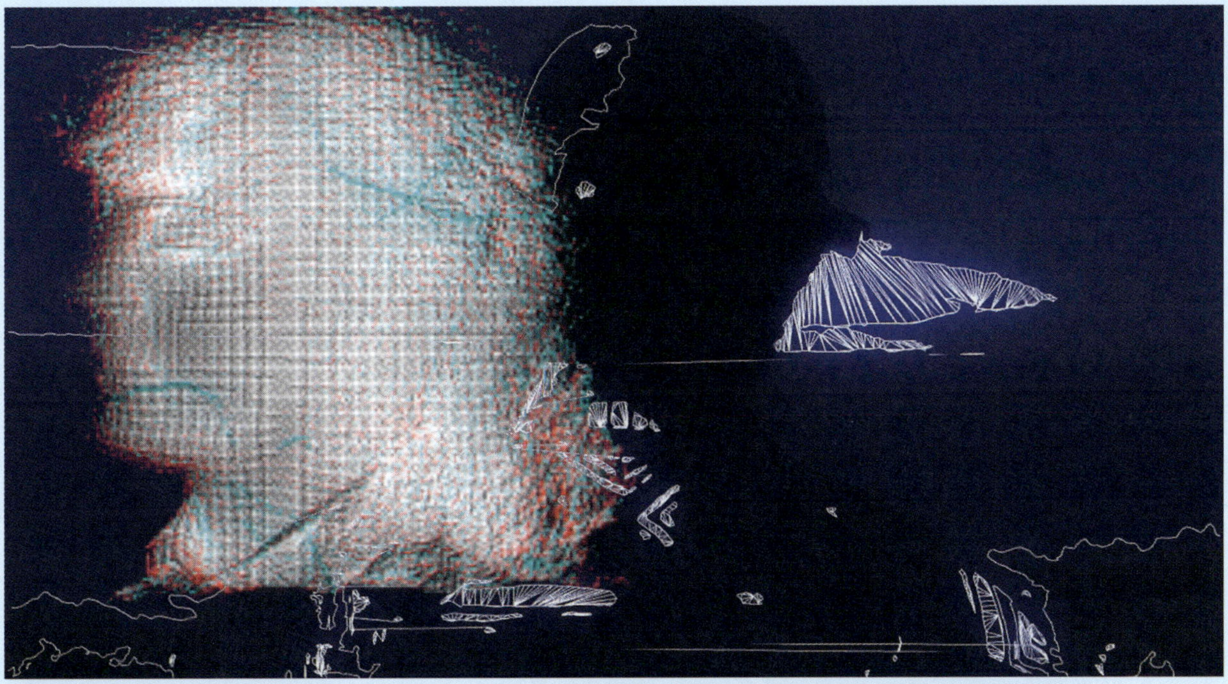

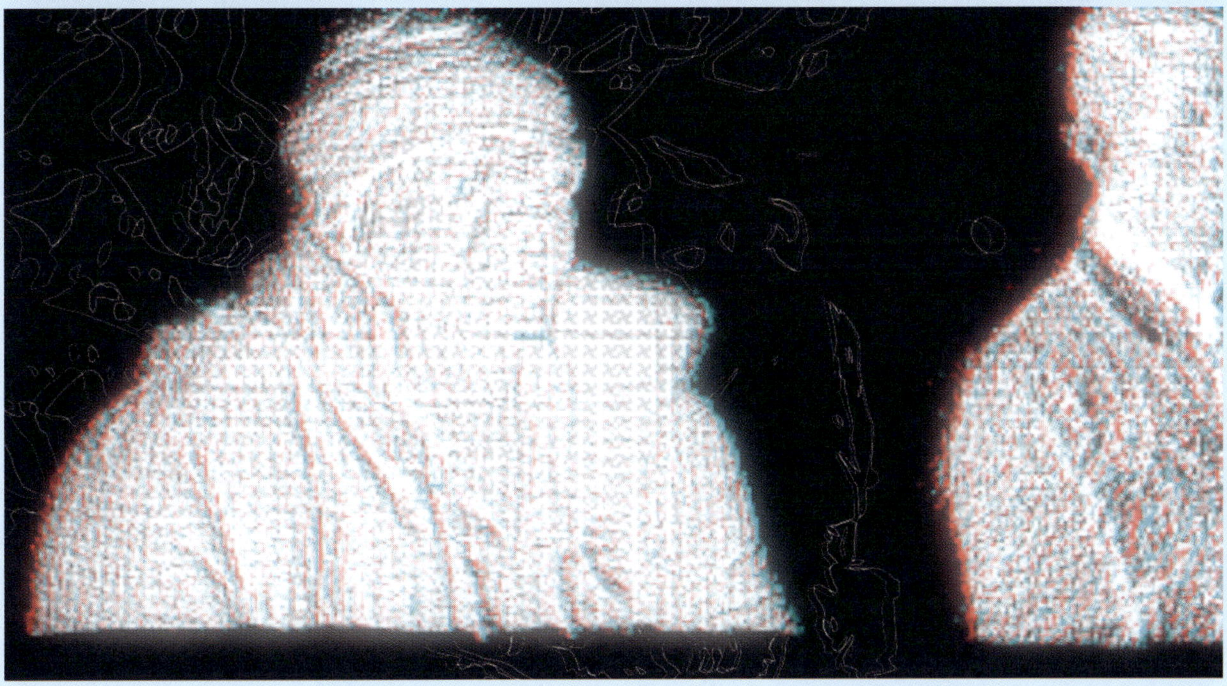

Transmigrations, 2025

Vivre ensemble

Togo

For the *Vivre ensemble* project (*Living Together*), artist Tesprit collaborated with musician Yao Bobby and eighty children to create artworks utilizing refuse collected from the streets and beaches of Lomé, Togo. The Togolese capital, which borders the Gulf of Guinea, produces more than 350,000 tons of household waste per year, a significant portion of which ends up in the sea. The discarded materials served as the foundational medium for their artistic productions.

"I am a Togolese artist. I was born and raised in Lomé, a small city in West Africa where everything revolves around the sea. When you cross the border from Ghana, it's the sight of the sea that tells you you've arrived in Lomé. The Aflao border, in particular, is closely connected to the ocean. My family comes from Aneho, one of the most enchanting places in Lomé, where the sea and the lagoon meet without mixing, creating a unique natural scene. My mother is a smoked fish seller, and I lost my father when was young. Her business helped pay for my education and that of my sisters. Aneho is known for its women who are experts in choosing and smoking the best fish in Togo. In Lomé, the sea is at the heart of everything. Music festivals, competitions, and recreational activities often take place there. On weekends, the whole city gathers on the beaches. Some people come to enjoy the little bars called *maquis* and the sea breeze, while others simply lie in on cloths under the shade of coconut trees. As a child, I used to play football on the beach and collect tin cans to make makeshift cars. I felt a deep sense of peace, relaxation, and freedom. I am never truly myself when I am far from the sea. But as I grew older, I saw the sea I loved fill with waste: plastic, old flip-flop soles, clothes, and many other things. Since the big market in Lomé is near the beach, a lot of trash ends up in the ocean. Faced with this disaster, I asked myself, what can we do to protect the ocean? Why doesn't my community, which relies so much on this sea for life, do more to protect it? That's how my motivation was born to create artworks from waste, especially from flip-flop soles, as a tribute to the sea that has always inspired me. It's a long process of collecting and cleaning, and I create very few works every year, but it's the one that's most in tune with myself. I like the story of the hummingbird that uses its beak to carry water to put out a forest fire. The other animals, much bigger, laugh at it, but the bird says it's doing its part. I know it's not much compared to the size of the problem, but I try to do my part, to give something back to the sea, just as it has given to me and my family."

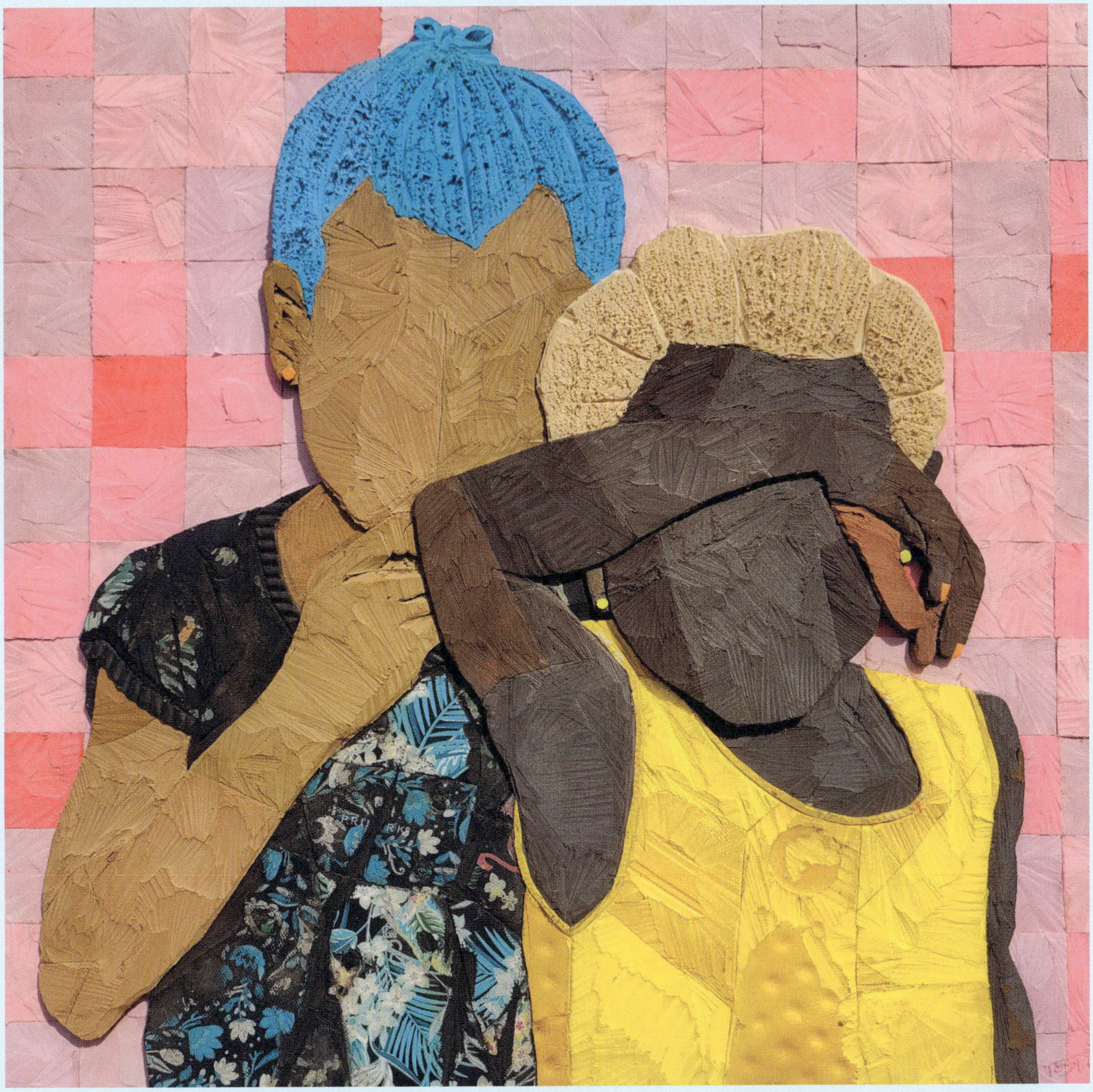

Sébla III, 2023, Flip flops on canvas, 60 × 60 cm

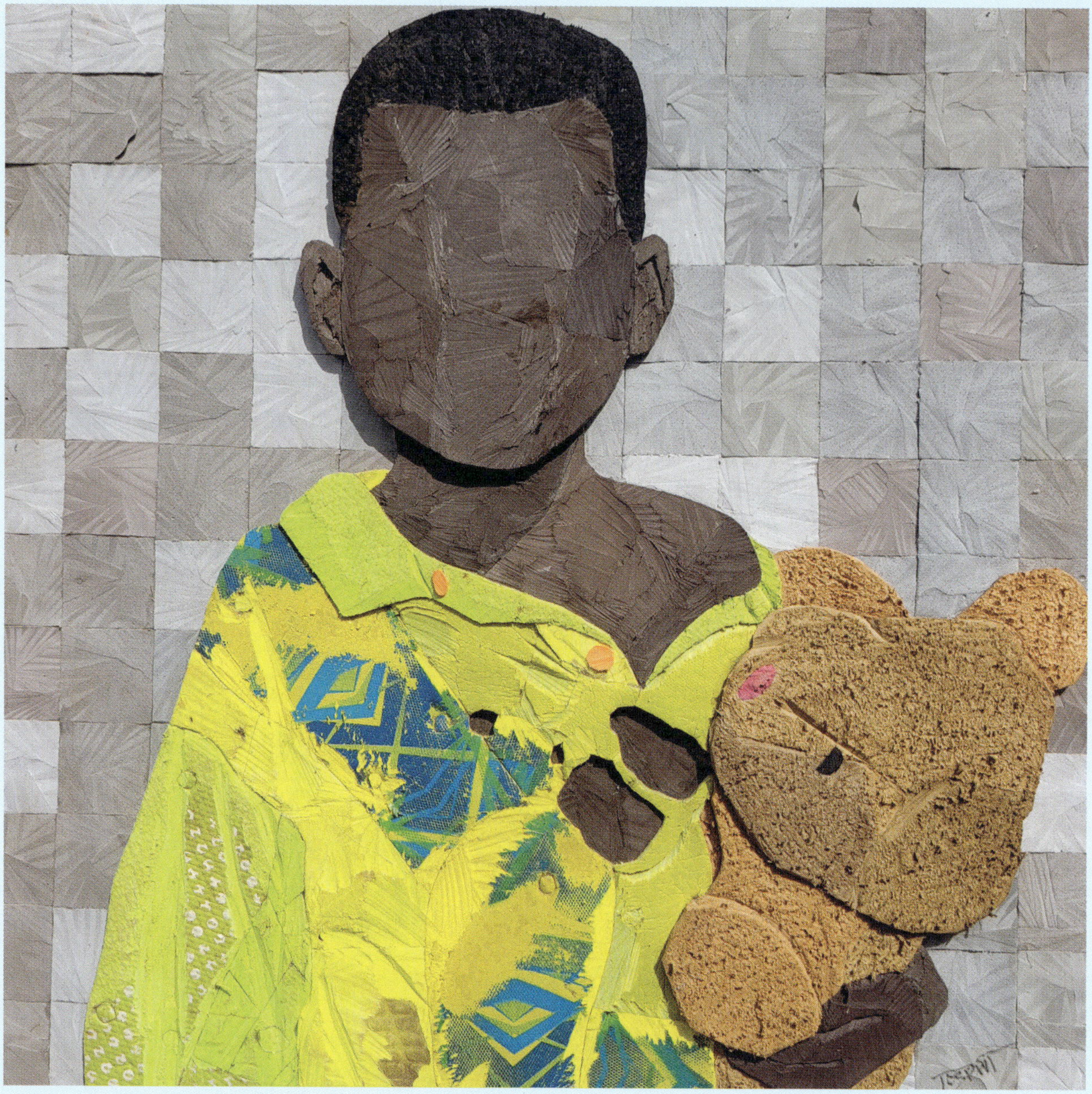

Lébénè, 2023, Flip flops on canvas, 60 × 60 cm

Credits

Cover:
The Goddess of the Deep Blue, 2022, Acrylic on canvas, 160 × 140 cm, © Arnold Fokam, Courtesy of the artist and Christophe Person Gallery.

Inner pages: pp.11-12: © Juria Toramae; p.22: © Shuyi Cao; pp.25-27: © Arnold Fokam, Courtesy of the artist and Christophe Person Gallery; p.30: © Claudia Müller Montes and Matías Labbé; p.31: © Claudia Müller Montes; p.33: © Maki Ohkojima, Photo by Akimi Ota; pp.35-37: © Nyaba Ouedraogo; p.39: © Ghizlane Sahli, Photo Nohan Ferreira; p.40: © Ghizlane Sahli; pp.42-43: © Katherine Takpannie; pp. 50-52: © Dineo Seshee Bopape; p.54-56: © Marina Gioti, Courtesy of the artist and Galerie Dominique Fiat; p.58-59: © Magali Grondin; pp.61-62: © Shitabile Mlotshwa, Courtesy of the artist, all rights reserved; p.64: © Enrique Ramírez; p.66: © Enrique Ramírez, Courtesy of the artist and the Parc naturel régional de la Narbonnaise en Méditerranée (France); p.68: © Lisa Reihana; pp.71-72: © Bao Vuong; pp.78-80: © Yoan Capote; p.83: © Maureen Gruben, © Tuk TV; p.84: © Maureen Gruben, Photo Kyra Kordoski; p.87: © Deborah Drisana Jack, Photo Courtesy of the artist; p.89: © Ange Arthur Koua, Photo Aurélien Mole; p.91: © Louisa Marajo, © ADAGP, Paris, 2025; p.93: © Barthélémy Toguo, © ADAGP, Paris, 2025; p.94-95: © Barthélémy Toguo. Courtesy Galerie Lelong & Co. et Bandjoun Station. © ADAGP, Paris, 2025; pp.96-97: © Zai Kuning, Courtesy of Ota Fine Arts Singapore/ Shanghai/ Tokyo; p.99: © Museo Nacional del Prado; p.104: © Achille Adonon; p.107: © Felipe Fittipaldi; p.108-109: © Naiza Khan; pp. 111-112: © Gideon Mendel, Courtesy of the artist; p.115: © Paribartana Mohanty; pp.118-119: ©Shneider Léon Hilaire. Courtesy Galerie MAGNIN-A, Paris; pp.121-122: © Judy Watson, Courtesy the Artist and Milani Gallery, Meeanjin/Brisbane. Photos Carl Warner; p.130: © Clay Apenouvon, Photo Alun B; p.132: © Clay Apenouvon, Photo Odile Pascale; p.135: © Edward Burtynsky; p.137: © Thaís Espaillat Ureña, Photo Jeremy García; p.138: © Heba Y. Amin; p.139: DR; p.140: © Andy Hughes; p.143: © Andy Hughes; p.145: © 2019 Denise Lira Ratinoff, © 2019 Patricio Aguilar Díaz, All Rights Reserved. Photo © 2019 Jorge Brantmayer; p.148: © Patrick Turkson; pp.156-158: © Emma Critchley; p.160: © Ana Teresa Fernández; p.162: © Inuk Silis Høegh / Ánorâk Film; p.165: © Lee Tzu-Tung, Photography by Wei Keng Hung / Provided by @Zonesoundcreative; p.166: © Lee Tzu-Tung; pp.168-169: © Charles Lim; p.171: © Uili Lousi, p.173: © Tita Salina and Irwan Ahmett, Photo Irwan Ahmett; p.174: © Tita Salina and Irwan Ahmett, Photo Yopie Nugraha; p.182: © Alimi Adewale, Courtesy of the Artist and Pearl Lam Galleries; p.185: © Soly Cissé; p.187-188: © Ale de la Puente, Courtesy of the artist; p.192: © Faris Ridzwan; p.194-195: © Rebecca Rutstein; p.197-198: © Kim Yip Tong; p.201: © Juria Toramae; p.206: © Ravi Agarwal; pp.210-212: © Martha Atienza; pp.214-217: © Nicolas Floc'h, ADAGP. Courtesy of the artist and Galerie Maubert; pp.219-220: © Jane Jin Kaisen; pp.223-224: © Mario Macilau, Courtesy of Ed Cross Fine Art; p.224: © Mario Macilau, Courtesy of Ed Cross Fine Art; p.226: © Manuel Ndoye; p.219: © Temandrota, Courtesy of the artist; pp.236-238: © Ursula Biemann; pp.242-243: © Nadia Huggins, Courtesy of the artist; p.247: © Kristina Pulejkova, Courtesy of the artist; p.249: © Robertina Šebjanič; p.252: © Shakil Solanki; p.255: © Lara Tabet; p.259: © Alberta Whittle; pp.266-268: © Amina Agueznay, Photo Khalil Nemmaoui. Courtesy CulturesInterface; p.271: © Ari Bayuaji, Courtesy of the artist; pp.273-274: © Megan Cope, Courtesy the Artist and Milani Gallery, Meeanjin/Brisbane, Photos Carl Warner; pp.276-268: © Jason deCaires Taylor; pp.282-284: © Jérémy Gobé; p.287: © Julia Lohmann, Photo Frankfurter Kunstverein; p.288: © Julia Lohmann; p.292: © Moffat Takadiwa, Photo Sebastiano Pellion di Persano. Courtesy Semiose, Paris; p.294: © Moffat Takadiwa, Photo A. Mole. Courtesy Semiose, Paris; pp. 302-303: © Ines Arámburo, Omar Kassimi, Rahul Ray, Mbal Tony, Mert Gökalp; p. 304: © Rahul Ray; p. 306: © Photo Rahul Ray; p. 309-311: © Mert Gökalp; p. 313: © Diego Esquivel; p. 315: Photo © Ines Arámburo; p. 318: © Mbal Tony / Renatura Congo; p. 321: © Youssef El Idrissi; pp. 323-324: © Tesprit.

Thanks

To the artists
Alimi Adewal
Achille Adonon
Amina Agueznay
Irwan Ahmett
Clay Apenouvon
Martha Atienza
Ari Bayuaji
Ursula Biemann
Dineo Seshee Bopape
Edward Burtynsky
Shuyi Cao
Yoan Capote
Soly Cissé
Megan Cope
Fabian Coto Chaves
Emma Critchley
Youssef El Idrissi
Thaís Espaillat Ureña
Ana Teresa Fernández
Felipe Fittipaldi
Nicolas Floc'h
Arnold Fokam
Marina Gioti
Jérémy Gobé
Mert Gökalp
Magalie Grondin
Maureen Gruben
Heba Y. Amin
Inuk Silis Høegh
Afroza Hossain (Sara)
Nadia Huggins
Andy Hughes
Deborah Drisana Jack
Jaddict
Jane Jin Kaisen
Naiza Khan
Ange Arthur Koua
Zai Kuning
Lee Tzu-Tung
Charles Lim
Denise Lira-Ratinoff
Julia Lohmann
Uili Lousi
Mario Macilau
Louisa Marajo
Gideon Mendel
Sithabile Mlotshwa
Paribartana Mohanty
Rémy-Gaylord Mongo-Etsion
Claudia Müller Montes
Manel Ndoye
Maki Ohkojima
Nyaba Léon Ouedraogo
Ale de la Puente
Kristina Pulejkova
Enrique Ramírez
Agarwal Ravi
Lisa Reihana
Faris Ridzwan
Rebecca Rutstein
Ghizlane Sahli
Tita Salina
Robertina Šebjanič
Shneider Léon Hilaire
Shakil Solanki
Winnie Soon
Lara Tabet
Moffat Takadiwa
Katherine Takpannie
Jason deCaires Taylor
Temandrota
Tesprit
Lhylian Therance Ralff
Barthélémy Toguo
Juria Toramae
Patrick Tagoe-Turkson
Bao Vuong
Judy Watson
Alberta Whittle
Kim Yip Tong

To the contributors
Sylvain Agostini
Hélène Artaud
Isabelle Autissier
André Hoffmann
Carlos Manuel Rodriguez
Jennifer Morris
Olivier Poivre d'Arvor
Rémy Rioux

To all the people who have been involved in this publication
Philippe Bauduin
Sylvie Boyer
Jean-François Chougnet
Jeremie Daussin-Charpantier
Anne Degroux
Emmanuelle Durand
Anne-Laure Durand-Servin
Hélène Gobert
Xiaoting Jin
Didier Kuhn
Yanns Laurans
Catherine Lecouffe
Damien Mermet
Sébastien Mosneron Dupin
Patrick Papazian
Axelle de Russé
Serge Snrech
Ronny Turiaf

The Ocean Manifesto

A Metis Arts and Development Fund Initiative
Tiphaine de Mombynes
Tanguy Abittan-Tenenbaum
Marie Diehl
Barbara Winkelried Quea
Raphaël Lemarchand
Violette Berthod

Texts by
Caroline Ha Thuc

Editorial Direction
David Desrimais

Editorial Coordination
Pierre-Édouard Couton
assisted by Lisa Valentin

Graphic Design
Emma Zampieri – Studio JBE
assisted by Axelle Perrais

Translations from French
Monique Gross (for texts by Olivier Poivre D'Arvor and Metis Fund)

Photoengraving
IGS – Print

Fonts
Atlantic (F37 Foundry)
Joly (Blaze Type)

This book is published on the occasion
of the third United Nations Ocean Conference (UNOC3)
organized in Nice (France), from June, 9 to 13, 2025.

ISBN: 978-2-36568-115-5
Legal deposit: June 2025
Printed in Lithuania in 2025

JBE Books
90 rue de la Folie-Méricourt
75011 Paris

jbe-books.com